Cognitive Technologies

Managing Editors: D. M. Gabbay J. Siekmann

For further volumes:
http://www.springer.com/series/5216

Johannes Fürnkranz
Dragan Gamberger
Nada Lavrač

Foundations of Rule Learning

Springer

Johannes Fürnkranz
FB Informatik
TU Darmstadt
Darmstadt
Germany

Nada Lavrač
Department of Knowledge Technologies
Jožef Stefan Institute
Ljubljana
Slovenia

Dragan Gamberger
Rudjer Bošković Institute
Zagreb
Croatia

Cognitive Technologies ISSN 1611-2482
ISBN 978-3-642-43046-6 ISBN 978-3-540-75197-7 (eBook)
DOI 10.1007/978-3-540-75197-7
Springer Heidelberg New York Dordrecht London

ACM Codes: I.2, H.2

Printed on acid-free paper

Springer is part of Springer Science+Business Media (www.springer.com)

Foreword

It was on a balmy spring day in Nanjing, while we were working on a paper on contrast discovery for the Journal of Machine Learning Research, that Nada Lavrač let slip that she was writing a book on rule learning with Johannes Fürnkranz and Dragan Gamberger. I must admit that I was initially skeptical. Rule learning is one of the core technologies in machine learning, but there is a good reason why nobody has previously had the audacity to write a book on it. The topic is large and complicated. There are a great variety of quite different machine learning activities that all use rules, in different ways, for different purposes. Many different machine learning communities have contributed to research in each of these areas, often inventing totally different terminologies. Certainly Fürnkranz, Lavrač, and Gamberger have been significant contributors to this large, diverse body of research, and hence were eminently qualified to take on the task, but was there any chance of making sense of this massive heterogeneous body of research within the confines of a book of manageable size?

Having now read the book, I am delighted to be able to answer this question with a resounding "Yes." The book you are reading provides a clear overview of the field. One secret to its success lies in the development of a clear unifying terminology that is powerful enough to cover the whole field. The many diverse sets of terminology developed within different research communities are then translated back into this unifying terminology, which acts as a lingua franca.

But this book is more than just a triumph of translation between multiple alternative sets of terminology. For the first time we have a consolidated detailed summary of the state of the art in rule learning. This book provides an excellent introduction to the field for the uninitiated, and is likely to lift the horizons of many working in subcommunities, providing a clear view of how the work in that subcommunity relates to work elsewhere.

It is notable that J. Ross Quinlan, famous as the developer of the quintessential decision tree learners C4.5 and C5/SEE5, is a strong advocate of rule learning. For example, in Quinlan (1993), with reference to his C4.5RULES software for converting decision trees into classification rules he concludes that the resulting rule set is "usually about as accurate as a pruned tree, but more easily understood by people."

Rule learning technologies provide an extensive toolkit of powerful learning techniques. This book, for the first time, makes the full extent of this toolkit widely accessible to both the novice and the initiate, and clearly maps the research landscape, from the field's foundations in the 1970s through to the many diverse frontiers of current research.

Geoffrey I. Webb

Preface

Rule learning is not only one of the oldest but also one of the most intensively investigated, most frequently used, and best developed fields of machine learning. In more than 30 years of intensive research, many rule learning systems have been developed for propositional and relational learning, and have been successfully used in numerous applications. Rule learning is particularly useful in intelligent data analysis and knowledge discovery tasks, where the compactness of the representation of the discovered knowledge, its interpretability, and the actionability of the learned rules are of utmost importance for successful data analysis.

The aim of this book is to give a comprehensive overview of modern rule learning techniques in a unifying framework which can serve as a basis for future research and development. The book provides an introduction to rule learning in the context of other machine learning and data mining approaches, describes all the essential steps of the rule induction process, and provides an overview of practical systems and their applications. It also introduces a feature-based framework for rule learning algorithms which enables the integration of propositional and relational rule learning concepts.

The book starts by introducing the necessary machine learning and data mining terminology. Chapter 1 provides the motivation for the research, development, and application of rule learning systems. Descriptions of the inductive rule learning task, basic propositional rule learning algorithms, and the basic rule quality evaluation measures are found in Chap. 2. The chapter ends with an overview of the best established rule learning systems. In Chap. 3, coverage space is defined, which enables qualitative and quantitative comparisons of predictive rule learning systems. Coverage space is extensively used in the rest of the book for the evaluation and illustration of various elements constituting the rule learning process.

The central part of the book consists of Chaps. 4–10. It starts by the presentation of features as unifying building blocks for propositional and relational rule learning. Additionally, Chap. 4 presents methods for detecting and eliminating irrelevant features as well as for handling missing and imprecise attribute values which are important constituents of the process of feature construction. Chapter 5 is devoted to relational rule learning: it describes relational feature construction as an upgrade of propositional feature construction, which enables the methodology described in the following chapters to be applied regardless of the propositional or relational nature of the features. Chapter 6 describes algorithms learning single rules by searching the hypothesis space for rules that cover many positive and few negative examples. These objectives are usually captured in search heuristics, which are extensively discussed in Chap. 7. Chapter 8 shows how individual rules can be combined into complete theories such as rule sets or decision lists. Chapter 9 addresses pruning of rules and rule sets with the goal of learning simple concepts in order to increase interpretability and to prevent overfitting. This part of the book ends with Chap. 10, which contains an extensive presentation of approaches used for classification into multiclass and structured output spaces, unsupervised learning, and regression.

The presentation of modern descriptive rule learning methods and their applications is in the final part of the book. Chapter 11 includes the presentation of contrast set mining, emerging pattern mining, and subgroup discovery approaches. It shows that these supervised descriptive rule learning approaches are compatible with standard rule learning definitions. This chapter ends by presenting subgroup visualization methods which are a useful ingredient of descriptive data analysis. A detailed description of four successful applications of rule learning is given in Chap. 12. These include a social science application in a demographic domain, a biological functional genomics analysis application concerning the analysis of gene expression, and two medical applications. The applications are used to illustrate the data analysis process, including data collection, data preparation, results evaluation, and results interpretation, and to illustrate the performance of selected rule learning algorithms described in the previous chapters of the book on real-world data. The book includes also a rich list of almost 300 references that will help the interested reader in further work.

The book is prepared as a monograph summarizing relevant achievements in the field of rule learning. The target audience is researchers looking for references for further study and teachers in need of detailed material for their machine learning courses. The book is structured as a self-contained introduction to the field of rule learning (Chaps. 1–3) followed by in-depth discussions of relevant subtopics. After reading the introductory part, a reader is prepared to directly skip to any of the following chapters. Jumping to Chap. 12 describing applications of rule learning may be a good motivation for studying the details of the methodology.

The book is written by authors who have been working in the field of rule learning for many years and who themselves developed several of the algorithms and approaches presented in the book. Although rule learning is assumed to be a well-established field with clearly defined concepts, it turned out that finding a unifying approach to present and integrate these concepts was a surprisingly difficult task. This is one of the reasons why the preparation of this book took more than 5 years of joint work.

A good deal of discussion went into the notation to use. The main challenge was to define a consistent notational convention to be used throughout the book because there is no generally accepted notation in the literature. The used notation is gently introduced throughout the book, and is summarized in Table I in a section on notational conventions immediately following this preface (pp. xi–xiii). We strongly believe that the proposed notation is intuitive. Its use enabled us to present different rule learning approaches in a unifying notation and terminology, hence advancing the theory and understanding of the area of rule learning.

Many researchers have substantially contributed to the content of this book, as most of the chapters are based on our original papers which were written in coauthorship. Most notably, the contributions of Peter Flach and Sašo Džeroski, with whom we have successfully collaborated in rule learning research for nearly a decade, are acknowledged. Other collaborators who have in several aspects contributed to our work include Francesco Billari, Bruno Crémilleux, Alexia Fürnkranz-Prskawetz, Marko Grobelnik, Eyke Hüllermeier, Frederik Janssen, Arno Knobbe, Antonija Krstačić, Goran Krstačić, Eneldo Loza Mencía, Petra Kralj Novak, Sang-Hyeun Park, Martin Scholz, Ashwin Srinivasan, Jan-Nikolas Sulzmann, Jakub Tolar, Geoffrey Webb, Lorenz Weizsäcker, Lars Wohlrab, and Filip Železný; we are grateful for their collaboration and contributions.

We are grateful to our institutions, Technical University of Darmstadt (Germany), Rudjer Bošković Institute (Zagreb, Croatia), and Jožef Stefan Institute (Ljubljana, Slovenia), for providing us with stimulating working environments, and to the national funding agencies who have supported and cofunded this work. We are also grateful to Springer, the publisher of this work, and especially to Ronan Nugent who has shown sufficient patience waiting for this book to be ready for publication, which took much longer than initially expected.

Our final thanks go to Geoffrey Webb for his inspiring Foreword to this book, as well as Richard Wheeler and Peter Ross, who proofread the book for language, consistency, and mathematical correctness. Richard's contribution was especially invaluable due to his passionate reading of the entire book, proving that the book contains sufficiently interesting material also for a very knowledgeable reader.

Darmstadt, Germany
Zagreb, Croatia
Ljubljana, Slovenia
January 2012

Johannes Fürnkranz
Dragan Gamberger
Nada Lavrač

Notational Conventions

In the following, we briefly note the notational conventions of this book. They are summarized at the end of this section in Table I.

Rule representation. Depending on the context, we use three different notations for rules:

Explicit: This notation explicitly marks the condition part (`IF`), the connection of the features in the conjunctive condition (via `AND`), and the conclusion (`THEN`). It is the most readable form and will be mostly used in application-oriented parts, where it is important to understand the semantics of the rule.

Example:
```
IF    Shape = triangle
  AND Color = red
  AND Size  = big
THEN  Class = Positive
```

Formal: This notation is based on formal (propositional) logic. The implication sign (←) is typically written from right to left (as in Prolog), but it may also appear in the other direction. It will be used if theoretical properties of rule learning algorithms are discussed.

Example: $\oplus \leftarrow$ Shape = triangle $\wedge$ Color = red $\wedge$ Size = big.

Prolog: This notation uses a PROLOG-like syntax. It will mostly be used on examples involving relational learning.

Example:
```
positive(X) :- shape(X,triangle),
               color(X,red),
               size(X,big).
```

Examples, attributes. Uppercase calligraphic font is used for denoting sets, and the corresponding letters in regular font are used for denoting the sizes of these sets. So, e.g., $\mathcal{E} = \{e_i\}, i = 1 \ldots E$ denotes a training set with E examples e_i. Examples may be represented with a set of attributes $\mathcal{A} = \{\mathbf{A}_j\}, j = 1 \ldots A$, and the class attribute is denoted as **C**. Attribute and class values are denoted as $v_{i,j}$ and c_i, respectively.

Features. The basic building blocks for rules are binary features **f**. The set of all features $\mathcal{F} = \{\mathbf{f}_k\}, k = 1 \ldots F$ is typically constructed from the examples and their attribute values. For example, $\mathbf{f}_1$ = `(Shape = triangle)` is a propositional feature, and $\mathbf{f}_2$ = `equal_shape(X,Y)` is a relational feature.

Rules and rule sets. The symbol **r** denotes a single rule, and $\mathcal{R}$ denotes a rule set consisting of R rules. Rules are built out of features. The number of such features that are included in the rule is L, the length of the rule. If it is clear from the context that the rule predicts the positive class, we may also describe a rule by a set of features.

Positive and negative examples. In a concept learning setting, the positive class will be denoted as $\oplus$, and the negative class as $\ominus$. $\mathcal{P}$ is the set of P positive examples, and $\mathcal{N} = \mathcal{E} \setminus \mathcal{P}$ is the set of N negative examples.

Coverage. Coverage is denoted by adding a hat or roof ($\hat{}$) on top of a letter (read as "covered"); noncoverage is denoted by a bar ($\bar{}$). Thus, $\hat{P}$ denotes the number of positive examples and $\hat{N}$ the number of negative examples covered by a rule. In case we need to compare these numbers for different rules, we introduce the argument **r**: $\hat{P}(\mathbf{r})$ and $\hat{N}(\mathbf{r})$ denote the numbers of positives and negatives covered by rule **r**, respectively. Their complements, i.e., the number of positive and negative examples not covered by a rule, are denoted as $\bar{P}$ and $\bar{N}$, respectively. Numbers $\hat{P}$ and $\hat{N}$ are also known as the numbers of true positives and false positives, $\bar{P}$ and $\bar{N}$ as the false negatives and true negatives, respectively. The corresponding rates are denoted as $\hat{\pi} = \frac{\hat{P}}{P}$ (true positive rate), $\hat{\nu} = \frac{\hat{N}}{N}$ (false positive rate), $\bar{\pi} = \frac{\bar{P}}{P}$ (false negative rate), and $\bar{\nu} = \frac{\bar{N}}{N}$ (true negative rate).

Multiple classes. In the multiclass case, we denote the C different classes with C_i, $i = 1, \ldots, C$. The number of examples in class C_i is P_i, and therefore $E = \sum_{i=1}^{C} P_i$. The number of examples that do not belong to class C_i is denoted as N_i. The corresponding sets are denoted as $\mathcal{P}_i$ and $\mathcal{N}_i$. Clearly, $\mathcal{N}_i = \mathcal{E} - \mathcal{P}_i = \bigcup_{j \neq i} \mathcal{P}_j$.

Refinements. Rules can be refined by adding features, a refinement with feature **f** may be described as $\mathbf{r}' = \mathbf{r} \cup \{\mathbf{f}\}$. Refinement of $\mathbf{r} = \mathrm{H} \leftarrow \mathrm{B}$ by feature **f** results in a rule with the same rule head (rule conclusion H) and the extended rule body $\mathrm{B} \wedge \mathbf{f}$. The prime ($'$) can also be used for denoting characteristics of the refined rule. For example, $\hat{P}'$ and $\hat{N}'$ are the positive and negative examples covered by the refinement $\mathbf{r}'$. The set of all refinements of rule **r** is denoted as $\rho(\mathbf{r})$. $\mathbf{r}^{\top}$ is the universal rule covering all examples, $\mathbf{r}_e$ is the bottom rule for example e, and $\mathbf{r}^{\perp}$ is the empty rule covering no examples.

Rule evaluation measures. Rules are evaluated using heuristic functions $H(\hat{P}, \hat{N})$ or $H(\mathbf{r})$, which are written in italic font. The equivalence of two metrics is denoted with $\sim$: $H_1 \sim H_2$ means that heuristics H_1 and H_2 sort rules in the same way.

Probabilities. The probability of an event x is denoted as $\Pr(x)$.

Complexities. For complexity analysis, we use the standard notation, where $g(x) = \Omega(f(x))$ denotes that g grows at least as fast as f, i.e., f forms a lower bound for g. Similarly, $g(x) = O(f(x))$ means that f forms an upper bound for g. We also use $g(x) = \Theta(f(x))$ to denote that f and g have the same asymptotic growth, i.e., there exist two constants c_1 and c_2 such that $c_1 f(x) \leq g(x) \leq c_2 f(x)$.

Table I Summary of notational conventions used throughout the book

$\mathcal{E} = \{e_1 \dots e_E\}$	Set of examples
$\mathcal{P} = \{p_i\}$	Set of positive examples
$\mathcal{N} = \{n_i\}$	Set of negative examples, $\mathcal{N} = \mathcal{E} \setminus \mathcal{P}$
$P = \lvert\mathcal{P}\rvert$	Number of positive examples
$N = \lvert\mathcal{N}\rvert$	Number of negative examples
$E = P + N$	Number of examples
$\hat{\mathcal{E}}$ $(\hat{E})$	Set (number) of examples covered by a rule
$\hat{\mathcal{P}}$ $(\hat{P})$	Set (number) of covered positive examples (*true positives*)
$\hat{\mathcal{N}}$ $(\hat{N})$	Set (number) of covered negative examples (*false positives*)
$\bar{\mathcal{E}}$ $(\bar{E})$	Set (number) of examples not covered by a rule
$\bar{\mathcal{P}}$ $(\bar{P})$	Set (number) of not covered positive examples (*false negatives*)
$\bar{\mathcal{N}}$ $(\bar{N})$	Set (number) of not covered negative examples (*true negatives*)
$\hat{\pi} = \hat{P}/P$	True positive rate (sensitivity or Precision)
$\bar{\nu} = \bar{N}/N$	True negative rate
$\hat{\nu} = \hat{N}/N = 1 - \bar{\nu}$	False positive rate (false alarm)
$\bar{\pi} = \bar{P}/P = 1 - \hat{\pi}$	False negative rate
$\mathcal{F} = \{\mathbf{f}_1, \dots \mathbf{f}_F\}$	Set of features
$\mathcal{A} = \{\mathbf{A}_1, \dots \mathbf{A}_A\}$	Set of attributes
$v_{i,j}$	The j th value of attribute $\mathbf{A}_i$
$\mathbf{C} = \{c_1, \dots c_C\}$	Class attribute
C	Number of class values in a multiclass problem
$\mathcal{P}_i, P_i, \hat{\pi}_i$	Set, number, rate of examples belonging to class c_i
$\mathcal{N}_i, N_i, \hat{\nu}_i$	Set, number, rate of examples that do not belong to class c_i
$\gamma_i = P_i/E$	Fraction of examples that belong to class c_i
$\mathbf{r} = \mathrm{H} \leftarrow \mathrm{B}$	A rule with body B and head H
$\rho(\mathbf{r})$	Rules resulting from applying refinement operator ρ to $\mathbf{r}$
$\sigma(\mathbf{r})$	Rules resulting from applying specialization operator σ to $\mathbf{r}$
$\gamma(\mathbf{r})$	Rules resulting from applying generalization operator γ to $\mathbf{r}$
$\mathbf{r}'$	Refinement of rule $\mathbf{r}$, typically of the form $\mathbf{r}' = \mathbf{r} \cup \{\mathbf{f}\}$
L	Length of a rule (number of features in the rule condition)
$\mathcal{R} = \{\mathbf{r}_1, \dots \mathbf{r}_R\}$	A rule-based theory (a set of rules)
$H(\hat{P}, \hat{N}) = H(\mathbf{r})$	Rule evaluation measure
$\sim$	Equivalence of rule evaluation measures
$\Pr(x)$	Probability of event x
$\square$	End of proof

Contents

Chapter 1
Machine Learning and Data Mining

Machine learning and data mining are research areas of computer science whose quick development is due to the advances in data analysis research, growth in the database industry and the resulting market needs for methods that are capable of extracting valuable knowledge from large data stores. This chapter gives an informal introduction to machine learning and data mining, and describes selected machine learning and data mining methods illustrated by examples. After a brief general introduction, Sect. 1.2 briefly sketches the historical background of the research area, followed by an outline of the knowledge discovery process and the emerging standards in Sect. 1.3. Section 1.4 establishes the basic terminology and provides a categorization of different learning tasks. Predictive and descriptive data mining techniques are illustrated by means of simplified examples of data mining tasks in Sects. 1.5 and 1.6, respectively. In Sect. 1.7, we highlight the importance of relational data mining techniques. The chapter concludes with some speculations about future developments in data mining.

1.1 Introduction

Machine learning (Mitchell, 1997) is a mature and well-recognized research area of computer science, mainly concerned with the discovery of models, patterns, and other regularities in data. Machine learning approaches can be roughly categorized into two different groups:

Symbolic approaches. Inductive learning of symbolic descriptions, such as rules (Clark & Niblett, 1989; Michalski, Mozetič, Hong, & Lavrač, 1986) decision

†This chapter is partly based on Lavrač & Grobelnik (2003).

DOI 10.1007/978-3-540-75197-7_1,

trees (Quinlan, 1986) or logical representations (De Raedt, 2008; Lavrač & Džeroski, 1994a; Muggleton, 1992). Textbooks that focus on this line of research include (Langley, 1996; Mitchell, 1997; Witten & Frank, 2005).

Statistical approaches. Statistical or pattern-recognition methods, including k-nearest neighbor or instance-based learning (Aha, Kibler, & Albert, 1991; Dasarathy, 1991), Bayesian classifiers (Pearl, 1988), neural network learning (Rumelhart & McClelland, 1986), and support vector machines (Schölkopf & Smola, 2001; Vapnik, 1995). Textbooks in this area include (Bishop, 1995; Duda, Hart, & Stork, 2000; Hastie, Tibshirani, & Friedman, 2001; Ripley, 1996).

Although the approaches taken in these fields are often quite different, their effectiveness in learning is often comparable (Michie, Spiegelhalter, & Taylor, 1994). Also, there are many approaches that cross the boundaries between the two approaches. For example, there are decision tree (Breiman, Friedman, Olshen, & Stone, 1984) and rule learning (Friedman & Fisher, 1999) algorithms that are firmly based in statistics. Similarly, ensemble techniques such as boosting (Freund & Schapire, 1997), bagging (Breiman, 1996) or random forests (Breiman, 2001a) may combine the predictions of multiple logical models on a sound statistical basis (Bennett et al., 2008; Mease & Wyner, 2008; Schapire, Freund, Bartlett, & Lee, 1998). This book is concerned only with the first group of methods, which result in symbolic, human-understandable patterns and models.

Due to the growth in the database industry and the resulting market needs for methods that are capable of extracting valuable knowledge from large data stores, *data mining* (DM) and *knowledge discovery in databases* (KDD) (Fayyad, Piatetsky-Shapiro, Smyth, & Uthurusamy, 1995; Han & Kamber, 2001; Piatetsky-Shapiro & Frawley, 1991) have recently emerged as a new scientific and engineering discipline, with separate workshops, conferences and journals. According to Witten and Frank (2005), data mining means "solving problems by analyzing data that already exists in databases". In addition to the mining of structured data stored in data warehouses—e.g., in the form of relational data tables—there has recently also been increased interest in the mining of unstructured data such as text and web.

Research areas related to machine learning and data mining include database technology and data warehouses, pattern recognition and soft computing, text and web mining, visualization, and statistics.

- *Database technology and data warehouses* are concerned with the efficient storage, access and manipulation of data.
- *Pattern recognition and soft computing* typically provide techniques for classifying data items.
- *Text and web mining* are used for web page analysis, text categorization, as well as filtering and structuring of text documents; natural language processing can provide useful tools for improving the quality of text mining results.
- *Visualization* concerns the visualization of data as well as the visualization of data mining results.
- *Statistics* is a classical data analysis discipline, mainly concerned with the analysis of large collections of numerical data.

As statistics already provides numerous data analysis tools (Breiman, 2001b; Friedman, 1998), a relevant question is whether machine learning and data mining are needed at all. There are several possible answers. First, as industry needs solutions for real-life problems, one of the most important issues is the problem solving speed: many data mining methods are able to deal with very large datasets in a very efficient way, while the algorithmic complexity of statistical methods may turn out to be prohibitive for their use on very large databases. Next, the results of the analysis need to be represented in an appropriate, usually human understandable way; apart from the analytical languages used in statistics, data mining methods also use other forms of formalisms, the most popular being decision trees and rule sets. The next important issue in a real-life setting concerns the assumptions about the data. In general one may claim that data mining deals with all sorts of structured tabular data (e.g., non-numeric, highly unbalanced, unclean data) as well as with non-structured data (e.g., text documents, images, multimedia), and does not make assumptions about the distribution of the data. Finally, while one of the main goals of statistics is hypothesis testing, one of the main goals of data mining is the construction of hypotheses.

1.2 Historical Background

Machine learning is a well-established research area of computer science. Early machine learning algorithms were perceptrons (later called neural networks, Rumelhart & McClelland, 1986), decision tree learners like ID3 (Quinlan, 1979, 1986) and CART (Breiman et al., 1984), and rule learners like AQ (Michalski, 1969; Michalski et al., 1986) and INDUCE (Michalski, 1980). These early algorithms were typically used to induce classifiers from a relatively small set of training examples (up to a thousand) described by a small set of attributes (up to a 100). An overview of early work in machine learning can be found in (Michalski, Carbonell, & Mitchell, 1983, 1986).

Data mining and knowledge discovery in databases appeared as a recognizable research discipline in the early 1990s (Piatetsky-Shapiro & Frawley, 1991), with the advent of a series of data mining workshops. The birth of this area was triggered by a need in the database industry to deliver solutions enhancing the traditional database management systems and technologies. At that time, these systems were able to solve the basic data management issues like how to deal with the data in transactional processing systems. In online transactional processing (OLTP) most of the processing scenarios were predefined. The main emphasis was on the stability and safety of solutions.

As the business emphasis changed from automation to decision support, limitations of OLTP systems in business support led to the development of the next-generation data management technology known as data warehousing. The motivation for data warehousing was to provide tools for supporting analytical operations for decision support that were not easily provided by the existing database query languages. Online analytical processing (OLAP) was introduced to enable inexpensive data access and insights which did not need to be defined in advance. However, the typical operations on data warehouses were similar to the ones from the traditional OLTP databases in that the user issued a query and received a data table as a result. The major difference between OLTP and OLAP is the average number of records accessed per typical operation. While a typical operation in OLTP affects only up to tens or hundreds of records in predefined scenarios, a typical operation in OLAP affects up to millions of records (sometimes all records) in the database in a non-predefined way.

The role of data mining in the above framework can be explained as follows. While typical questions in OLTP and OLAP are of the form: '*What is the answer to the given query?*', data mining—in a somewhat simplified and provocative formulation—addresses the question '*What is the right question to ask about this data?*'. The following explanation can be given. Data warehousing/OLAP provides analytical tools enabling only user-guided analysis of the data, where the user needs to have enough advance knowledge about the data to be able to raise the right questions in order to get the appropriate answers. The problem arises in situations when the data is too complex to be appropriately understood and analyzed by a human. In such cases data mining can be used, providing completely different types of operations for handling the data, aimed at hypothesis construction, and providing answers to questions which—in most cases—cannot be formulated precisely.

1.3 Knowledge Discovery Process and Standardization

Data mining is the core stage of the knowledge discovery process that is aimed at the extraction of interesting—nontrivial, implicit, previously unknown and potentially useful—information from data in large databases (Fayyad, Piatetsky-Shapiro, & Smyth, 1996). Data mining projects were initially carried out in many different ways with each data analyst finding their own way of approaching the problem, often through trial-and-error. As the data mining techniques and businesses evolved, there was a need for data analysts to better understand and standardize the knowledge discovery process, which would—as a side effect—demonstrate to prospective customers that data mining was sufficiently mature to be adopted as a key element of their business. This led to the development of the *cross-industry standard process for data mining* (CRISP-DM; Chapman et al., 2000), which is intended to be independent of the choice of data mining tools, industry segment, and the application/problem to be solved.

The CRISP-DM methodology defines the crucial steps of the knowledge discovery process. Although in most data mining projects, several iterations of individual steps or step sequences need to be performed, these basic guidelines are very useful both for the data analyst and the client. Below is a list of CRISP-DM steps.

1. *Business understanding:* understanding and defining of business goals and the actual goals of data mining.
2. *Data understanding:* familiarization with the data and the application domain, by exploring and defining the relevant prior knowledge.
3. *Data preparation through data cleaning and preprocessing*: creating the relevant data subset through data selection, as well as finding of useful properties/attributes, generating new attributes, defining appropriate attribute values and/or value discretization.
4. *Data mining*: the most important step of this process, which is concerned with choosing the most appropriate data mining tools—from the available tools for summarization, classification, regression, association, clustering—and searching for patterns or models of interest.
5. *Evaluation and interpretation of results*: aided by pattern/model visualization, transformation, and removal of redundant patterns.
6. *Deployment:* the use of the discovered knowledge.

A terminological note needs to be made at this point. While data mining is considered to be the core step of the knowledge discovery process, in this book—as with most industrial applications—we use the term data mining interchangeably with knowledge discovery.

In addition to the CRISP-DM standardized methodology for building data mining applications, standards covering specific phases of the process are also emerging. These standards include:

- The XML-based Predictive Modeling Markup Language (PMML) (Pechter, 2009) standard for storing and sharing data mining results,
- A standard extending the Microsoft analysis server with new data mining functionality (OLE DB for data mining, using a customized SQL language),
- Part of the ISO effort to define multimedia and application-specific SQL types and their methods, including support for data mining functionality (SQL/MM), and
- The emerging Java API for data mining (JDM).

The standardization efforts and numerous tools available (IBM Intelligent Miner, SAS Enterprise Miner, SPSS Clementine, and many others), including the publicly available academic data mining platforms WEKA (Hall et al., 2009; Witten & Frank, 2005), RAPID-I (formerly YALE; Mierswa, Wurst, Klinkenberg, Scholz, & Euler, 2006), the Konstanz Information Miner KNIME (Berthold et al., 2009), ORANGE (Demšar, Zupan, & Leban, 2004), and the statistical data analysis package R (Everitt & Hothorn, 2006; Torgo, 2010) demonstrate that data mining has made progress towards becoming a mature and widely used technology for analytical practices.

Most of the available tools are capable of mining data in tabular format, describing a dataset in terms of a fixed collection of attributes (properties), as is the case with transactional databases. More sophisticated tools are available for data mining from relational databases, data warehouses and stores of text documents.

Methods and tools for the mining of advanced database systems and information repositories (object-oriented and object-relational databases, spatial databases, time-series data and temporal data, multimedia data, heterogeneous and legacy data, World-Wide Web) still lack commercial deployment.

1.4 Terminology and Categorization of Learning Tasks

In the simplest case, data mining techniques operate on a single data table. Rows in the data table correspond to objects (*training examples*) to be analyzed in terms of their properties (*attributes*) and the concept (*class*) to which they belong. There are two main approaches:

Supervised learning. Supervised learning assumes that training examples are classified (labeled by class labels)

Unsupervised learning. Unsupervised learning concerns the analysis of unclassified examples.

In both cases, the goal is to induce a *model* for the entire dataset, or to discover one or more *patterns* that hold for some part of the dataset.

In supervised learning, data is usually formed from examples (records of given attribute values) which are labeled by the class to which they belong (Kotsiantis, Zaharakis, & Pintelas, 2006). The task is to find a model (a classifier) that will enable a newly encountered instance to be classified. Examples of discrete classification tasks are classification of countries based on climate, classification of cars based on gas consumption, or prediction of a diagnosis based on patient's medical condition.

Let us formulate a classification/prediction task, and illustrate it by a simplified example. As described above, we are given a database of observations described with a fixed number of attributes $\mathbf{A}_i$, and a designated class attribute $\mathbf{C}$. The learning task is to find a mapping f that is able to compute the class value $\mathbf{C} = f(\mathbf{A}_1, \ldots, \mathbf{A}_n)$ from the attribute values of new, previously unseen observations.

Table 1.1 shows a very small, artificial sample database.[1] The database contains the results of a survey on 14 individuals, concerning the approval or disapproval of a certain issue. Each individual is characterized by four attributes—Education (with possible values primary school, secondary school, or university), MaritalStatus (with possible values single, married, or divorced), Sex (male or female), and HasChildren (yes or no)—that encode rudimentary information about their sociodemographic background. The last column, Approved, is the class attribute, encoding whether the individual approved or disapproved of the issue.

The task is to use the information in this *training set* to derive a model that is able to predict whether a person is likely to approve or disapprove the issue, based on the four demographic characteristics. While there are statistical techniques that are able

[1] The dataset is adapted from the well-known dataset Quinlan (1986).

Table 1.1 A sample database

No.	Education	Marital status	Sex	Has children	Approved
1	Primary	Single	Male	No	No
2	Primary	Single	Male	Yes	No
3	Primary	Married	Male	No	Yes
4	University	Divorced	Female	No	Yes
5	University	Married	Female	Yes	Yes
6	Secondary	Single	Male	No	No
7	University	Single	Female	No	Yes
8	Secondary	Divorced	Female	No	Yes
9	Secondary	Single	Female	Yes	Yes
10	Secondary	Married	Male	Yes	Yes
11	Primary	Married	Female	No	Yes
12	Secondary	Divorced	Male	Yes	No
13	University	Divorced	Female	Yes	No
14	Secondary	Divorced	Male	No	Yes

to solve particular instances of this problem, mainly focusing on the analysis of numeric data, machine learning and data mining techniques focus on the analysis of categorical, non-numeric data, and on the interpretability of the result.

Typical data mining approaches find patterns or models in a single data table, while some, like most of the relational data mining approaches, (Džeroski & Lavrač, 2001; Lavrač & Džeroski, 1994a) find patterns/models from data stored in multiple tables, e.g., in a given relational database.

Propositional learning. Data mining approaches that find patterns/models in a given single table are referred to as *attribute-value* or *propositional* learning approaches, as the patterns/models they find can be expressed in propositional logic.

Relational learning. *First-order learning* approaches are also referred to as *relational data mining* (RDM) (Džeroski & Lavrač, 2001), *relational learning* (RL) (Quinlan, 1990) or *inductive logic programming* (ILP) (Lavrač & Džeroski, 1994a; Muggleton, 1992), as the patterns/models they find are expressed in relational formalisms of first-order logic.

We further distinguish between predictive and descriptive data mining. In the example above, a predictive data mining approach will aim at building a predictive classification model for classifying new instances into one of the two class values (yes or no). On the other hand, in descriptive data mining the input data table will typically not contain a designated class attribute and will aim at finding patterns describing the relationships between other attribute values.

Predictive data mining. Predictive data mining methods are supervised. They are used to induce models or theories (such as decision trees or rule sets) from class-labeled data. The induced models can be used for prediction and classification.

Descriptive data mining. Descriptive data mining methods are typically unsupervised. They are used to induce interesting patterns (such as association rules) from unlabeled data. The induced patterns are useful in exploratory data analysis.

While there is no clear distinction in the literature, we will generally use the term *pattern* for results of a descriptive data mining process, whereas we will use the terms *model*, *theory*, or *hypothesis* for results of a predictive data mining task.

The next two sections briefly introduce the two main learning approaches, predictive and descriptive induction.

1.5 Predictive Data Mining: Induction of Models

This data analysis task is concerned with the induction of models for classification and prediction purposes, and is referred to as *predictive induction*. Two symbolic data mining methods that result in classification/prediction models are outlined in this section: decision tree induction and rule set induction.

1.5.1 Decision Tree Induction

A *decision tree* is a classification model whose structure consists of a number of *nodes* and *arcs*. In general, a node is labeled by an attribute name, and an arc by a valid value of the attribute associated with the node from which the arc originates. The top-most node is called the *root* of the tree, and the bottom nodes are called the *leaves*. Each leaf is labeled by a class (value of the class attribute). When used for classification, a decision tree is traversed in a top-down manner, following the arcs with attribute values satisfying the instance that is to be classified. The traversal of the tree leads to a leaf node and the instance is assigned the class label of the leaf. Figure 1.1 shows a decision tree induced from the training set shown in Table 1.1.

A decision tree is constructed in a top-down manner, starting with the most general tree consisting of only the root node, and then refining it to a more specific tree structure. A small tree consisting only of the root node is *too general*, while the most specific tree which would construct a leaf node for every single data instance would be *too specific*, as it would *overfit* the data. The art of decision tree construction is to construct a tree at the right 'generality level' which will adequately generalize the training data to enable high predictive accuracy on new instances.

The crucial step in decision tree induction is the choice of an attribute to be selected as a node in a decision tree. Typical attribute selection criteria use a function that measures the *purity* of a node, i.e., the degree to which the node contains only examples of a single class. This purity measure is computed for a node and

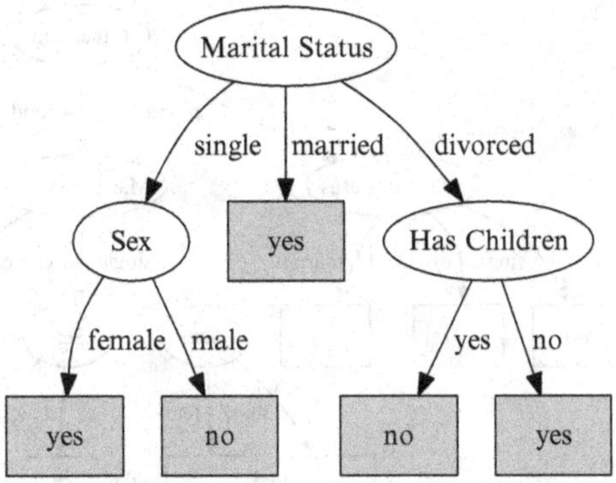

Fig. 1.1 A decision tree describing the dataset shown in Table 1.1

all successor nodes that result from using an attribute for splitting the data. In the well-known C4.5 decision tree algorithm, which uses information-theoretic entropy as a purity measure (Quinlan, 1986), the difference between the original purity value and the sum of the purity values of the successor nodes weighted by the relative sizes of these nodes, is used to estimate the utility of this attribute, and the attribute with the largest utility is selected for expanding the tree.

To see the importance of this choice, consider a procedure that constructs decision trees simply by picking the next available attribute instead of the most informative attribute. The result is a much more complex and less comprehensible tree (Fig. 1.2). Most leaves cover only a single training example, which means that this tree is overfitting the data. Consequently, the labels that are attached to the leaves are not very reliable. Although the trees in Figs. 1.1 and 1.2 both classify the training data in Table 1.1 correctly, the former appears to be more trustworthy, and it has a higher chance of correctly predicting the class values of new data.[2]

Note that some of the attributes may not occur at all in the tree; for example, the tree in Fig. 1.1 does not contain a test on Education. Apparently, the data can be classified without making a reference to this variable. In addition, the attributes in the upper parts of the tree (near the root) have a stronger influence on the value of the target variable than the nodes in the lower parts of the tree, in the sense that they participate in the classification of a larger number of instances.

As a result of the recursive partitioning of the data at each step of the top-down tree construction process, the number of examples that end up in each node decreases steadily. Consequently, the reliability of the chosen attributes decreases with increasing depths of the tree. As a result, overly complex models are generated, which explain the training data but do not generalize well to unseen data. This is known as *overfitting*. This is the main reason why the state-of-the-art decision tree learners employ a post-processing phase in which the generated tree is simplified

[2]The preference for simpler models is a heuristic criterion known as *Occam's razor*, which appears to work well in practice. It is often addressed in the literature on model selection, but its utility has been the subject of discussion (Domingos, 1999; Webb, 1996).

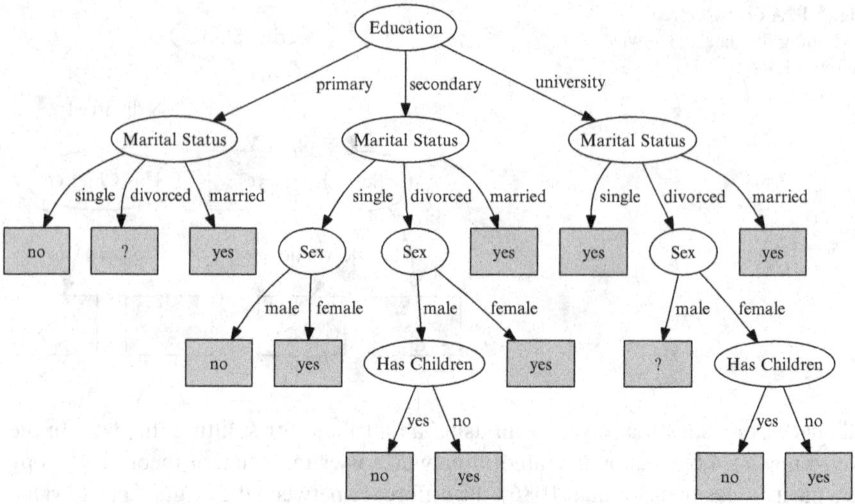

Fig. 1.2 A bad decision tree describing the dataset shown in Table 1.1

by *pruning* branches and nodes near the leaves, which results in replacing some of the interior nodes of the tree with a new leaf, thereby removing the subtree that was rooted at this node. It is important to note that the leaf nodes of the new tree are no longer pure nodes, containing only training examples of the same class labeling the leaf; instead the leaf will bear the label of the most frequent class at the leaf.

Many decision tree induction algorithms exist, the most popular being C4.5 and its variants: a commercial product SEE5, and J48, which is available in the WEKA workbench (Witten & Frank, 2005), as open source.

1.5.2 Rule Set Induction

Another important machine learning technique is the induction of rule sets. The learning of rule-based models has been a main research goal in the field of machine learning since its beginning in the early 1960s. Recently, rule-based techniques have also received increased attention in the statistical community (Friedman & Fisher, 1999).

A rule-based classification model consists of a set of if–then rules. Each *rule* has a conjunction of attribute values (which will in the following be called *features*) in the conditional part of the rule, and a class label in the rule consequent. As an alternative to such logical rules, *probabilistic rules* can be induced; in addition to the predicted class label, the consequent of these rules consists also of a list of probabilities or numbers of covered training instances for each possible class label (Clark & Boswell, 1991).

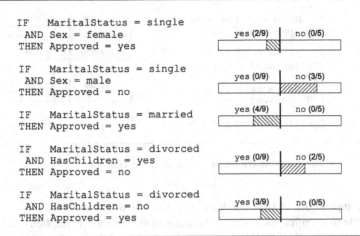

```
IF   MaritalStatus = single
 AND Sex = female
THEN Approved = yes

IF   MaritalStatus = single
 AND Sex = male
THEN Approved = no

IF   MaritalStatus = married
THEN Approved = yes

IF   MaritalStatus = divorced
 AND HasChildren = yes
THEN Approved = no

IF   MaritalStatus = divorced
 AND HasChildren = no
THEN Approved = yes
```

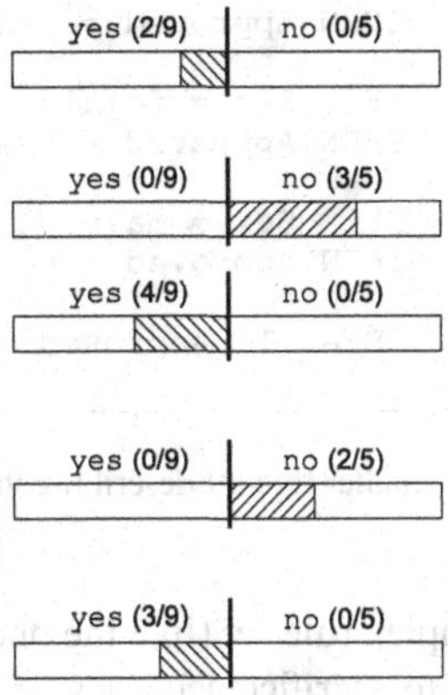

Fig. 1.3 A rule set describing the dataset shown in Table 1.1

Rule sets are typically simpler and more comprehensible than decision trees. To see why, note that a decision tree can also be interpreted as a set of if—then rules. Each leaf in the tree corresponds to one rule, where the conditions encode the path that is taken from the root to this particular leaf, and the conclusion of the rule is the label of that leaf. Figure 1.3 shows the set of rules that corresponds to the tree in Fig. 1.1. Note the rigid structure of these rules. For example, the first condition always uses the same attribute, namely, the one used at the root of the tree. Next to each rule, we show the proportion of covered examples for each class value.

The main difference between the rules generated by a decision tree and the rules generated by a rule learning algorithm is that the former rule set consists of nonoverlapping rules that span the entire instance space (i.e., each possible combination of attribute values will be covered by exactly one rule). Relaxing this constraint—by allowing for potentially overlapping rules that need not span the entire instance space—may often result in smaller rule sets. However, in this case, we need mechanisms for tie breaking (which rule to choose when more than one covers the example to be classified) and default classifications (what classification to choose when no rule covers the given example). Typically, one prefers rules with a higher ratio of correctly classified examples from the training set.

Figure 1.4 shows a particularly simple rule set which uses two different attributes in its first two rules. Note that these two rules are overlapping: several examples will be covered by more than one rule. For instance, examples 3 and 10 are covered by both the first and the third rule. These conflicts are typically resolved by using the more accurate rule, i.e., the rule that covers a higher proportion of examples that support its prediction (the first one in our case). Also note that this rule set makes two mistakes (the last two examples). These might be resolved by resorting to a

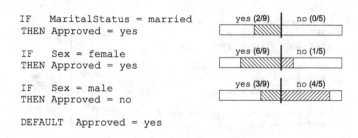

```
IF   MaritalStatus = married
THEN Approved = yes

IF   Sex = female
THEN Approved = yes

IF   Sex = male
THEN Approved = no

DEFAULT  Approved = yes
```

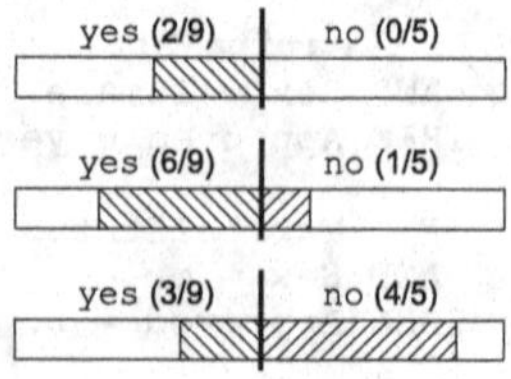

Fig. 1.4 A smaller rule set describing the dataset shown in Table 1.1

more complex rule set (like the one in Fig. 1.3) but, as stated above, it is often more advisable to sacrifice accuracy on the training set for model simplicity in order to avoid overfitting the training data. Finally, note the *default rule* at the end of the rule set. This is added for the case when certain regions of the data space are not represented in the training set.

The key ideas for learning such rule sets are quite similar to the ideas used in decision tree induction. However, instead of recursively partitioning the dataset by optimizing the purity measure over all successor nodes (in the literature, this strategy is also known as *divide-and-conquer* learning), rule learning algorithms only expand a single successor node at a time, thereby learning a complete rule that covers part of the training data. After a complete rule has been learned, all examples that are covered by this rule are removed from the training set, and the procedure is repeated with the remaining examples (this strategy is also known as *separate-and-conquer* learning). Again, pruning is a good idea for rule learning, which means that the rules only need to cover examples that are *mostly* from the same class. It turns out to be advantageous to prune rules immediately after they have been learned, that is before successive rules are learned (Fürnkranz, 1997).

1.5.3 *Rule Sets Versus Decision Trees*

There are several aspects which make rule learning attractive. First of all, decision trees are often quite complex and hard to understand. Quinlan (1993) has noted that even pruned decision trees may be too cumbersome, complex, and inscrutable to provide insight into the domain at hand and has consequently devised procedures for simplifying decision trees into pruned production rule sets (Quinlan, 1987a, 1993). Additional evidence for this comes from Rivest (1987), showing that decision lists (ordered rule sets) with at most k conditions per rule are strictly more expressive than decision trees of depth k. A similar result has been proved by Boström (1995).

Moreover, the restriction of decision tree learning algorithms to nonoverlapping rules imposes strong constraints on learnable rules. One problem resulting from this constraint is the *replicated subtree problem* (Pagallo & Haussler, 1990); it often happens that identical subtrees have to be learned at various places in a decision tree, because of the fragmentation of the example space imposed by the restriction to nonoverlapping rules. Rule learners do not make such a restriction and are thus less susceptible to this problem. An extreme example for this problem has been provided by Cendrowska (1987), who showed that the minimal decision tree for the concept x defined as

```
IF A = 3 AND B = 3 THEN Class = x
IF C = 3 AND D = 3 THEN Class = x
```

has 10 interior nodes and 21 leafs assuming that each attribute A ...D can be instantiated with three different values.

Finally, propositional rule learning algorithms extend naturally to the framework of *inductive logic programming* framework, where the goal is basically the induction of a rule set in first-order logic, e.g., in the form of a Prolog program.[3] First-order background knowledge can also be used for decision tree induction (Blockeel & De Raedt, 1998; Kramer, 1996; Lavrač, Džeroski, & Grobelnik, 1991; Watanabe & Rendell, 1991), but once more, Watanabe and Rendell (1991) have noted that first-order decision trees are usually more complex than first-order rules.

1.6 Descriptive Data Mining: Induction of Patterns

While a decision tree and a set of rules represent a model (a theory) that can be used for classification and/or prediction, the goal of data analysis may be different. Instead of model construction, the goal may be the discovery of individual patterns/rules describing regularities in the data. This form of data analysis is referred to as *descriptive induction* and is frequently used in exploratory data analysis.

As opposed to decision tree and rule set induction, which result in classification models, *association rule learning* is an unsupervised learning method, with no class labels assigned to the examples. Another method for unsupervised learning is *clustering*, while *subgroup discovery*—aimed at finding descriptions of interesting population subgroups—is a descriptive induction method for pattern learning, but is at the same time a form of supervised learning due to a defined property of interest acting as a class.

[3]Prolog is a programming language, enabling knowledge representation in first-order logic (Lloyd, 1987; Sterling & Shapiro, 1994). We will briefly return to learning in first-order logic in Sect. 1.7; a systematic treatment of relational rule learning can be found in Chap. 5.

```
IF    Education = university
THEN Sex = female

IF    Approved = no
THEN Sex = male

IF    Education = secondary
 AND MaritalStatus = divorced
THEN HasChildren = no
 AND Approved = yes
```

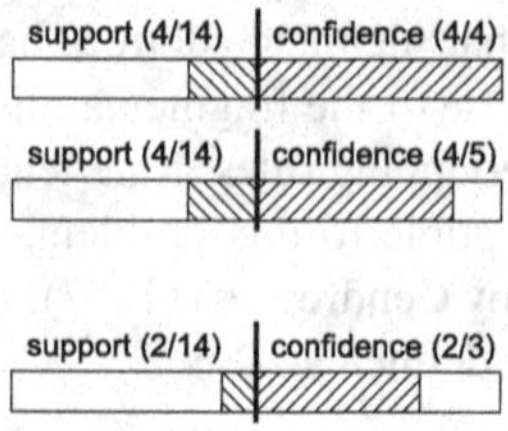

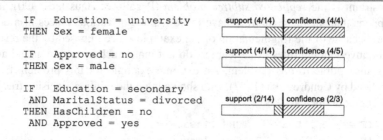

Fig. 1.5 Three rules induced by an association rule learning algorithm

1.6.1 Association Rule Learning

The problem of inducing *association rules* (Agrawal, Mannila, Srikant, Toivonen, & Verkamo, 1995) has received much attention in the data mining community. It is defined as follows: given a set of transactions (examples), where each transaction is a set of items, an association rule is an expression of the form B $\rightarrow$ H, where B and H are sets of items, and B $\rightarrow$ H is interpreted as IF B THEN H, meaning that the transactions in a database which contain B tend to contain H as well.

Figure 1.5 shows three examples for association rules that could be discovered in the dataset of Table 1.1. The first rule states that in this dataset, all people with a university education were female. This rule is based on four observations in the dataset. The fraction of entries in the database that satisfy all conditions (both in body and head) is known as the *support* of the rule. Thus, the support of the rule is the ratio of the number of records having true values for all items in B and H to the number of all records in the database. As 4 of a total of 14 persons are both female and have university education, the support of the first rule is $4/14 \approx 0.286$.

The second rule also has a support of $4/14$, because four people in the database do not approve and are male. However, in this case, the strength of the rule is not as strong as in the previous case, because only $4/5 = 0.8$ of all persons that do not approve were actually male. This value is called the *confidence* of the rule. It is calculated as the ratio of the number of records having true values for all items in B and H to the number of records having true values for all items in B.

Unlike with classification rules, the head of an association rule may also contain a conjunction of conditions. This is illustrated by the third rule, which states that divorced people with secondary education typically have no children and approve.

In all rules there is no distinction between the class attribute and all other attributes: the class attribute may appear on any side of the rule or not at all. In fact, typically association rules are learned from databases with binary features (called *items*) without any dedicated class attribute. Thus association rule discovery is an unsupervised learning task. Most algorithms, such as the well-known APRIORI

```
IF   MaritalStatus = single
 AND Sex = male
THEN Approved = no

IF   MaritalStatus = single
THEN Approved = no

IF   Sex = female
THEN Approved = yes
```

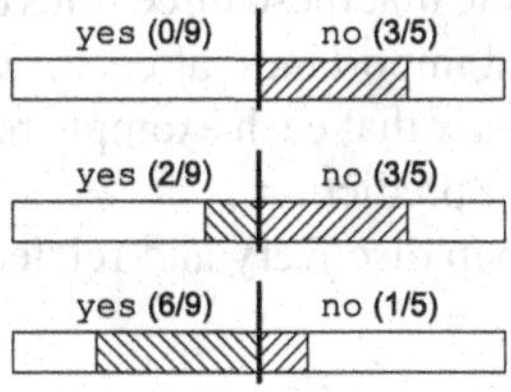

Fig. 1.6 Three subgroup descriptions induced by a subgroup discovery algorithm

algorithm (Agrawal et al., 1995), find all association rules that satisfy minimum support and minimum confidence constraints.

An in-depth survey of association rule discovery is beyond the scope of this book, and, indeed, the subject has already been covered in other monographs (Adamo, 2000; Zhang & Zhang, 2002). We will occasionally touch upon the topic when it seems appropriate (e.g., the level-wise search algorithm, which forms the basis of APRIORI and related techniques, is briefly explained in Sect. 6.3.2), but for a systematic treatment of the subject we refer the reader to the literature.

1.6.2 Subgroup Discovery

In subgroup discovery the task is to find sufficiently large population subgroups that have a significantly different class distribution than the entire population (the entire dataset). Subgroup discovery results in individual rules, where the rule conclusion is a class (the property of interest). The main difference between learning of classification rules and subgroup discovery is that the latter induces single rules (subgroups) of interest, which aim at revealing interesting properties of groups of instances, not necessarily at forming a rule set used for classification.

Figure 1.6 shows three subgroup descriptions that have been induced with the MAGNUM OPUS descriptive rule learning system (Webb, 1995).[4] While the first and third rules could also be found by classification rule algorithms (cf. Fig. 1.4), the second rule would certainly not be found because it has a comparably low predictive quality. There are almost as many single persons that approve than there are singles that do not approve. Nevertheless, this rule can be considered to be an interesting subgroup because the class distribution of covered instances (2 yes and 3 no) is significantly different than the distribution in the entire dataset (9 yes and 5 no). Conversely, a classification rule algorithm would not find the first rule because if we accept the second rule for classification, adding the first one does not improve

[4]The rules are taken from Kralj Novak, Lavrač, and Webb (2009).

classification performance, i.e., it is *redundant* with respect to the second rule. Finally, note that these three rules do not cover all the examples. While it is typically considered important that each rule covers a significant number of examples, it is not necessary that each example be covered by some rule, because the rules will not be used for prediction.

Subgroup discovery and related techniques are covered in depth in Chap. 11 of this book.

1.7 Relational Data Mining

Both predictive and descriptive data mining are usually performed on a single database relation, consisting of examples represented with values for a fixed number of attributes. However, in practice, the data miner often has to face more complex scenarios. Suppose that data is stored in several tables, e.g., it has a *relational database* form. In this case the data has to be transformed into a single table in order to be able to use standard data mining techniques. The most common data transformation approach is to select one table as the main table to be used for learning, and try to incorporate the contents of other tables by aggregating the information contained in the tables into summary attributes, which are added to the main table. The problem with such transformations is that some information may be lost while the aggregation may also introduce artifacts, possibly leading to inappropriate data mining results. What one would like to do is to leave data conceptually unchanged and rather use data mining tools that can deal with multirelational data.

Integrating data from multiple tables through joins or aggregation can cause loss of meaning or information. Suppose we are given two relations: `customer(CID,Name,Age,SpendALot)` encodes the ID, name, and age of a customer, and the information whether this customer spends a lot, and `purchase(CID,ProdID,Date,Value,PaymentMode)` encodes a single purchase by a customer with a given ID. Each customer can make multiple purchases, and we are interested in characterizing customers that spend a lot. Integrating the two relations via a natural join will result in a relation `purchase1(CID,Name,Age,SpendALot,ProdID,Date,Value, PaymentMode)`. However, this is problematic because now each row corresponds to a purchase and not to a customer, and we intend to analyze our information with respect to customers. An alternative would be to aggregate the information contained in the `purchase` relation. One possible aggregation could be the relation `customer1(CID,Name,Age,NofPurchases, TotalValue,SpendALot)`, which aggregates the number of purchases and their total value into new attributes. Naturally, some information has been lost during the aggregation process.

The following pattern can be discovered by a relational rule learning system if the relations `customer` and `purchase` are considered together.

```
customer(CID,Name,Age,yes) :-
        Age > 30,
        purchase(CID,PID,D,Value,PM),
        PM = creditcard,
        Value > 100.
```

This pattern, written in a Prolog-like syntax, says: *'a customer spends a lot if she is older than 30, has purchased a product of value more than 100 and paid for it by credit card.'* It would not be possible to induce such a pattern from either of the relations `purchase1` and `customer1` considered on their own.

We will return to relational learning in Chap. 5, where we take a feature-based view on the problem.

1.8 Conclusion

This chapter briefly described several aspects of machine learning and data mining, aiming to provide the background and basic understanding of the topics presented in this book. To conclude, let us make some speculations about future developments in data mining.

With regard to data mining research, every year the research community addresses new open problems and new problem areas, for many of which data mining is able to provide value-added answers and results. Because of the interdisciplinary nature of data mining, there is a big inflow of new knowledge, widening the spectrum of problems that can be solved by the use of this technology. Another reason why data mining has a scientific and commercial future was given by Friedman (1998): "Every time the amount of data increases by a factor of 10, we should totally rethink how we analyze it."

To achieve its full commercial exploitation, data mining is still lacking the standardization to the degree of, for example, the standardization available for database systems. There are initiatives in this direction, which will diminish the monopoly of expensive closed-architecture systems. For data mining to be truly successful it is important that data mining tools become available in major database products as well as in standard desktop applications (e.g., spreadsheets). Other important recent developments are open source data mining services, tools for online construction of data mining workflows, as well as the terminology and ingredients of data mining through the development of a data mining ontology (Lavrač, Kok, de Bruin, & Podpečan, 2008; Lavrač, Podpečan, Kok, & de Bruin, 2009).

In the future, we envisage intensive development and increased usage of data mining in specific domain areas, such as bioinformatics, multimedia, text and web data analysis. On the other hand, as data mining can be used for building surveillance systems, recent research also concentrates on developing algorithms for mining databases without compromising sensitive information (Agrawal & Srikant, 2000). A shift towards automated use of data mining in practical systems is also expected to become very common.

Chapter 2
Rule Learning in a Nutshell

This chapter gives a brief overview of inductive rule learning and may therefore serve as a guide through the rest of the book. Later chapters will expand upon the material presented here and discuss advanced approaches, whereas this chapter only presents the core concepts. The chapter describes search heuristics and rule quality criteria, the basic covering algorithm, illustrates classification rule learning on simple propositional learning problems, shows how to use the learned rules for classifying new instances, and introduces the basic evaluation criteria and methodology for rule-set evaluation.

After defining the learning task in Sect. 2.1, we start with discussing data (Sect. 2.2) and rule representation (Sect. 2.3) for the standard propositional rule learning framework, in which training examples are represented in a single table, and the outputs are if–then rules. Section 2.4 outlines the rule construction process, followed by a more detailed description of its parts: the induction of individual rules is presented as a search problem in Sect. 2.5, and the learning of rule sets in Sect. 2.6. One of the classical rule learning algorithms, CN2, is described in more detail in Sect. 2.7. Section 2.8 shows how to use the induced rule sets for the classification of new instances, and the subsequent Sect. 2.9 discusses evaluation of the classification quality of the induced rule sets and presents cross-validation as a means for evaluating the predictive accuracy of rules. Finally, Sect. 2.10 gives a brief historical account of some influential rule learning systems.

[†]This chapter is partly based on (Flach & Lavrač, 2003).

J. Fürnkranz et al., *Foundations of Rule Learning*, Cognitive Technologies,
DOI 10.1007/978-3-540-75197-7_2, © Springer-Verlag Berlin Heidelberg 2012

Given:

- a *data description language*, defining the form of data,
- a *hypothesis description language*, defining the form of rules,
- a *coverage function* COVERED(r, e), defining whether rule r covers example e,
- a *class* attribute **C**, and
- a set of *training examples* $\mathcal{E}$, instances for which the class labels are known, described in the data description language.

Find:

a *hypothesis* in the form of a rule set $\mathcal{R}$ formulated in the hypothesis description language which is

- *complete*, i.e., it covers all the examples, and
- *consistent*, i.e., it predicts the correct class for all the examples.

Fig. 2.1 Definition of the classification rule learning task

2.1 Problem Definition

Informally, we can define the problem of learning classification rules as follows:

> Given a set of training examples, find a set of classification rules that can be used for prediction or classification of new instances.

Note that we distinguish between the terms *examples* and *instances*. Both are usually described by *attribute values*. Examples refer to instances labeled by a *class label*, whereas instances themselves bear no class label. An instance is *covered* by a rule if its description satisfies the rule conditions, and it is *not covered* if its description does not satisfy the rule conditions. An example is *correctly covered* by the rule if it is covered and the class of the rule equals the class label of the example, or *incorrectly covered* if its description satisfies the rule conditions, but the class label of the rule is not equal to the class label of the example.

The above informal definition leaves out several details. A more formal definition is shown in Fig. 2.1. It includes important additional preliminaries for the learning task, such as the representation formalism used for describing the data (*data description language*) and for describing the induced set of rules (*hypothesis description language*). We use the term *hypothesis* to denote the output of learning because of the hypothetical nature of induction, which can never guarantee that the output of inductive learning will not be falsified by new evidence presented to the learner. However, we will also often use the terms *model* or *theory* as synonyms for hypothesis. Finally, we also need a *coverage function*, which connects the hypothesis description with the data description. The restrictions imposed by the languages defining the format and scope of data and knowledge representation are also referred to as the *language bias* of the learning problem.

Note that the definition of the classification rule learning task of Fig. 2.1 describes an idealistic scenario with no errors in the data where a complete and consistent

Given:

- a *data description language*, imposing a bias on the form of data,
- a *target concept*, typically denoted with $\oplus$,
- a *hypothesis description language*, imposing a bias on the form of rules,
- a *coverage function* COVERED($\mathbf{r}, e$) defining whether rule $\mathbf{r}$ covers example e,
- a set of *positive examples* $\mathcal{P}$, instances for which it is known that they belong to the target concept
- a set of *negative examples* $\mathcal{N}$, instances for which it is known that they do not belong to the target concept

Find:

a *hypothesis* as a set of rules $\mathcal{R}$ described in the hypothesis description language, providing the definition of the target concept which is

- *complete*, i.e., it covers all examples that belong to the concept, and
- *consistent*, i.e., it does not cover any example that does not belong to the concept.

Fig. 2.2 Definition of the concept learning task

hypothesis can be induced. However, in realistic situations, completeness and consistency have to be replaced with less strict criteria for measuring the quality of the induced rule set.

Propositional rules. This chapter focuses on *propositional rule induction* or *attribute-value rule learning*. Representatives of this class of learners are CN2 (Clark & Boswell, 1991; Clark & Niblett, 1989) and RIPPER (Cohen, 1995). An example of rule learning from the statistics literature is PRIM (Friedman & Fisher, 1999). In this language, a classification rule is an expression of the form:

IF *Conditions* THEN *c*

where *c* is the *class label*, and the *Conditions* are a conjunction of simple logical tests describing the properties of instances that have to be satisfied for the rule to 'fire'. Thus, a rule essentially corresponds to an implication *Conditions* $\rightarrow c$ in propositional logic, which we will typically write in the opposite direction of the implication sign ($c \leftarrow$ *Conditions*).

Concept learning. Most rule learning algorithms assume a *concept learning* task, a special case of the classification learning problem, shown in Fig. 2.2. Here the task is to learn a set of rules that describe a single *target class c* (often denoted as $\oplus$), also called the *target concept*. As training information, we are given a set of *positive* examples, for which we know that they belong to the target concept, and a set of *negative* examples, for which we know that they do *not* belong to the concept. In this case, it is typically sufficient to learn a theory for the target class only. All instances that are not covered by any of the learned rules will be classified as negative. Thus, a *complete* hypothesis is one that covers all positive examples, and

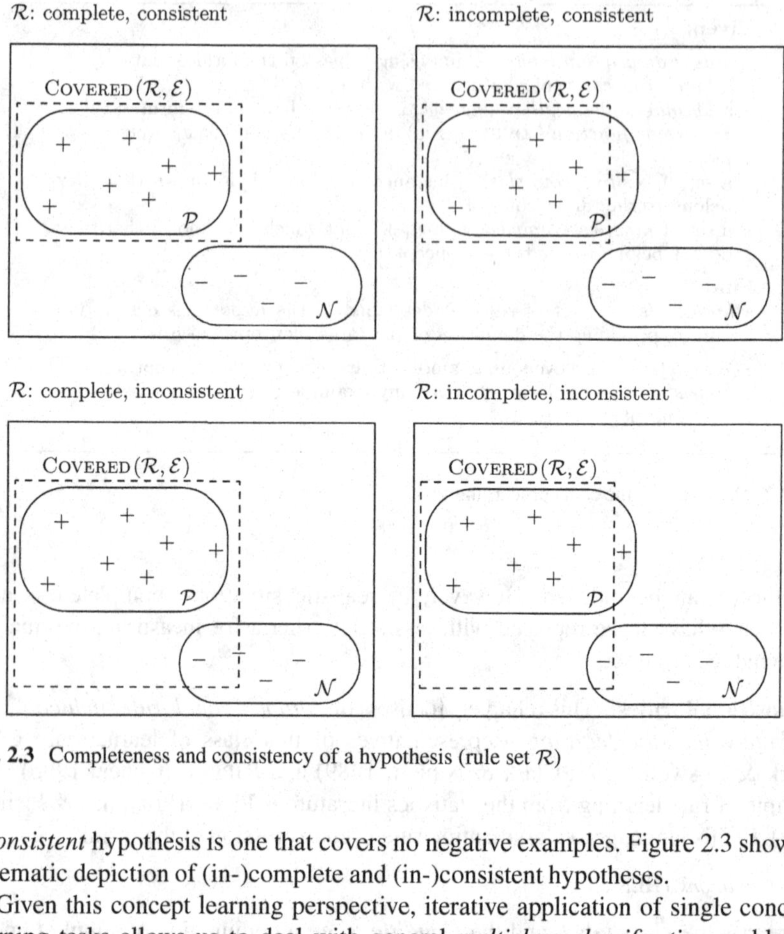

Fig. 2.3 Completeness and consistency of a hypothesis (rule set $\mathcal{R}$)

a *consistent* hypothesis is one that covers no negative examples. Figure 2.3 shows a schematic depiction of (in-)complete and (in-)consistent hypotheses.

Given this concept learning perspective, iterative application of single concept learning tasks allows us to deal with general *multiclass classification* problems. Suppose that training instances are labeled with three class labels: c_1, c_2, and c_3. The above definition of the learning task can be applied if we form three different learning tasks. In the first task, instances labeled with class c_1 are treated as the positive examples, and instances labeled c_2 and c_3 are the negative examples. In the next run, class c_2 will be considered as the positive class, and finally, in the third run, rules for class c_3 will be learned. Due to this simple transformation of a multiclass learning problem into a number of concept learning tasks, concept learning is a central topic of inductive rule learning. This type of transformation of multiclass problems to two-class concept learning problems is also known as *one-against-all* class binarization. Alternative ways for handling multiple classes are discussed in Chap. 10.

Overfitting. Generally speaking, consistency and completeness—as required in the task definition of Fig. 2.1—are very strict conditions. They are unrealistic in learning from large, *noisy* datasets, which contain random errors in the data, either due to incorrect class labels or errors in instance descriptions. Learning a complete and consistent hypothesis is undesirable in the presence of noise, because the hypothesis will try to explain the errors as well. This is known as *overfitting* the data.

It is also possible that the data description language or the hypothesis description language are not expressive enough to allow a complete and consistent hypothesis, in which case the target class needs to be approximated. Another complication is caused by target classes that are not strictly disjoint. To deal with these cases, the consistency and completeness requirements need to be relaxed and replaced with some other evaluation criteria, such as sufficient *coverage* of positive examples, high *predictive accuracy* of the hypothesis or its *significance* above the requested, predefined threshold. These measures can be used both as heuristics to guide rule construction and as measures to evaluate the quality of induced hypotheses. Some of these measures and related issues will be discussed in more detail in Sect. 2.7 and, subsequently, in Chaps. 7 and 9.

Background knowledge. The above definition of the learning task assumes that the learner has no prior knowledge about the problem and that it learns exclusively from training examples. However, difficult learning problems typically require a substantial body of prior knowledge. We refer to declarative prior knowledge as *background knowledge*. Using background knowledge, the learner may express the induced hypotheses in a more natural and concise manner. In this chapter we mostly disregard background knowledge, except in the process of constructing features (attribute values) used as ingredients in forming rule conditions. However, background knowledge plays a crucial role in relational rule learning, addressed in Chap. 5.

2.2 Data Representation

In classification tasks as defined in Fig. 2.1, the input to a classification rule learner consists of a set of training examples, i.e., instances with known class labels. Typically, these instances are described in a so-called attribute-value representation: An *instance* description has the form $(v_{1,j}, \ldots, v_{n,j})$, where each $v_{i,j}$ is the value of *attribute* $\mathbf{A}_i$, $i \in \{1, \ldots, A\}$. An attribute can either have a finite set of values (*discrete*) or take real numbers as values (*continuous* or *numerical*). An *example* e_j is a vector of attribute values labeled by a class label $e_j = (v_{1,j}, \ldots, v_{n,j}, c_j)$, where each $v_{i,j}$ is a value of attribute $\mathbf{A}_i$, and $c_j \in \{c_1, \ldots, c_C\}$ is one of the C possible values of class attribute $\mathbf{C}$. The class attribute is also often called the *target attribute*. A *dataset* is a set of examples. We will normally organize a dataset in tabular form, with columns for the attributes and rows or *tuples* for the examples.

As an example, consider the dataset in Table 2.1.[1] Like the dataset of Table 1.1, it characterizes a number of individuals by four attributes: EducationMaritalStatus,

[1]The dataset is adapted from the well-known contact lenses dataset (Cendrowska, 1987; Witten & Frank, 2005).

Table 2.1 A sample three-class dataset

No.	Education	Marital status	Sex	Has children	Car
1	Primary	Married	Female	No	Mini
2	Primary	Married	Male	No	Sports
3	Primary	Married	Female	Yes	Mini
4	Primary	Married	Male	Yes	Family
5	Primary	Single	Female	No	Mini
6	Primary	Single	Male	No	Sports
7	Secondary	Married	Female	No	Mini
8	Secondary	Married	Male	No	Sports
9	Secondary	Married	Male	Yes	Family
10	Secondary	Single	Female	No	Mini
11	Secondary	Single	Female	Yes	Mini
12	Secondary	Single	Male	Yes	Mini
13	University	Married	Male	No	Mini
14	University	Married	Female	Yes	Mini
15	University	Single	Female	No	Mini
16	University	Single	Male	No	Sports
17	University	Single	Female	Yes	Mini
18	University	Single	Male	Yes	Mini

Sex, and HasChildren. However, the target value is now not a binary decision (whether a certain issue is approved or not), but a three-valued attribute, which encodes what car the person is driving. For ease of reference, we have numbered the examples from 1 to 18.

The reader may notice that the set of examples is *incomplete* in the sense that not all possible combinations of attribute values are present. This situation is typical for real-world applications where the training set consists only of a small fraction of all possible examples. The task of a rule learner is to learn a rule set that serves a twofold purpose:

1. The learned rule set should help to uncover the hidden relationship between the input attributes and the class value, and
2. it should generalize this relationship to new, previously unseen examples.

Table 2.2 shows the remaining six examples in this domain, for which we do not know their classification during training, indicated by question marks in the last column. However, the class labels can, in principle, be determined, and their values are shown in square brackets. If these classifications are known, such a dataset is also known as a *test set*, if its purpose is to evaluate the predictive quality of the learned theory, or a *validation set*, if its purpose is to provide an internal evaluation that the learning algorithm may use to improve its performance.

In the following, we will use the examples from Table 2.1 as the training set, and the examples of Table 2.2 as the test set of a rule learning algorithm.

Table 2.2 A test set for the database of Table 2.1

No.	Education	Marital status	Sex	Has children	Car
19	Primary	Single	Female	Yes	? [mini]
20	Primary	Single	Male	Yes	? [family]
21	Secondary	Married	Female	Yes	? [mini]
22	Secondary	Single	Male	No	? [sports]
23	University	Married	Male	Yes	? [family]
24	University	Married	Female	No	? [mini]

2.3 Rule Representation

Given a set of preclassified objects (called *examples*), usually described by attribute values, a rule learning system constructs one or more rules of the form:

$$\texttt{IF } \mathbf{f}_1 \texttt{ AND } \mathbf{f}_2 \texttt{ AND} \ldots \texttt{AND } \mathbf{f}_L \texttt{ THEN } Class = c_i$$

The condition part of the rule is a logical conjunction of features (also called *conditions*), where a *feature* $\mathbf{f}_k$ is a test that checks whether the example to classify has the specified property or not. The number L of such features (or conditions) is called the *rule length*.

In the attribute-value framework that we sketched in the previous section, features $\mathbf{f}_k$ typically have the form $\mathbf{A}_i = v_{i,j}$ for discrete attributes, and $\mathbf{A}_i < v$ or $\mathbf{A}_i \geq v$ for continuous attributes (here, v is a threshold value that does not need to correspond to a value of the attribute observed in examples). The conclusion of the rule contains a class value c_i. In essence, this means that for all examples that satisfy the body of the rule, the rule predicts the class value c_i.

The condition part of a rule $\mathbf{r}$ is also known as the *antecedent* or the *body* (B) of the rule, and the conclusion is also known as the *consequent* or the *head* (H) of the rule. The terms 'head' and 'body' have their origins in common notation in clausal logic, where an implication is denoted as B → H, or equivalently, H ← B, of the form

$$c_i \leftarrow \mathbf{f}_1 \wedge \mathbf{f}_2 \wedge \ldots \wedge \mathbf{f}_L$$

We will also frequently use this formal syntax, as well as the equivalent Prolog-like syntax

```
ci :- f1, f2, ..., fL.
```

In logical terminology, the body consists of a conjunction of *literals*, and the head is a single literal. Such rules are also known as *determinate clauses*. General clause may have a disjunction of literals in the head. More on the logical foundations can be found in Chap. 5.

An example set of rules that could have been induced in our sample domain is shown in Fig. 2.4a. The numbers between square brackets indicate the number of covered examples from each class. All the rules, except for the second, cover only examples from a single class, i.e., these rules are *consistent*. On the other hand, the second rule is inconsistent because it misclassifies one training example (#13). Note that the fourth and fifth rule would each misclassify one example from the test set (#20 and #23), but this is not known to the learner. The first rule is *complete* with regard to the class family (covers all the examples of this class), the second is complete with regard to the class sports. Again, this only refers to the training examples that are known to the learner, the first rule would not be complete for class family with respect to the entire domain because it does not cover example #20 of the test set.

Collectively, the rules classify all the training examples, i.e., the learned theory is complete for the given training set (and, in fact, for the entire domain). The theory is not consistent, because it misclassifies one training example. However, we will see later that this is not necessarily bad due to a phenomenon called *overfitting* (cf. Sect. 2.7).

Also note that the counts for the class mini add up to 16 examples, while there are only 12 examples from this class. Thus, some examples must be covered by more than one rule. This is possible, because the rules are *overlapping*. For example, example 13 is covered by the second and by the fifth rule. As both rules make contradicting predictions, there must be a procedure for determining the final prediction (cf. Sect. 2.8).

This is not the case for the *decision list*, shown in Fig. 2.4b. Here the rules are tried from top to bottom, and the first rule that fires is used to assign the class label to the instance to be classified. Thus, the class counts of each rule only show the examples that are not covered by previous rules. Moreover, the rule set ends in a *default rule* that will be used for class assignment when none of the previous rules fire.

The numbers that show the class distribution of the examples covered by a rule are not necessary. If desired, we can simply ignore them and interpret the rule categorically. However, the rules also give an indication about the reliability of a rule. Generally speaking, the more biased the distribution is towards a single class, and the more examples are covered by the rule, the more reliable is the rule. For example, intuitively the third rule in Fig. 2.4a is more reliable than the second rule, because it covers more examples, and it also covers only examples of a single class. Rules one, four, and five are also consistent, but they cover fewer examples. Indeed, it turns out that rules four and five misclassify examples in the test set. This intuitive understanding of rule reliability will be formalized in Sect. 2.5.3, where it is used for choosing among a set of candidate rules.

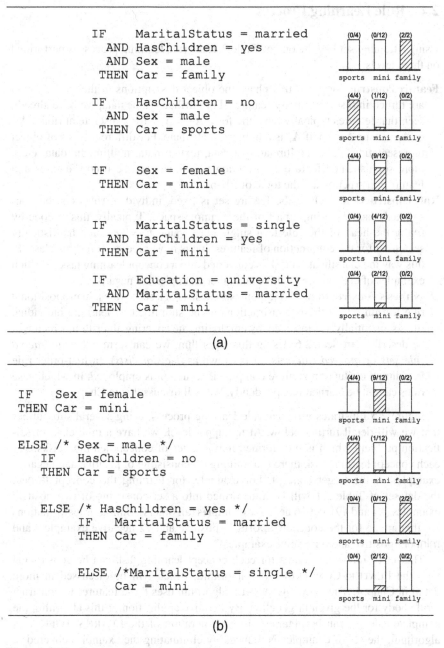

```
IF    MaritalStatus = married
  AND HasChildren = yes
  AND Sex = male
 THEN Car = family

IF    HasChildren = no
  AND Sex = male
 THEN Car = sports

IF    Sex = female
 THEN Car = mini

IF    MaritalStatus = single
  AND HasChildren = yes
 THEN Car = mini

IF    Education = university
  AND MaritalStatus = married
THEN  Car = mini
```

(a)

```
IF   Sex = female
THEN Car = mini

ELSE /* Sex = male */
   IF   HasChildren = no
   THEN Car = sports

   ELSE /* HasChildren = yes */
        IF   MaritalStatus = married
        THEN Car = family

        ELSE /*MaritalStatus = single */
             Car = mini
```

(b)

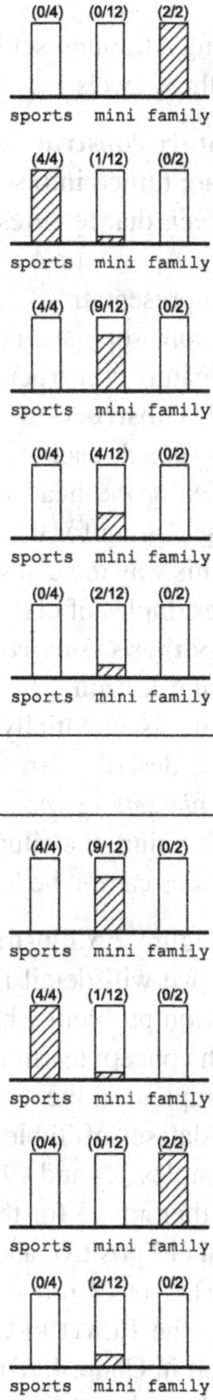

Fig. 2.4 Different types of rule-based theories induced from the car dataset. **(a)** Rule set. **(b)** Decision list

2.4 Rule Learning Process

Using a training set like the one of Table 2.1, the rule learning process is performed on three levels:

Feature construction. In this phase the object descriptions in the training data are turned into sets of binary features. For attribute-value data, we have already seen that features typically have the form $\mathbf{A}_i = v_{i,j}$ for a discrete attribute $\mathbf{A}_i$, or $\mathbf{A}_i < v$ or $\mathbf{A}_i \geq v$ if $\mathbf{A}_i$ is a numerical attribute. For different types of object representations (e.g., multirelational data, textual data, multimedia data, etc.), more sophisticated feature construction techniques can be used. Features and feature construction are the topic of Chap. 4.

Rule construction. Once the feature set is fixed, individual rules can be constructed, each covering a part of the example space. Typically, this is done by fixing the head of the rule to a single class value $\mathbf{C} = c_j$, and heuristically searching for the conjunction of features that is most predictive for this class. In this way the classification task is converted into a concept learning task in which examples of class c_i are positive and other examples are negative.

Hypothesis construction. A hypothesis consists of a set of rules. In propositional rule learning, hypothesis construction can be simplified by learning individual rules sequentially, for instance, by employing the covering algorithm, which will be described in Sect. 2.6. Using this algorithm, we can form either *unordered rule sets* or *ordered rule sets* (also known as *decision lists*). In first-order rule learning, the situation is more complex if recursion is employed, in which case rules cannot be learned independently. We will discuss this in Chap. 5.

Figure 2.5 illustrates a typical rule learning process, using several subroutines that we will detail further below. At the upper level, we have a multiclass classification problem which is transformed into a series of concept learning tasks. For each concept learning task there is a training set consisting of positive and negative examples of the target concept. For example, for learning the concept family, the dataset of Table 2.1 will be transformed into a set consisting of two positive examples (#4 and #9) and 16 negative examples (all others). Similar transformations are then made for the concepts sports (4 positive and 12 negative examples) and mini (12 positive and 6 negative examples).

The set of relevant features for each concept learning task can be constructed with the FEATURECONSTRUCTION algorithm, which will be discussed in more detail in Chap. 4. The LEARNONERULE algorithm uses these features to construct a rule body for the given target class. By iterative application of this algorithm the complete rule set can be obtained. In each iteration of the LEARNSETOFRULES algorithm, the set of examples is reduced by eliminating the examples covered in the previous iteration. When all positive examples have been covered, or some other stopping criterion is satisfied, the concept learning task is completed. The set of rules describing the target class is returned to the LEARNRULEBASE algorithm and included into the set of rules for classification.

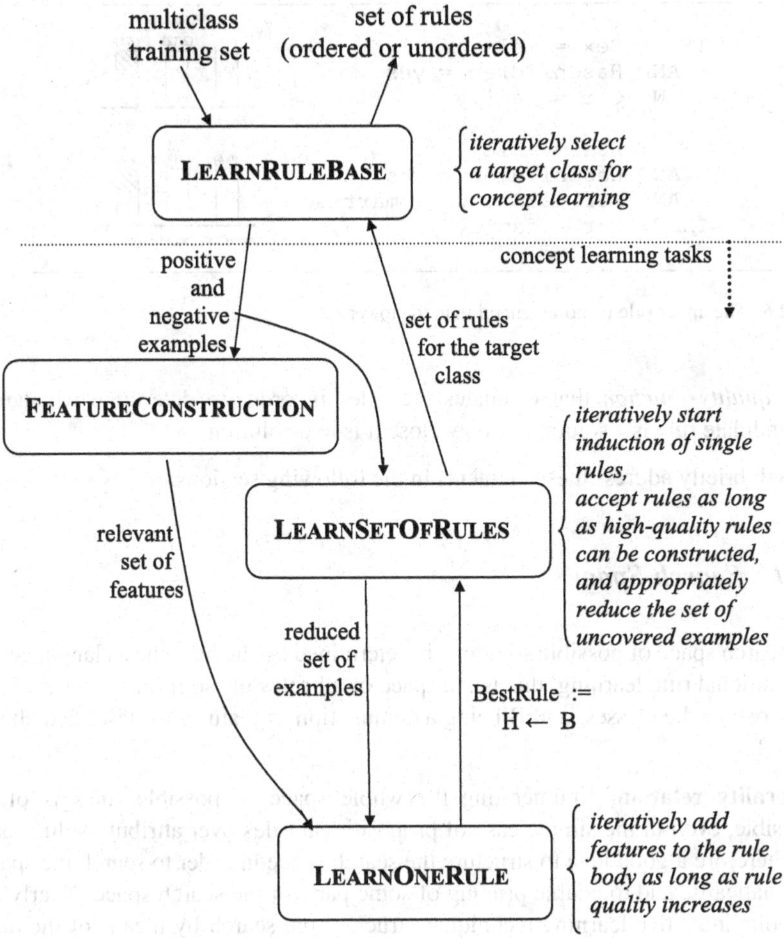

Fig. 2.5 Rule learning process

In the following sections, we will take a closer look at the key subroutines of this process, learning a single rule from data, and assembling multiple rules to a hypothesis in the form of a rule-based theory.

2.5 Learning a Single Rule

Learning of individual rules can be regarded as a search problem (Mitchell, 1982). To formulate the problem in this way, we have to define

- An appropriate *search space*
- A *search strategy* for searching through this space

```
IF      Sex = male
  AND HasChildren = yes
THEN    Car = family

IF      Sex = male
  AND HasChildren = yes
  AND MaritalStatus = married
THEN    Car = family
```

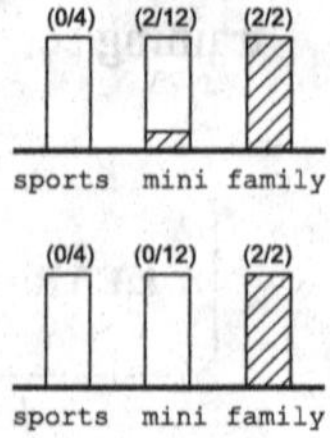

Fig. 2.6 The upper rule is more general than the lower rule

– A *quality function* that evaluates the rules in order to determine whether a candidate rule is a solution or how close it is to a solution.

We will briefly address these elements in the following sections.

2.5.1 Search Space

The search space of possible solutions is determined by the hypothesis language. In propositional rule learning, this is the space of all rules of the form $c \leftarrow \text{B}$, with c being one of the classes, and B being a conjunction of features as described above (Sect. 2.3).

Generality relation. Enumerating the whole space of possible rules is often infeasible, even in the simple case of propositional rules over attribute-value data. It is therefore a good idea to structure the search space in order to search the space systematically, and to enable pruning of some parts of the search space. Nearly all symbolic inductive learning techniques structure the search by means of the dual notions of *generalization* and *specialization* (Mitchell, 1997).

Generality is most easily defined in terms of coverage. Let $\text{COVERED}(\mathbf{r}, \mathcal{E})$ stand for the subset of examples in $\mathcal{E}$ which are covered by rule $\mathbf{r}$.

Definition 2.5.1 (Generality). A rule $\mathbf{r}$ is said to be *more general than* rule $\mathbf{r}'$, denoted as $\mathbf{r}' \subseteq \mathbf{r}$, iff

– Both $\mathbf{r}$ and $\mathbf{r}'$ have the same consequent, and
– $\text{COVERED}(\mathbf{r}', \mathcal{E}) \subseteq \text{COVERED}(\mathbf{r}, \mathcal{E})$.

We also say that $\mathbf{r}'$ is *more specific than* $\mathbf{r}$.

As an illustration, consider the two rules shown in Fig. 2.6.
The second rule has more features in its body and thus imposes more constraints on the examples it covers than the first. Thus, it will cover fewer examples and is therefore more specific than the first. In terms of coverage, the first rule covers four instances of Table 2.1 (examples 4, 9, 12, and 18), whereas the second rule covers

only two of them (4 and 9). Consequently, the first rule is more general than the second rule.

In case of continuous attributes, conditions involving inequalities are compared in the obvious way: e.g., a condition like `Age < 25` is more general than `Age < 20`. On the other hand, condition `Age = 22` would be less general than the first, but is incomparable to the second because it is neither a subset nor a superset of this rule.

The above definition of generality is sometimes called *semantic* generality because it is concerned with the semantics of the rules reflected in the examples they cover. However, computing this generality relation requires us to evaluate rules against a given dataset, which is costly. For learning conjunctive rules, a simple *syntactic* criterion can be used instead: given the same rule consequent, rule $\mathbf{r}$ is more general than rule $\mathbf{r}'$ if the antecedent of $\mathbf{r}'$ imposes at least the same constraints as the antecedent of $\mathbf{r}$, i.e., when $\textsc{Conditions}(\mathbf{r}) \subseteq \textsc{Conditions}(\mathbf{r}')$. For example, in Fig. 2.6, the lower rule is also a syntactic specialization of the upper rule, because the latter can be transformed into the former by deleting the condition `MaritalStatus = married`.

It is easy to see that syntactic generality defines a sufficient, but not necessary condition for semantic generality. For example, specialization could also operate over different attribute values (e.g., `Vienna` $\subseteq$ `Austria` $\subseteq$ `Europe`) or over different attributes (e.g., `Pregnancy = yes` $\subseteq$ `Sex = female`).

Structuring the search space. The generality relation can be used to structure the hypothesis space by ordering rules according to this relation. It is easily seen that the relation of generality between rules is reflexive, antisymmetric, and transitive, hence a partial order.

The search space has a unique most general rule, the *universal rule* $\mathbf{r}^{\top}$, which has the body true and thus covers all examples, and a unique most specific rule, the *empty rule* $\mathbf{r}_{\perp}$, which has the body false and thus covers no examples. All other rules are more specific than the universal rule and more general than the empty rule. Thus, the universal rule is also called the *top rule*, and the empty rule is also called the *bottom rule* of the hypothesis space, which is indicated by the symbols $\top$ and $\perp$. However, the term bottom rule is also often used to refer to the most specific rule $\mathbf{r}_e$ that covers a given example e. Such a bottom rule typically consists of a conjunction of all features that are true for this particular example. We will use the terms universal rule and empty rule for the unique most general and most specific rules in the hypothesis space, and reserve the term bottom rule for the most specific rule relative to a given example.

The syntactic generality relation can be used to define a so-called *refinement operator* that allows navigation in this ordered space. A rule can be *specialized* by conjunctively adding a condition to the rule, or it can be *generalized* by deleting one of its conditions. Figure 2.7 shows the space of all generalizations of the conjunction `MaritalStatus = married, HasChildren=yes, Sex=male`. This rule could be reached by six different paths that start from the universal rule at the top. Each step on this path consists of refining the rule in the

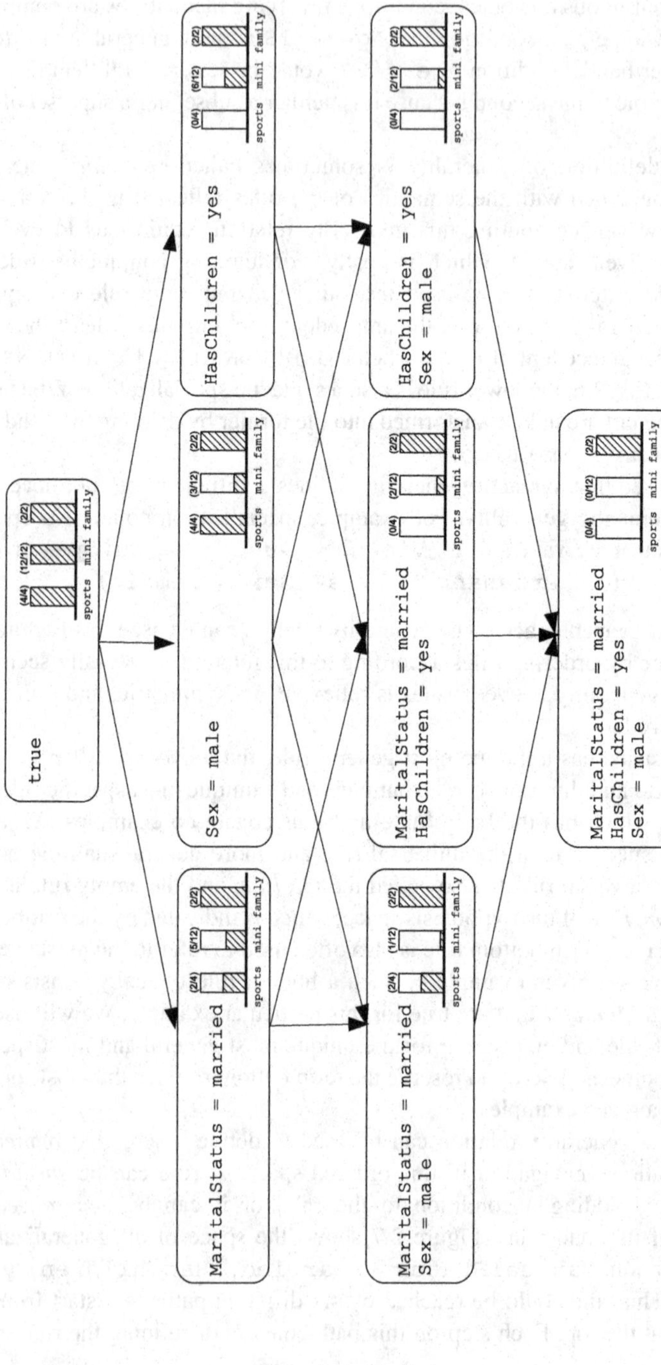

Fig. 2.7 All generalizations of MaritalStatus = married, HasChildren = yes, Sex = male, shown as a generalization hierarchy

current node by adding a condition, resulting in a more specific rule that covers fewer examples. Thus, since a more specific rule will cover (the same or) a subset of the already covered examples, making a rule more specific (or *specializing it*) is a way to obtain consistent (pure) rules which cover only examples of the target class. In this case, each path successively removes examples of all classes other than family, eventually resulting in a rule that covers all examples of this class and no examples from other classes.

Note, however, that Fig. 2.7 only shows a small snapshot of the actual search space. In principle, the universal rule could be refined into nine rules with a single condition (one for each possible value of each of the four attributes), which in turn can be refined into 30 rules with 2 conditions, 44 rules with 3 conditions, and 24 rules with 4 conditions before we arrive at the empty rule. Thus, the total search space has $1 + 9 + 30 + 44 + 24 + 1 = 109$ rules. The number of paths through this graph is $24 \times 4! = 576$.

Thus it is important to avoid searching unpromising branches and to avoid searching parts of the graph multiple times. By exploiting the monotonicity of the generality relation, the partially ordered search space can be searched efficiently because

- When generalizing rule $\mathbf{r}'$ to $\mathbf{r}$ all training examples covered by $\mathbf{r}'$ will also be covered by $\mathbf{r}$,
- When specializing rule $\mathbf{r}$ to $\mathbf{r}'$ all training examples not covered by $\mathbf{r}$ will also not be covered by $\mathbf{r}'$.

Both properties can be used to prune large parts of the search space of rules. The second property is often used in conjunction with positive examples. If a rule does not cover a positive example, all specializations of that rule can be pruned, as they also cannot cover the example. Similarly, the first property is often used with negative examples: if a rule covers a negative example, all its generalizations can be pruned since they will cover that negative example as well.

Searching through such a *refinement graph*, i.e., a graph which has rules as its nodes and applications of a refinement operator as edges, can be seen as a balancing act between rule coverage (the proportion of examples covered by a rule) and rule precision (the proportion of examples correctly classified by a rule). We will address the issue of rule quality evaluation in Sect. 2.5.3.

2.5.2 Search Strategy

For learning a single rule, most learners use one of the following search strategies.

- *General-to-specific* or top-down learners start from the most general rule and repeatedly specialize it as long as the found rules still cover negative examples. Specialization stops when a rule is consistent. During the search, general-to-specific learners ensure that the rules considered cover at least one positive example.

function LEARNONERULE(c_i,$\mathcal{P}_i$,$\mathcal{N}_i$)

Input:
c_i: a class value
$\mathcal{P}_i$: a set of positive examples for class c_i
$\mathcal{N}_i$: a set of negative examples for class c_i
$\mathcal{F}$: a set of features

Algorithm:
$\mathbf{r} := (c_i \leftarrow \text{B})$, where $\text{B} \leftarrow \emptyset$
repeat
build refinements $\rho(\mathbf{r}) \leftarrow \{\mathbf{r}' \mid \mathbf{r}' = (c_i \leftarrow \text{B} \wedge \mathbf{f})\}$ for all $\mathbf{f} \in \mathcal{F}$
evaluate all $\mathbf{r}' \in \rho(\mathbf{r})$ according to a quality criterion
$\mathbf{r} :=$ the best refinement $\mathbf{r}'$ in $\rho(\mathbf{r})$
until r satisfies a quality threshold
or covers no examples from $\mathcal{N}_i$

Output:
learned rule **r**

Fig. 2.8 A general-to-specific hill-climbing algorithm for single rule learning

– *Specific-to-general* or bottom-up learners start from a most specific rule (either the empty rule or a bottom rule for a given example), and then generalize the rule until it cannot further be generalized without covering negative examples.

The first approach generates rules from the top of the generality ordering downwards, whereas the second proceeds from the bottom of the generality ordering upwards. Typically, top-down search will find more general rules than bottom-up search, and is thus less cautious and makes larger inductive leaps. General-to-specific search is very well suited for learning in the presence of noise because it can easily be guided by heuristics.

Specific-to-general search strategies, on the other hand, seem better suited for situations where fewer examples are available and for interactive and incremental processing. These learners are, however, quite susceptible to noise in the data, and cannot be used for hill-climbing searches, such as a bottom-up version of the LEARNONERULE algorithm introduced below. Bottom-up algorithms must therefore be combined with more elaborate refinement operators. Even though bottom-up learners enjoyed some popularity in inductive logic programming, most practical systems nowadays use a top-down strategy.

Using a refinement operator, it is easy to define a simple general-to-specific search algorithm for learning individual rules. A possible implementation of this algorithm, called LEARNONERULE, is sketched in Fig. 2.8. The algorithm repeatedly refines the current best rule, and selects the best of all computed refinements according to some quality criterion. This amounts to a *top-down hill-climbing*[2]

[2]If the term 'top-down hill-climbing' sounds contradictory: hill-climbing refers to the process of greedily moving towards a (local) optimum of the evaluation function, whereas top-down refers to the fact that the search space is searched by successively specializing the candidate rules, thereby moving downwards in the generalization hierarchy induced by the rules.

search strategy. LEARNONERULE is, essentially, equivalent to the algorithm used in the PRISM learning system (Cendrowska, 1987). It is straightforward to modify the algorithm to return not only one but a beam of the b best rules, using the so-called *beam search* strategy.[3] This strategy is, for example, used in the CN2 learning algorithm.

The LEARNONERULE algorithm contains several heuristic choices. For example, it uses a heuristic quality function for selecting the best refinement, and it stops rule refinement either when a stopping criterion is satisfied or when no further refinement is possible. We will briefly discuss these options in the next section, but refer to Chaps. 7 and 9 for more details.

2.5.3 Evaluating the Quality of Rules

A key issue in the LEARNONERULE algorithm of Fig. 2.8 is how to evaluate and compare different rules, so that the search can be focused on finding the best possible rule refinement. Numerous measures are used for rule evaluation in machine learning and data mining. In classification rule induction, frequently used measures include *precision*, *information gain*, *correlation*, the *m-estimate*, the *Laplace estimate*, and others. In this section, we focus on the basic principle underlying these measures, namely a simultaneous optimization of consistency and coverage, and present a few simple measures. Two more measures will be presented in Sect. 2.7, but a detailed discussion of rule learning heuristics will follow in Chap. 7.

Terminological and notational conventions. In concept learning, examples are either positive or negative examples of a given target class $\oplus$, and they are covered (predicted positive) or not covered (predicted negative) by a rule $\mathbf{r}$ or set of rules $\mathcal{R}$. Positive examples correctly predicted to be positive are called *true positives*, correctly predicted negative examples are called *true negatives*, positives incorrectly predicted as negative are called *false negatives*, and negatives predicted as positive are called *false positives*. This situation can be plotted in the form of a 2×2 table, as shown in Table 2.3.

In the following, we will briefly introduce some of our notational conventions. A summary can be found in Table I in a separate section in the frontmatter (pp. xi–xiii). We will use the letters $\mathcal{E}$, $\mathcal{P}$, and $\mathcal{N}$ to refer to all examples, the positive examples, and the negative examples, respectively. Calligraphic font is used for denoting sets, and the corresponding uppercase letters E, P, and N are used for denoting the sizes of these sets. Table 2.3 thus shows the four possible subsets into which the example set $\mathcal{E}$ can be divided, depending on whether the example is positive or negative, and

[3] Beam search is a heuristic search algorithm that explores a graph by expanding just a limited set of the most promising nodes (cf. also Sect. 6.3.1).

Table 2.3 Confusion matrix depicting the notation for sets of covered and uncovered positive and negative examples (in *calligraphic font*) and their respective absolute numbers (in *parantheses*)

Examples	Covered	Not Covered	
Positive	$\hat{\mathcal{P}}$ ($\hat{P}$) (true positives)	$\bar{\mathcal{P}}$ ($\bar{P}$) (false negatives)	$\mathcal{P}$ (P)
Negative	$\hat{\mathcal{N}}$ ($\hat{N}$) (false positives)	$\bar{\mathcal{N}}$ ($\bar{N}$) (true negatives)	$\mathcal{N}$ (N)
Total	$\hat{\mathcal{E}}$ ($\hat{E}$)	$\bar{\mathcal{E}}$ ($\bar{E}$)	$\mathcal{E}$ (E)

whether it is covered or not covered by rule **r**. Coverage is denoted by adding a hat (ˆ) on top of a letter; noncoverage is denoted by a bar (¯).

Goals of rule learning heuristics. The goal of a rule learning algorithm is to find a simple set of rules that explains the training data and generalizes well to unseen data. This means that individual rules have to simultaneously optimize two criteria:

- *Coverage:* the number of positive examples that are covered by the rule ($\hat{P}$) should be maximized, and
- *Consistency:* the number of negative examples that are covered by the rule ($\hat{N}$) should be minimized.

Thus, we have a multi-objective optimization problem, namely to simultaneously maximize $\hat{P}$ and minimize $\hat{N}$. Equivalently, one can minimize $\bar{P} = P - \hat{P}$ and maximize $\bar{N} = N - \hat{N}$. Thus, the quality of a rule can be characterized by four of the entries in the confusion matrix. As P and N are constant for a given dataset, the heuristics effectively only differ in the way they trade off completeness (maximizing $\hat{P}$) and consistency (minimizing $\hat{N}$). Thus they may be viewed as functions $H(\hat{P}, \hat{N})$.

What follows is a very short selection of rule quality measures. All of them are applicable for a single rule **r** but, in principle, they can also be used for evaluating a set of rules constructed for the positive class (an example is covered by a rule set if it is covered by at least one rule from the set). The presented selection does not aim for completeness or quality, but is meant to illustrate the main problems and principles. An exhaustive survey and analysis of rule evaluation measures is presented in Chap. 7.

Selected rule learning heuristics. As discussed above, the two key values that characterize the quality of a rule are $\hat{P}$, the number of covered positive examples, and $\hat{N}$, the number of covered negative examples. Optimizing each one individually is insufficient, as it will either neglect consistency or completeness.

A simple way to trade these values off is to form a linear combination, in the simplest case

$$\boxed{\mathit{CovDiff}(\mathbf{r}) = \hat{P} - \hat{N}}$$

which gives equal weight to both components. One can also normalize the two components and use the difference between the *true positive rate* ($\hat{\pi}$) and the *false positive rate* ($\hat{\nu}$).

$$\boxed{\mathit{RateDiff}(\mathbf{r}) = \frac{\hat{P}}{P} - \frac{\hat{N}}{N} = \hat{\pi} - \hat{\nu}}$$

Instead of taking the difference, one can also compute the relative frequency of positive examples in all the covered examples:

$$\boxed{\mathit{Precision}(\mathbf{r}) = \frac{\hat{P}}{\hat{P} + \hat{N}} = \frac{\hat{P}}{\hat{E}}}$$

Essentially, this measure estimates the probability $\Pr(\oplus \mid \mathrm{B})$ that an example that is covered by (the body of) a rule $\mathbf{r}$ is positive. This measure is known under several names, including *precision*, *confidence*, and *rule accuracy*. We will stick with the first term.

These are only three simple examples that are meant to illustrate how a trade-off between consistency and coverage is achieved. They are not among the best-performing heuristics. Later in this chapter (in Sect. 2.7.2), we will introduce two more heuristics that are commonly used to fight overfitting.

2.5.4 Example

We will now look at a concrete example of a rule learning algorithm at work. We again use the car database from Table 2.1, and, for the moment, rule precision as a measure of rule quality. Consider calling LEARNONERULE to learn the first rule for the class `Car = family`. The rule is initialized with an empty body, so that it classifies all examples into class `family`.

```
IF     true
THEN   Car = family
```

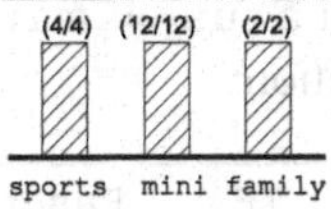

This rule covers all four examples of class `sports`, all two examples of class `family`, and all 12 examples of class `mini`. Given 2 true positives and 16 false positives, it has precision $\frac{2}{18} = 0.11$.

In the next run of the `repeat` loop, the algorithm of Fig. 2.8 will need to select the most promising refinement by conjunctively adding the best feature to the currently empty rule body. In this case there are as many refinements as there are values for all attributes; there are $3 + 2 + 2 + 2 = 9$ possible refinements in the car domain. Shown below are the two possible refinements that concern the attribute `HasChildren`:

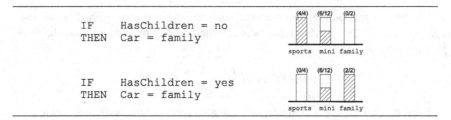

```
IF    HasChildren = no
THEN  Car = family
```

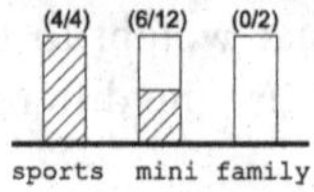


```
IF    HasChildren = yes
THEN  Car = family
```

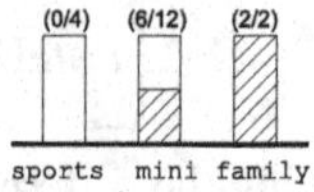

Clearly the second refinement is better than the first for predicting the class family. Its precision is estimated at $\frac{2}{8} = 0.25$. As it turns out, this rule is the best one in this iteration, and we proceed to refine it further.

Table 2.4 presents all seven possible refinements in the second iteration. Next to *Precision*, heuristic values for *CovDiff*, *RateDiff*, and *Laplace* are presented.[4] In bold are the best refinements for each evaluation measure. It can be noticed that for *CovDiff*, *Precision*, and *Laplace* there are three best solutions, while for *RateDiff* there are only two. Selecting at random among optimal solutions and using, for example, *Precision*, it can happen that we select the first refinement `HasChildren = yes AND Education = primary`, which is not an ideal solution according to *RateDiff*. The example demonstrates a common fact that different heuristics may result in different refinement selections and consequently also in different final solutions.

This is confirmed by the third iteration. If refinement `HasChildren = yes AND Education = primary` is used, then the final solution is:

```
IF    HasChildren = yes
  AND Education = primary
  AND Sex = male
THEN  Car = family
```

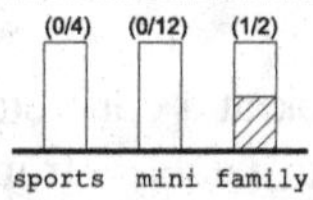

This rule covers one example of class `family` and no examples of other classes. In contrast to that, if we start with `HasChildren = yes AND MaritalStatus = married` then all heuristics will successfully find the optimal solution:

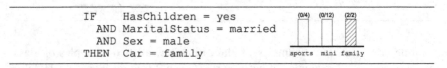

```
IF    HasChildren = yes
  AND MaritalStatus = married
  AND Sex = male
THEN  Car = family
```

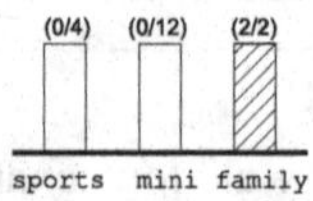

[4] *Laplace* will be defined in Sect. 2.7.

Table 2.4 All possible refinements of the rule `IF HasChildren = yes THEN Car = family` in the second iteration step of LEARNONERULE. Shown is the feature that is added to the rule, the number of covered examples of each of the three classes, and the evaluation of four different heuristics

	Covered examples of class			Heuristic evaluation			
Added feature	Sports	Mini	Family	*CovDiff*	*RateDiff*	*Precision*	*Laplace*
`Education = primary`	0	1	1	**0**	0.437	**0.5**	**0.5**
`Education = secondary`	0	2	1	−1	0.375	0.333	0.4
`Education = university`	0	3	0	−3	−0.187	0.0	0.2
`MaritalStatus = married`	0	2	2	**0**	**0.875**	**0.5**	**0.5**
`MaritalStatus = single`	0	4	0	−4	−0.25	0.0	0.167
`Sex = male`	0	2	2	**0**	**0.875**	**0.5**	**0.5**
`Sex = female`	0	4	0	−4	−0.25	0.0	0.167

2.6 Learning a Rule-Based Model

Real-world hypotheses can only rarely be formulated with a single rule. Thus, both general-to-specific learners and specific-to-general learners repeat the procedure of single rule learning on a reduced example set, if the constructed rule by itself does not cover all positive examples. They use thus an iterative process to compute disjunctive hypotheses consisting of more than one rule.

In this section, we briefly discuss methods that repeatedly call the LEARNONE RULE algorithm to learn multiple rules and combine them into a rule set. We will first discuss the covering algorithm, which forms the basis of most rule learning algorithms, and then discuss how we can deal with multiclass problems.

2.6.1 The Covering Algorithm

The *covering* or *separate-and-conquer* strategy has its origins in the AQ family of algorithms (Michalski, 1969). The term *separate-and-conquer* has been coined by Pagallo and Haussler (1990) because of the way of developing a theory that characterizes this learning strategy: learn a rule that covers a part of the given training examples, remove the covered examples from the training set (the *separate* part), and recursively learn another rule that covers some of the remaining examples (the *conquer* part) until no examples remain. The terminological choice is a matter of personal taste; both terms can be found in the literature.

The basic covering algorithm shown in Fig. 2.9 learns a set of rules $\mathcal{R}_i$ for a given class c_i. It starts to learn a rule by calling the LEARNONERULE algorithm. After the found rule is added to the hypothesis, examples covered by that rule are deleted from the current set of examples, so that they will not influence the generation of subsequent rules. This is done via calls to COVERED$(\mathbf{r}, \mathcal{E})$, which returns the subset of examples in $\mathcal{E}$ that are covered by rule $\mathbf{r}$. This cycle of adding rules and removing covered examples is repeated until no more examples of the given class remain. In this case, all examples of this class are covered by at least one rule. We will see later (Sect. 2.7) that sometimes it may be advisable to leave some examples uncovered, i.e., no more rules will be added as soon as some external *stopping criterion* is satisfied.

2.6.2 Learning a Rule Base for Classification Problems

The basic LEARNSETOFRULES algorithm can only learn a rule set for a single class. In a concept learning setting, this rule set can be used to predict whether an example is a member of the class c_i or not. However, many real-world problems are multiclass, i.e., it is necessary to learn rules for more than one class.

```
function LEARNSETOFRULES(c_i, P_i, N_i)
  Input:
    c_i: a class value
    P_i: a set of positive examples for class c_i
    N_i: a set of negative examples for class c_i, where N_i = E \ P_i

  Algorithm:
    P_i^cur := P_i, N_i^cur := N_i
    R_i := ∅
    repeat
        r := LEARNONERULE(c_i, P_i^cur, N_i^cur)
        R_i := R_i ∪ {r}
        P_i^cur := P_i^cur \ COVERED(r, P_i^cur)
        N_i^cur := N_i^cur \ COVERED(r, N_i^cur)
    until R_i satisfies a quality threshold or P_i^cur is empty

  Output:
    R_i the rule set learned for class c_i
```

Fig. 2.9 The covering algorithm for rule sets

```
function LEARNRULEBASE(E)
  Input:
    E set of training examples

  Algorithm:
    R := ∅
    for each class c_i, i = 1 to C do
        P_i := {subset of examples in E with class label c_i}
        N_i := {subset of examples in E with other class labels}
        R_i := LEARNSETOFRULES(c_i, P_i, N_i)
        R := R ∪ R_i
    endfor
    R := R ∪ {default rule (c_max ← true)}
        where c_max is the majority class in E.

  Output:
    R the learned rule set
```

Fig. 2.10 Constructing a set of rules in a multiclass learning setting

A straightforward way to tackle such problems is to learn a *rule base* $\mathcal{R} = \bigcup_i \mathcal{R}_i$ that consists of a rule set $\mathcal{R}_i$ for each class. This can be learned with the algorithm LEARNRULEBASE, shown in Fig. 2.10, which simply iterates calls to LEARNSETOFRULES over all the C classes c_i. In each iteration the current positive class will be learned against the negatives provided by all other classes.

At the end, we need to learn a *default rule*, which simply predicts the majority class in the data set. This rule is necessary in order to make sure that new examples that may not be covered by any of the learned rules, can nevertheless be assigned a class value.

This strategy of repeatedly learning one rule set for each class is also known as the *one-against-all* learning strategy. We note in passing that other strategies are possible. This, and several other learning strategies (including strategies for learning decision lists) are the subject of Chap. 10.

2.7 Overfitting Avoidance

Most top-down rule learners can be fit into the high-level description provided in the previous sections. For doing so, we need to configure the LEARNONERULE algorithm of Fig. 2.8 with appropriate heuristics for

- Evaluating the quality of a single rule,
- Deciding when to stop refining a rule, and
- Deciding when to stop adding rules to a rule set for a given class.

So far, we have defined very simple rule evaluation criteria, and used consistency and completeness as stopping criteria. However, these choices are appropriate only in idealistic situations. For practical applications, one has to deal with the problem of *overfitting*, which is a common phenomenon in data analysis (cf. also Sect. 2.1). Essentially, the problem is that rule sets that exactly fit the training data often do not generalize well to unseen data. In such cases, heuristics are needed to trade off the quality of a rule or rule set with other factors, such as their complexity.

In the following, we will briefly discuss the choices that are made by the CN2 learning algorithm. More elaborate descriptions of rule evaluation criteria can be found in Chap. 7, and stopping criteria are discussed in more detail in Chap. 9.

2.7.1 *Rule Evaluation in* CN2

Rules are evaluated on a training set of examples, but we are interested in estimates of their performance on the whole example set. In particular for rules that cover only a few examples, their evaluation values may not be representative for the entire domain. For simplicity, we illustrate this problem by estimating the precision heuristic, but in principle the argument applies to any function where a population probability is to be estimated from sample frequencies.

A key problem with precision is that for very low numbers of $\hat{N}$ and $\hat{P}$, this measure is not very robust. If both $\hat{P}$ and $\hat{N}$ are low, one extra covered positive or negative example may significantly change the evaluation value. Compare, e.g., two rules $\mathbf{r}_1$ and $\mathbf{r}_2$, both covering no negative examples ($\hat{N}_1 = \hat{N}_2 = 0$), but the first one covers 1 positive ($\hat{P}_1 = 1$), and the second one covers 99 positive examples ($\hat{P}_2 = 99$). Both have a precision of 1.0. However, if it turns out that each rule covers one additional negative example ($\hat{N}_1 = \hat{N}_2 = 1$), the evaluation of $\mathbf{r}_1$ drops to $\frac{1}{1+1} = 0.5$, while the evaluation of $\mathbf{r}_2$ is still very high ($\frac{99}{1+99} = 0.99$).

The Laplace estimate addresses this problem by adding two 'virtual' covered examples, one for each class, resulting in the formula

$$Laplace(\mathbf{r}) = \frac{\hat{P}+1}{(\hat{P}+1)+(\hat{N}+1)} = \frac{\hat{P}+1}{\hat{E}+2}$$

In the above example, the estimate for $\mathbf{r}_1$ would be $\frac{2}{3}$, rather than 1, which is much closer to the value $\frac{1}{2}$ that results from covering one additional negative example. The estimate asymptotically approaches 1 if the number of true positives increases, but with finite amounts of data the probability estimate will never be exactly 1 (or 0). In fact, the Laplace correction of the relative frequency of covered positive examples is an unbiased estimate for the probability $\Pr(\oplus \mid B)$ of the positive class given the body of the rule.[5] Example calculations for the Laplace values can be seen in Table 2.4.

2.7.2 *Stopping Criteria in* CN2

CN2 can use a significance measure to enforce the induction of reliable rules. If CN2 is used to induce sets of unordered rules, the rules are usually required to be highly *significant* (at the 99 % level), and thus reliable, representing a regularity unlikely to have occurred by chance. To test significance, CN2 uses the *likelihood ratio statistic* (Clark & Niblett, 1989) that compares the class probability distributions in the set of covered examples with the distribution over the training set, i.e., the distribution of examples covered by the rule compared to the prior distribution of examples in the training set. A rule is deemed reliable if these two distributions are significantly different. For a two-class problem, rule significance is measured as follows:

$$LRS(\mathbf{r}) = 2 \cdot \left(\hat{P} \cdot \log_2 \frac{\frac{\hat{P}}{\hat{P}+\hat{N}}}{\frac{P}{P+N}} + \hat{N} \cdot \log_2 \frac{\frac{\hat{N}}{\hat{P}+\hat{N}}}{\frac{N}{P+N}} \right)$$

[5] If $C > 2$ classes are used, the relative frequencies for each class should be estimated with $\frac{\hat{P}_i+1}{\sum_{j=1}^{C} \hat{P}_j + C}$, where $\hat{P}_i$ is the number of examples of class i covered by the rule and C is the number of classes. However, if we estimate the probability whether an example that is covered by the body of a rule is also covered by its head or not, we have a binary distinction even in multiclass problems.

For a multiclass problem, the value of the likelihood ratio statistic is computed as follows[6]:

$$LRS(\mathbf{r}) = 2 \cdot \sum_{i=1}^{C} \hat{P}_i \log_2 \frac{\hat{\pi}_i}{\gamma_i}$$

where $\hat{\pi}_i = \frac{\hat{P}_i}{\hat{E}}$ is the proportion of covered examples of class c_i, and $\gamma_i = \frac{P_i}{E}$ is the overall proportion of examples of class c_i. This statistic is distributed as χ^2 with $C - 1$ degrees of freedom. The rule is only considered significant if its likelihood ratio is above a specified significance threshold.

2.8 Making a Prediction

After having learned a rule set, how do we use it for classifying new instances? Classification with a decision list is quite straightforward. To classify a new instance, the rules are tried in order. The first rule that covers the instance is used for classification/prediction. If no induced rule fires, a default rule is invoked, which typically predicts the majority class of uncovered training examples.

In the case of unordered rule sets, the situation is more complicated because all the rules are tried and predictions of those that cover the example are collected. Two problems have to be dealt with:

- *Conflict resolution:* multiple overlapping rules may make possibly conflicting predictions for the same example
- *Uncovered examples:* no rule may cover the example.

As we have seen in the previous section, the second case is handled by adding a default rule to the rule set.

The typical solution for the first problem is to use a voting mechanism to obtain the final prediction. Conflicting decisions are resolved by taking into account the number of examples of each class (from the training set) covered by each rule. CN2, for example, sums up the class distributions attached to the rules to determine the most probable class. For example, in the rule set of Fig. 2.4a, example 13 is covered by rules nos. 2 and 5. If we sum up the class counts for each rule, we obtain 4 counts for `sports`, 3 counts for `mini`, and none for `family`. As a result, the person would still be assigned to the `sports` car class. These counts can also be

[6]Clark and Niblett (1989, p. 269) define the likelihood ratio statistic in the form $LRS(\mathbf{r}) = 2 \cdot \sum_{i=1}^{C} \hat{P}_i \cdot \log_2 \frac{\hat{P}_i}{\mathbb{E}\hat{P}_i}$, where $\mathbb{E}\hat{P}_i = \gamma_i \cdot \hat{E}$ is the expected *number* of examples of class c_i that the rule would cover if the covered examples were distributed with the same relative class frequencies as in the original dataset. A simple transformation gives our formulation with ratios of observed relative frequencies and expected relative frequencies.

used for probabilistic estimation (Džeroski, Cestnik, & Petrovski, 1993). In this case, we would predict class `sports` with a probability of 4/7 and class `mini` with a probability of 3/7. Class `family` would be ruled out.

As an alternative to standard CN2 classification, where the rule class distribution is computed in terms of the numbers of examples covered, class distribution can be given also in terms of probabilities (computed by the relative frequencies $\hat{\pi}$ of each class). In the above example, rule 2 predicts `sports` with a probability of 4/5, and class `mini` with a probability of 1/5, whereas rule 5 predicts `mini` with a probability of 1. Now the probabilities are averaged (instead of summing the numbers of examples), resulting in the final probability distribution [0.4, 0, 0.6], and now the correct class `mini` is predicted. By using this voting scheme the rules covering a small number of examples are not so heavily penalized (as is the case in CN2) when classifying a new example. However, this may also become a problem when some of the rules overfit the data. Thus, often a Laplace-correction is used for estimating these probabilities from the data, as discussed in the previous section.

One may also use different aggregations than averaging the class probability distributions. RIPPER (Cohen, 1995), for example, resolves ties by using the rule that has the largest (Laplace-corrected) probability estimate for its majority class. Note that this approach is equivalent to converting the rule set into a decision list by sorting its rules according to the above-mentioned probability estimates and using the first one that fires.

In the literature on rule learning, we can find a number of more elaborate approaches for handling conflicting predictions and for predictions in the case when no rule covers an example. We will discuss a few of them in Sect. 10.2.

2.9 Estimating the Predictive Accuracy of Rules

Classification quality of the rule set is measured by the classification accuracy, which is defined as the percentage of the total number of correctly classified examples in all classes relative to the total number of tested examples. As a special case, for a binary classification problem, rule set accuracy is computed as follows:

$$Accuracy(\mathcal{R}) = \frac{\hat{P} + \bar{N}}{P + N}$$

Note that accuracy measures the classification accuracy of the whole rule set on both positive and negative examples. It should not be confused with *rule accuracy*, for which we use the term *precision*. The coverage difference measure *CovDiff* (Sect. 2.5.3) is equivalent to evaluating the accuracy of a single rule, when we consider that $\bar{N} = N - \hat{N}$, and that P and N are constant and can be omitted when comparing the accuracy of different rules in the same domain.

Instead of accuracy, results are often presented with *classification error*, which is simply

$$Error(\mathcal{R}) = 1 - Accuracy(\mathcal{R}) = \frac{\bar{P} + \hat{N}}{P + N}$$

Obviously, both measures are equivalent, and the choice is a matter of personal taste.

2.9.1 Hold-Out Set Evaluation

It is not valid to estimate the predictive accuracy of a theory on the same dataset that was used for learning. As an illustration, consider a rule learning algorithm that simply constructs one rule for each positive training example by formulating the body of the rule as a conjunction of all attribute values that appear in the definition of the example, and classifying everything else as negative. Such a set of rules would have 100 % accuracy[7] if estimated on the training data, but in reality, this theory would misclassify all examples that belong to the positive class except those that it has already seen during training. Thus, no generalization takes place, and the rule is worthless for practical purposes.

In order to solve this problem, predictive accuracy should be estimated on a separate part of the data, a so-called *test set*, that was removed (held out) from the training phase. Recall that Table 2.2 shows such a test set for the car domain in Table 2.1. If we classify these examples with the decision list of Fig. 2.4a, example 20 will be misclassified as `mini` (by the default rule), while all other examples will be classified correctly. Thus, the estimated predictive accuracy of this rule set would be $5/6 \approx 83.3\,\%$. On the training data, the decision list made only one mistake in 18 examples, resulting in an accuracy of $17/18 \approx 94.4\,\%$. Even though these measures are computed from a very small data sample and are thus quite unreliable, the result that the test accuracy is lower than the training accuracy is typical.

2.9.2 Cross-Validation

While hold-out set evaluation produces more accurate estimates for the predictive accuracy, it also results in a waste of training data because only part of the available data can be used for learning. It may also be quite unreliable if only a single, small test set is used, as in the above example.

Cross-validation is a common method for solving this problem: while all of the data are used for learning, the accuracy of the resulting theory is estimated by performing k hold-out experiments as described above. For this purpose, the data is divided into n parts. In each experiment, $n-1$ parts are combined into a training set,

[7] We assume that there are no contradictory examples in the training set, an assumption that does not always hold in practice, but the basic argument remains the same.

function CROSSVALIDATION($\mathcal{E}$,k)

Input:
$\mathcal{E}$ set of training examples
k the number of folds

Algorithm:
randomly partition $\mathcal{E}$ into f disjoint folds $\mathcal{E}_j$ of approximately the same size
for $j = 1$ to f **do**
$\mathcal{E}_{test} := \mathcal{E}_j$
$\mathcal{E}_{train} := \mathcal{E} \setminus \mathcal{E}_{test}$
$\mathcal{R} :=$ LEARNRULEBASE($\mathcal{E}_{train}$)
$v_j :=$ EVALUATE($\mathcal{R}, \mathcal{E}_{test}$)
endfor
$v := \frac{1}{k}\sum_{j=1}^{k} v_j$

Output:
v the estimated quality

Fig. 2.11 Estimation of the prediction quality of a rule set induced by LEARNRULEBASE via cross-validation

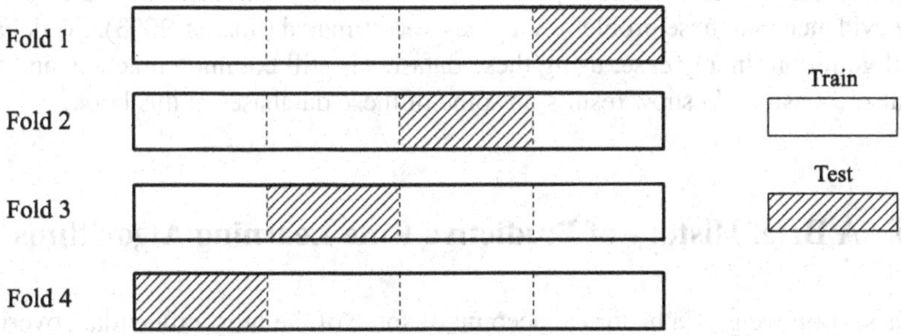

Fig. 2.12 A visual representation of fourfold cross-validation, where predictive accuracy is estimated as the average rule set accuracy of four hold-out experiments

and the remaining part is used for testing. A theory is then learned on the training set and evaluated on the test set. This is repeated until each part (and thus each training example) has been used for testing once. The final accuracy is then estimated as an average of the accuracy estimates computed in each such hold-out experiment. The cross-validation algorithm is shown in Fig. 2.11. Note that the procedure can be used for estimating any aspect of the quality of the learned rules. It is thus shown with a generic function EVALUATE, which can be instantiated with any common evaluation measure, such as accuracy, recall and precision, area under the ROC curve, etc.

Figure 2.12 shows a schematic depiction of a fourfold cross-validation. In practice, tenfold cross-validation is most commonly used. An interesting special case is *leave-one-out cross-validation*, where in each iteration only a single example is held out from training and subsequently used for testing. Because of the large number of theories to be learned (one for each training example), this is only feasible for small datasets.

2.9.3 Benchmark Datasets

Results on a single dataset are typically not very meaningful. Therefore, machine learning techniques are often evaluated on a large set of benchmark datasets. There are several collections of benchmark datasets available; the best-known is the *UCI machine learning repository* (Frand & Asuncion, 2010).[8]

This approach has several advantages, including that tests on a large collection of databases reduce the risk of getting spurious results, and that published results on these datasets are, in principle, reproducible. Some researchers are even working on experiment databases that collect experimental results on these benchmark datasets for various algorithms under various experimental setups in order to increase the repeatability and the reliability of reported results (Blockeel & Vanschoren, 2007).

On the other hand, this approach also has disadvantages. Most notably, it is unclear how representative these datasets are for real-world applications because many datasets have been donated because someone has (typically successfully) presented them in a publication. Thus there is a certain danger that the collection is biased against hard problems, and that algorithms that are *overfitting the UCI repository* may actually not be applicable to real-world problems. Although there is some evidence that these problems may be overestimated (Soares, 2003), the debate is still going on. In any case, using these datasets is still common practice, and we will also occasionally show results on some of these databases in this book.

2.10 A Brief History of Predictive Rule Learning Algorithms

In this section we give a historical account of some of the most influential covering algorithms. Most of them are still in use and are regularly cited in the rule learning literature.

2.10.1 AQ

AQ can be considered as the original covering algorithm. Its original version was conceived by Ryszard Michalski in the 1960s (Michalski, 1969). Over the years, numerous versions and variants of the algorithm appear in the literature (Bergadano, Matwin, Michalski, & Zhang, 1992; Kaufman & Michalski, 2000; Michalski, 1980; Michalski & Larson, 1978; Michalski, Mozetič, Hong, & Lavrač, 1986). A very good summary of the early versions of AQ is given in (Clark & Niblett, 1987, 1989).

[8] At the time of this writing, the collection contains 177 datasets in a great variety of different domains, including bioinformatics, medical applications, financial prognosis, game playing, politics, and more.

The basic algorithm features many typical components of the covering algorithm, but it also has a few interesting particularities. The algorithm uses a top-down beam search for finding the best rule.[9] However, contrary to most successors, AQ does not search all possible refinements of a rule; it only considers refinements that cover a particular example, the so-called *seed example*. In each step, it looks for refinements (*extensions* in AQ) of rules that cover the seed example, but do not cover a randomly chosen negative example. All possible refinements that meet this constraint are evaluated with rule learning heuristics (in the simplest version it was positive coverage p), and the best b rules are maintained in the current beam. Refinements are iterated until one or more rules are found that cover only positive examples, and the best rule among those is added to the current theory.

2.10.2 *PRISM*

PRISM (Cendrowska, 1987) was the first algorithm that used a conventional top-down search without being constrained by a particular, randomly selected pair of positive and negative examples. While this is less efficient than AQ's method, it results in a stable algorithm that does not depend on random choices. Cendrowska (1987) also realized the main advantage of rule sets over decision trees, namely that decision trees are constrained to find a theory with nonoverlapping rules, which can be more complex than a corresponding theory with overlapping rules that could be discovered by PRISM. Although the system has not received wide recognition in the literature, mostly due to its inability to address the problem of noise handling, it can be considered as an important step. A version of the algorithm, with all its practical limitations, is implemented in the WEKA toolbox.

2.10.3 *CN2*

CN2 (Clark & Niblett, 1989), named after the initials of its inventors, tried to combine ideas from AQ with the then-popular decision tree learning algorithm ID3 (Quinlan, 1983). The key observation was that learning a single rule corresponds to learning a single branch in a decision tree. Thus, Clark and Niblett (1989) proposed to evaluate each possible condition (the first version of CN2 used ID3's information gain) instead of focusing only on conditions that separate a randomly selected pair of positive and negative examples. This is basically the same idea that was also discovered independently in the PRISM learner. However, CN2's contributions extend much further. Most importantly, CN2 was the first rule learning system that recognized the overfitting problem, and proposed first measures to counter it. First,

[9]The beam is called a *star* in AQ's terminology, and the beam width is called the *star size*.

AQ could select a mislabeled negative example as a positive seed example, and therefore be forced to learn a rule that covers this example. CN2's batch selection of conditions is less susceptible to this problem, as it is independent of random selections of particular examples. Second, CN2 included a prepruning method based on a statistical significance test. In particular, it filtered out all rules that were insignificant according to the likelihood ratio statistics described in Sect. 9.2.4. Finally, in a later version, Clark and Boswell (1991) also proposed the use of the Laplace measure as an evaluation heuristic (cf. Sect. 7.3.6). Finally, following AQ's example, CN2 used a beam search, which is less likely to suffer from search myopia than hill-climbing approaches.

Another innovation introduced by CN2 was the ability to handle multiple classes. In the original version of the algorithm, Clark and Niblett (1989) proposed to handle multiple classes just as in ID3, by using information gain as a search heuristic and by picking the majority class among the examples that are covered by the final rule. In effect, this approach learns a decision list, a concept that was previously introduced by Rivest (1987). Clark and Boswell (1991) later suggested an alternative approach, which treats each class as a separate concept, thereby using the examples of this class as the positive examples and all other examples as the negative examples. The main advantage of this approach is that it allows the search for a rule that targets one particular class, instead of optimizing the distribution over all classes, as decision tree algorithms have to do, and as the first version of CN2 did as well. This approach, which they called *unordered*, as opposed to the *ordered* decision list approach, is nowadays better known as one-against-all, and is still widely used (cf. Chap. 10).

Section 2.7 gave a detailed account of CN2 at work. The system found many successors, including *m*FOIL (Džeroski & Bratko, 1992) and ICL (De Raedt & Van Laer, 1995), which upgraded it to first-order logic, or BEXA (Theron & Cloete, 1996), which used a more expressive hypothesis language by including disjunctions of attribute values in the search space.

2.10.4 FOIL

FOIL (First-Order Inductive Learner; Quinlan, 1990) was the first relational learning algorithm that received attention beyond the field of *relational data mining* and *Inductive Logic Programming* (ILP). It basically works like CN2 and PRISM discussed above; it learns a concept with the covering loop and learns individual concepts with a top-down refinement operator.

A minor difference between these algorithms and FOIL is that FOIL does not evaluate the quality of a rule, but its *information gain* heuristic (cf. Sect. 7.5.3) evaluates the improvement of a rule with respect to its predecessor. Thus, the quality measure of the same rule may be different, depending on the order of the conditions in the rule. For this reason, FOIL can only use hill-climbing search, and not a beam search like CN2 does. Another minor difference is the Minimum Description Length (MDL)-based stopping criterion that is employed by FOIL (cf. Sect. 9.3.4).

The main difference, however, is that FOIL allowed the use of first-order background knowledge. Instead of only being able to use tests on single attributes, FOIL could employ tests that compute relations between multiple attributes, and could also introduce new variables in the body of a rule. Unlike most other ILP algorithms, FOIL was not directly built on top of a Prolog-engine (although it could be), but it implemented its own reasoning module in the form of a simple tuple calculus. A *tuple* is a valid instantiation of all variables that are used in the body of a rule. Adding a literal that introduces a new variable results in extending the set of valid tuples with new variables. As a new variable can often be bound to more than one value, one tuple can be extended to several tuples after a new variable is introduced. Effectively, counting the covered tuples amounts to counting the number of possible proofs for each (positive or negative) covered example. We provide a more detailed description of the algorithm along with an example that illustrates the operation with tuples in Sect. 6.5.1.

New variables may be used to formulate new conditions, i.e., adding a literal that extends the current tuples may be a good idea, even if the literal on its own does not help to discriminate between positive and negative examples, and would therefore not receive a high evaluation by conventional evaluation measures. Quinlan (1991) proposed a simple technique for addressing this problem: whenever the information gain is below a certain percentage of the highest achievable gain, all so-called *determinate literals* are added to the body of a rule. These are literals that add a new variable, but have at most one binding for this variable, so that they will not lead to an increase in the number of tuples (cf. also Sect. 5.5.5).

FOIL was very influential, and numerous successor algorithms built upon its ideas (Quinlan & Cameron-Jones, 1995a). Its influence was not confined to ILP itself, but also propositional variants of the algorithm were developed and used (e.g., Mooney, 1995).

2.10.5 RIPPER

The main problem with all above-mentioned algorithms was overfitting. Even algorithms that employed various techniques for overfitting avoidance turned out to be ineffective. For example, Fürnkranz (1994a) showed that FOIL's MDL-based stopping criterion is not effective: in experiments in the king-rook-king chess endgame domain, with a controlled level of noise in the training data, the size of the theory learned by FOIL grows with the number of training examples. This means that the algorithm overfits the noise in the training data. Similar experiments with CN2 showed the same weakness.

RIPPER (Repeated Incremental Pruning to Produce Error Reduction; Cohen, 1995) was the first rule learning system that effectively countered the overfitting problem. It is based on several previous works, most notably the *incremental reduced error pruning* idea that we describe in detail in Sect. 9.4.1, and added various ideas of its own (cf. also Sect. 9.4.3). Most notably, Cohen (1995) proposed an interesting post-processing phase for optimizing a rule set in the context of other rules. The key idea is to remove one rule out of a previously learned rule set and try to relearn it not only in the context of previous rules (as would be the case in the regular covering rule), but also in the context of subsequent rules. The resulting algorithm, RIPPER, has been shown to be competitive with C4.5RULES (Quinlan, 1993) without losing I-REP's efficiency (Cohen, 1995).

All in all, RIPPER provided a powerful rule learning system that was used in several practical applications. In fact, RIPPER is still state of the art in inductive rule learning. An efficient C implementation of the algorithm can be obtained from its author, and an alternative implementation named JRIP can be found in the WEKA data mining library (Witten & Frank, 2005). In several independent studies, RIPPER and JRIP have proved to be among the most competitive rule learning algorithms available today. Many authors have also tried to improve RIPPER. A particularly interesting approach is the FURIA algorithm, which incorporates some ideas from fuzzy rule induction into RIPPER (Hühn & Hüllermeier, 2009b).

2.10.6 *PROGOL*

PROGOL (Muggleton, 1995), another ILP system, was interesting for several reasons. First, it adopted ideas of the AQ algorithm to inductive logic programming. Like AQ, it selects a seed example and computes a minimal generalization of the example with respect to the available background knowledge and a given maximum inference depth. The resulting *bottom rule* is then in a similar way used for constraining the search space for a subsequent top-down search for the best generalization.

Second, unlike AQ, PROGOL does not use a heuristic search for the best generalization of the seed example, but it uses a variant of best-first A^* search in order to find the best possible generalization of the given seed example. This proves to be feasible because of the restrictions of the search space that are imposed by the randomly selected seed example.

PROGOL proved very successful in various applications (Bratko & Muggleton, 1995), some of which were published in scientific journals of their respective application areas (Cootes, Muggleton, & Sternberg, 2003; King et al., 2004; Sternberg & Muggleton, 2003). An efficient implementation of the algorithm (Muggleton & Firth, 2001) is available from http://wwwhomes.doc.ic.ac.uk/~shm/Software/.

2.10.7 *ALEPH*

To our knowledge, the most widely used ILP system is ALEPH (A Learning Engine for Proposing Hypotheses). Unfortunately, there are no descriptions of this system in the formal literature, and references have been restricted to an extensive user manual

(Srinivasan, 1999). Some of ALEPH's popularity seems to be due to two aspects: that it is in just one physical file,[10] and that it is written in Prolog. Except for that, there is, in fact, one other aspect of the program that makes it different to almost any other ILP system. ALEPH was conceived as a workbench for implementing and testing—under one umbrella—concepts and procedures from a variety of different ILP systems and papers. It is possibly the only ILP system that can construct rules, trees, constraints and features; invent abnormality predicates; perform classification, regression, clustering, and association-rule mining; allow changes in search strategy (general-to-specific (top-down), specific-to-general (bottom-up), bidirectional, and so on); search algorithm (hill-climbing, exhaustive search, stochastic search, etc.); and evaluation functions. It even allows the user to specify their own search procedure, proof strategies, and visualization of hypotheses.

Given this plethora of options, it is not surprising that each user has her own preferred way to use this program, and we will not attempt to prescribe one over the other. It would, however, be amiss if we did not mention two uses of ALEPH that seem to be most popular in the ILP literature.

In the first, ALEPH is configured to identify a set of classification rules, using a randomized version of the covering algorithm. Individual rules in the set are identified using a general-to-specific, compression-based heuristic search guided by most-specific bottom clauses. This is sufficiently similar to the procedure followed by PROGOL (Muggleton, 1995) to be of interest to ILP practioners seeking a Prolog-based approximation to that popular system.

Another interesting use of ALEPH is to identify Boolean features (Joshi, Ramakrishnan, & Srinivasan, 2008; Ramakrishnan, Joshi, Balakrishnan, & Srinivasan, 2008; Specia, Srinivasan, Joshi, Ramakrishnan, & das Graças Volpe Nunes, 2009), that are subsequently used by a propositional learner to construct models (in the cases cited, this learner was a support vector machine). The key idea is to construct bottom clauses that cover individual examples, and to use the resulting rules as features in a subsequent induction phase. The quality of a rule or feature is defined via some syntactic and semantic constraints (such as minimum recall and precision). In this way, ALEPH may also be viewed as a system for first-order subgroup discovery (cf. Chap. 11).

This specific use of rules constructed by an ILP system as Boolean features appears to have been demonstrated first in (Srinivasan & King, 1997). It has since proved to be an extremely useful way to employ an ILP system (Kramer, Lavrač, & Flach, 2001). The idea of constructing propositional representations from first-order ones goes back at least to 1990, with the LINUS system (Lavrač & Džeroski, 1994a; Lavrač, Džeroski, & Grobelnik, 1991), and perhaps even earlier to work in the mid-1980s done by Michalski and coworkers. We will describe such approaches in more detail in Chap. 5.

[10] http://www.comlab.ox.ac.uk/oucl/research/areas/machlearn/Aleph/

2.10.8 OPUS

OPUS (Optimized Pruning for Unordered Search; Webb, 1995) was the first rule learning system that demonstrated the feasibility of a full exhaustive search through all possible rule bodies for finding a rule that maximizes a given quality criterion (or heuristic function). The key idea is the use of *ordered search* that prevents a rule being generated multiple times. This means that even though there are $l!$ different orders of the conditions of a rule of length l, only one of them can be taken by the learner for finding this rule. In addition, OPUS uses several techniques that allow it to prune significant parts of the search space, so that this search method becomes feasible. Follow-up work (Webb, 2000; Webb & Zhang, 2005) has shown that this technique is also an efficient alternative for association rule discovery, provided that the database to mine fits into the memory of the learning system.

2.10.9 CBA

Exhaustive search is also used by association rule discovery algorithms like APRIORI or its successors (cf. Sect. 6.3.2). The key difference is the search technique employed. Contrary to OPUS, association rule learners typically employ a memory-intensive level-wise breadth-first search. They return not a single best rule, but all rules with a given minimum support and a given minimum confidence.

In general, association rules can have arbitrary combinations of features in the head of a rule. However, the head can also be constrained to hold only features related to the class rule. In this case, an association rule algorithm can be used to discover all classification rules with a given minimum support and confidence. The learner's task is then to combine a subset of these rules into a final theory.

One of the first and best-known algorithms that employed this principle for learning predictive rules is CBA (Liu, Hsu, & Ma, 1998; Liu, Ma, & Wong, 2000). In its simplest version, the algorithm selects the final rule sets by sorting all classification rules according to confidence and incrementally adding rules to the final set until all examples are covered or the quality of the rule set decreases.

There is a variety of variations of this basic algorithm that have been discussed in the literature (e.g., Bayardo Jr., 1997; Jovanoski & Lavrač, 2001; Li, Han, & Pei, 2001; Yin & Han, 2003). Mutter, Hall, and Frank (2004) performed an empirical comparison between some of them. A good survey of pattern-based classification algorithms can be found in (Bringmann, Nijssen, & Zimmermann, 2009).

2.11 Conclusion

This chapter provided a gentle introduction to rule learning, mainly focusing on supervised learning of predictive rules. In this setting, the goal is to learn understandable models for predictive induction. We have seen a broad overview of the main components that are common to all rule learning algorithms, such as the covering loop for learning rule sets or decision lists, the top-down search for single rules, evaluation criteria for rules, and techniques for fighting overfitting. In the following chapters, we will return to these issues and discuss them in more detail.

However, other techniques for generating rule sets are possible. For example, rules can be generated from induced decision trees. As we have seen in Chap. 1, nodes in a decision tree stand for attributes or conditions on attributes, arcs stand for values of attributes or outcomes of those conditions, and leaves assign classes. A decision list like the one in Fig. 2.4b can be seen as a right-branching decision tree. Conversely, each path from the root to a leaf in a decision tree can be seen as a rule.

Standard algorithms for learning decision trees (such as C4.5) are quite similar to the CN2 algorithm for learning decision lists in that the aim of extending a decision tree with another split is to reduce the class impurity in the leaves (usually measured by entropy). However, as discussed in Sect. 1.5, rule sets are often more compact than decision trees. Consequently, a rule set can be considerably simplified during the conversion of a decision tree to a set of rules (Quinlan, 1987a, 1993). For example, Frank and Witten (1998) suggested the PART algorithm, which tries to integrate this simplification into the tree induction process by focusing only on a single branch of a tree.

Chapter 3
Formal Framework for Rule Analysis

This chapter provides a formal framework which enables the analysis of the key elements of rule learning: features, search heuristics for feature and rule evaluation, and heuristics/constraints for rule overfitting avoidance. After a brief introduction, Sect. 3.2 discusses *receiver operating characteristic* (ROC) analysis, the basis of coverage spaces which will serve as an analytic tool for visualizing the behavior of rule learning algorithms and their key elements. The coverage space will then be formally introduced in Sect. 3.3, and its relation to the ROC space will be discussed. Finally, in Sect. 3.4, we show how typical rule learning algorithms behave in coverage space.

3.1 Introduction

The goal of supervised rule induction is the discovery of rules that reflect relevant dependencies between classes and attribute values describing the given data. The induced rules can be used for classifying new instances or they can be a starting point for expert interpretation, potentially leading to the formulation of interesting and relevant knowledge that generalizes the information contained in the training examples.

Recall that rule learning has two complementary goals: the ideal classification rule will cover all examples of the target class (*completeness*) and will not cover any other examples (*consistency*). Note that these constraints should be satisfied for all examples in the domain, not only for the examples which were used for training the classifier. In practice, the completeness and consistency constraints have to be relaxed, so that a good rule will cover many (but not necessarily all) target class examples and possibly also a few examples from other classes. Typically, there is a trade-off between these two objectives: if we try to construct a rule

†Parts of this chapter are based on Fürnkranz and Flach (2005).

J. Fürnkranz et al., *Foundations of Rule Learning*, Cognitive Technologies,
DOI 10.1007/978-3-540-75197-7_3, © Springer-Verlag Berlin Heidelberg 2012

that correctly classifies many target class (positive) cases, the rule will typically also cover many non-target class (negative) examples. Conversely, minimizing the number of covered negative examples will typically also reduce the number of covered positive examples, although hopefully to a lesser degree. Feature and rule quality measures, which will be the subject of later chapters, trade off these two objectives into a joint measure to be optimized in rule construction and rule post-processing.

This chapter provides a framework which will enable us to study the different biases of feature and rule quality measures in terms of their trade-off between the covered positive and negative examples. More specifically, it will allow us to study covering properties of rules in a two-dimensional space, called the *coverage space*, in which the number of positive examples correctly covered by the rule (*true positives* $\hat{P}$) is plotted on the y-axis, and the number of covered negative examples (*false positives* $\hat{N}$) is plotted on the x-axis (Fürnkranz & Flach, 2005). Alternatively, instead of studying the $\hat{P}/\hat{N}$ trade-off of rules in the coverage space, some researchers prefer to analyze the trade-off between the true and false positive rates $\hat{\pi}$ and $\hat{\upsilon}$, which results in the so-called *ROC space*. For analytical purposes within the same problem domain, the coverage space introduced in this chapter is more convenient. In this book, it will be used as the basic analytic tool for evaluating the quality of features, rules, and rule sets. We start with a brief summary of the fundamental concepts of ROC spaces.

3.2 ROC Space

Originally, the use of ROC spaces was proposed for the presentation and comparison of the covering quality of complete classifiers. However, because covering properties of individual rules and individual features and their logical combinations are defined in the same way as they are defined for complete rule sets, ROC spaces can also be used for presenting covering qualities of individual rules and features. For simplicity, we will use the term 'rules' in the following, but note that it is meant to be a representative for rules, classifiers (rule sets), or features.

ROC analysis has its origins in signal detection theory. A ROC curve is a two-dimensional plot for analyzing the operating characteristics of the receiving station in terms of its hit and false alarm rates (cf., e.g., Egan, 1975). It can be straightforwardly adapted to binary classification as a framework for selecting the best classifier for unknown classification costs (Provost & Fawcett, 2001). In the concept learning framework (Fig. 2.2), a point in ROC space shows the performance of a classifier in terms of *false positive rate* $\hat{\upsilon} = \frac{\hat{N}}{N}$ (plotted on the x-axis), and sensitivity or *true positive rate* $\hat{\pi} = \frac{\hat{P}}{P}$ (plotted on the y-axis).

A point $(\hat{\upsilon}, \hat{\pi})$ depicting rule $\mathbf{r}$ in ROC space is determined by the covering properties of the rule. This allows us to plot rules in a two-dimensional space, one dimension for each of these two measurements. The *perfect rule* is at the upper left corner, at point $(0, 1)$, covering only positive and no negative examples, i.e., a complete and consistent rule. The origin $(0, 0)$ represents the empty rule $\mathbf{r}_{\perp}$, which classifies no examples as positive, whereas the point $(1, 1)$ corresponds to the universal rule $\mathbf{r}^{\top}$, which classifies all examples as positive, respectively. If all rules are plotted in this ROC space, the best rule is the one that is closest to $(0, 1)$, where the used distance measure depends on the relative misclassification costs of the two classes.

The ROC space is also appropriate for evaluating the quality of a learned rule, since rules near the diagonal can be discarded as insignificant. The reason is that the points on the diagonal represent rules whose true positive rate equals their false positive rate: their success rate on the positive examples equals their error rate on the negative examples and vice versa. One can imagine these points to be rules that classify an example as positive with the probability $p_{\mathbf{r}}^{\oplus} := \Pr(\oplus \mid \mathrm{B}_{\mathbf{r}})$. For $p_{\mathbf{r}}^{\oplus} = 0$, we have the empty rule, whereas $p_{\mathbf{r}}^{\oplus} = 1$ represents the universal rule. The points with $0 < p_{\mathbf{r}}^{\oplus} < 1$ form the diagonal. For example, the point that throws a fair coin for deciding whether an example is predicted as positive or negative ($p_{\mathbf{r}}^{\oplus} = 1/2$) is the center of the ROC space.

The distance of the rules from the diagonal can thus be used as a quality measure. The further a rule is away from the diagonal, the better is its ability to discriminate between positive and negative examples. The perfect rule in the upper-left corner of the ROC space has a maximum distance from the diagonal. Note that the same also holds for the rule in the lower-right corner of the ROC space. This point represents a rule that is maximally wrong, i.e., it makes a mistake on every example.[1]

One can also show that all rules that are optimal under some (linear) cost model must lie on the *convex hull* of a set of points in the ROC space (Provost & Fawcett, 2001). Essentially, the convex hull starts in $\mathbf{r}_{\perp}$ and ends in $\mathbf{r}^{\top}$, and encloses all the points so that no straight line connecting two points will lie outside of the hull (the thick lines in Fig. 3.1). This hull is also known as the *ROC curve*. A rule that lies below this convex hull can never be optimal under any linear cost model and can thus be removed from further consideration.[2] Figure 3.1 shows seven optimal rules including $\mathbf{r}_1$ and $\mathbf{r}_2$ on the convex hull, while two rules $\mathbf{r}_3$ and $\mathbf{r}_4$ below the convex hull are suboptimal.

ROC curves can also be used for evaluating algorithms that rank the instances according to their probability of being positive. The key idea behind this is to draw a ROC curve for which each point corresponds to a threshold in the predicted ranking, i.e., to a classifier that classifies all examples that are ranked above this threshold as positive and all others as negative. The ideal ranker would rank all positive examples before all negative examples, which corresponds to a ROC curve that goes from the

[1]Note that this rule carries a lot of information because one only needs to invert its predictions to have a perfect rule.

[2]Strictly speaking this would involve the construction of confidence bands as the ROC curve has to be estimated from samples. This is, however, outside the scope of this chapter. We refer the reader to Macskassy et al. (2005).

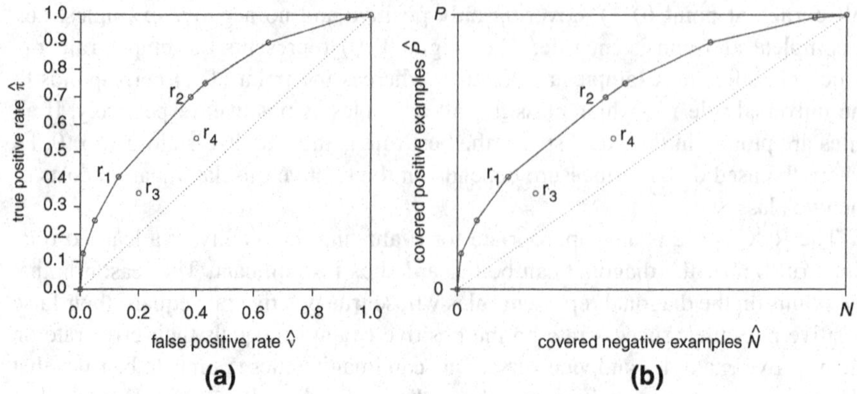

Fig. 3.1 A set of rules in (**a**) ROC space and (**b**) coverage space. Seven rules including $\mathbf{r}_1$ and $\mathbf{r}_2$ (marked with *full circles*) lie on the convex hull and are thus optimal for some cost trade-off. The two rules $\mathbf{r}_3$ and $\mathbf{r}_4$ (marked with *empty circles*) can never be optimal

empty rule $\mathbf{r}_\perp$ to the perfect rule and from there to the universal rule $\mathbf{r}^\top$. The worst ranker would rank the instances randomly, which corresponds to a ROC curve that moves along the diagonal. For evaluating the quality of a ranker, one can compute the *area under the ROC curve (AUC)*, which would be 1 for the ideal ranker and $\frac{1}{2}$ for a random ranker. The AUC corresponds to the probability that a randomly selected positive example is ranked before a randomly selected negative example.

For a more detailed account of ROC analysis for classification and ranking we refer to Fawcett (2006).

3.3 Coverage Space

While the ROC space is well-known from the literature, we believe that in the context of rule learning, it is more convenient to think in terms of absolute numbers of covered examples. Thus, instead of plotting $\hat{\pi}$ over $\hat{\nu}$, we plot the absolute number of covered positive examples $\hat{P}$ over the absolute number of covered negative examples $\hat{N}$. Following Fürnkranz and Flach (2005), we call this space the *coverage space*.[3] The coverage space is equivalent to the ROC space when comparing the quality of rules induced in a single domain. For example, see the right part of Fig. 3.1, where the false positive rate $\hat{\nu}$ on the x-axis is replaced by $\hat{N}$, the number of negatives covered by rule $\mathbf{r}$, and the true positive rate $\hat{\pi}$ on the y-axis is replaced by $\hat{P}$, the number of positives covered by rule $\mathbf{r}$.

[3]In previous work, Gamberger and Lavrač (2002) used the term TP/FP space, and Fürnkranz and Flach (2003, 2004) used the term PN-space. However, we prefer the term coverage space, which is used consistently in this book.

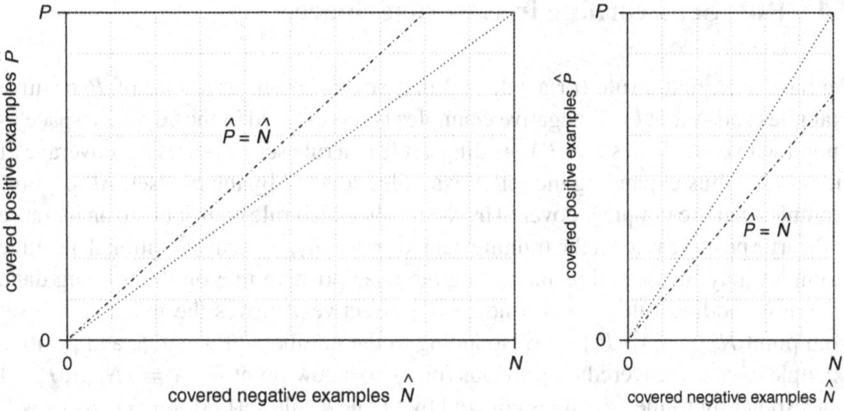

Fig. 3.2 The coverage space is a ROC space based on the absolute numbers of covered examples

Table 3.1 ROC spaces vs. coverage spaces

Property	ROC space	Coverage space
x-axis	$\hat{\nu} = \frac{\hat{N}}{N}$	$\hat{N}$
y-axis	$\hat{\pi} = \frac{\hat{P}}{P}$	$\hat{P}$
Empty rule $\mathbf{r}_\perp$	$(0, 0)$	$(0, 0)$
Perfect rule $\mathbf{r}^*$	$(0, 1)$	$(0, P)$
Universal rule $\mathbf{r}^\top$	$(1, 1)$	(N, P)
Resolution	$(\frac{1}{N}, \frac{1}{P})$	$(1, 1)$
Slope of diagonal	1	$\frac{P}{N}$
Slope of $\hat{P} = \hat{N}$ line	$\frac{N}{P}$	1

More formally, the coverage space is a two-dimensional space of points $(\hat{N}, \hat{P})$, where $\hat{N}$ denotes the number of negative examples covered by a rule (false positives) and $\hat{P}$ denotes the number of positive examples covered by a rule (true positives). Thus, contrary to the ROC space, the y- and x-axes are not necessarily of equal length, but the dimensions are $0 \leq \hat{N} \leq N$ and $0 \leq \hat{P} \leq P$, respectively. Figure 3.2 shows two examples, one for the case where the number of negative examples N exceeds the number of positive examples P (left), and one for the opposite case (right). We will typically assume $P \leq N$ as in the left graph of Fig. 3.2 because this is visually more pleasing.

A coverage graph can be turned into a ROC graph by simply normalizing the P and N-axes to the scale $[0, 1] \times [0, 1]$. Table 3.1 compares some of the properties of the coverage space to those of the ROC space. Nevertheless, the presentation of rules in coverage space has several properties that may be of interest depending on the purpose of the visualization. Particularly, the absolute numbers of covered positive and negative examples allow one to map the covering strategy into a sequence of nested coverage spaces, which we will briefly illustrate in the next section.

3.4 Rule Set Learning in Coverage Space

We have seen that a rule (or a rule set) that covers $\hat{P}$ out of a total of P positive examples and $\hat{N}$ out of N negative examples is represented in the coverage space as a point with coordinates $(\hat{N}, \hat{P})$. Adding a rule to a rule set increases the coverage of the rule set, thus expanding the set of examples covered by the rule set. All positive examples that are uniquely covered by the newly added rule contribute to an increase of the true positive rate on the training data. Conversely, covering additional negative examples may be viewed as increasing the false positive rate on the training data. Therefore, adding rule $\mathbf{r}_{i+1}$ to rule set $\mathcal{R}_i$ effectively moves the induced rule set from point $\mathcal{R}_i = (\hat{N}_i, \hat{P}_i)$ (corresponding to the numbers of negative and positive examples that are covered by previous rules), to a new point $\mathcal{R}_{i+1} = (\hat{N}_{i+1}, \hat{P}_{i+1})$ (corresponding to the examples covered by the new rule set). Moreover, $\mathcal{R}_{i+1}$ will typically be closer to (N, P) and farther away from $(0, 0)$ than $\mathcal{R}_i$.

Consequently, learning a rule set one rule at a time may be viewed as a path through coverage space, where each point on the path corresponds to the addition of a rule to the theory. Such a *coverage path* starts at $(0, 0)$, which corresponds to the empty rule set which does not cover any examples. Figure 3.3 shows the coverage path for a theory with three rules. Each point $\mathcal{R}_i = \bigcup_{j=1}^{i}\{\mathbf{r}_j\}$ represents the rule set consisting of the first i rules. Adding a rule moves the induced rule set to the next point $\mathcal{R}_{i+1} = \mathcal{R}_i \cup \{\mathbf{r}_{i+1}\}$.

Removing the covered (positive and negative) examples has the effect of switching to a subspace of the original coverage space, using the last learned rule as the new origin.[4] Thus the path may also be viewed as a sequence of nested coverage spaces CS_i. After the final rule has been learned, one can imagine adding yet another rule with a body that is always true. Adding such a rule has the effect that the theory now classifies *all* examples as positive, i.e., it will take us to the point $\mathcal{R}^{\top} = (N, P)$. Note that even this universal theory might be optimal under some cost assumptions.

As discussed in the previous chapter, the vast majority of algorithms use a heuristic top-down hill-climbing or beam search strategy for finding individual rules: rules are successively specialized by greedily adding the most promising feature to the rule body (cf. Fig. 2.8). Just as with adding rules to a rule set, successive refinements of a rule describe a path through coverage space (Fig. 3.3, right). However, in this case, the path starts at the upper right corner (covering all positive and negative examples), and successively proceeds towards the origin, which corresponds to the empty rule $\mathbf{r}_{\perp}$, a rule that is too specific to cover any example.

---

[4]Indeed, in every step of the covering algorithm, covered examples are removed from the current training set. However, depending on the implementation, the algorithm either removes all the covered examples (as depicted in Fig. 3.3) or removes only the covered positive examples. This will be discussed in more detail in Sect. 8.2.

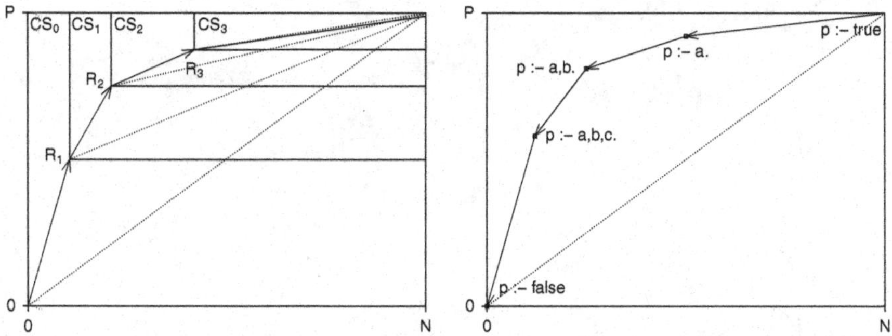

Fig. 3.3 Schematic depiction of the paths in the coverage space for (*left*) the covering strategy of learning a rule set by adding one rule at a time and (*right*) top-down specialization of a single rule

3.5 Conclusion

This chapter introduced the coverage space as a framework that allows us to study covering properties of features, rules, and rule sets in terms of the trade-off between the number of positive examples covered by the rule (true positives, $\hat{P}$) and the number of covered negative examples (false positives, $\hat{N}$). It will be used as the basic analytic tool for analyzing covering properties of features, which is discussed in Chap. 4, and for visualizing the behavior of rule quality evaluation measures by looking at their *isometrics*—lines that connect rules that are evaluated equally by the heuristic—which is the topic of Chap. 7.

Chapter 4
Features

Rule learning systems use features as the main building blocks for rules. A feature can be a simple attribute-value test or a test for the validity of a complex domain knowledge relationship. Most existing concept learning systems generate features in the rule construction process. In contrast, this chapter shows that the separation of the feature construction and rule construction process has several theoretical and practical advantages. In particular, explicit usage of features enables a unifying framework of both propositional and relational rule learning. We demonstrate procedures for generating a set of simple features that—in domains with no contradictory examples—enable the construction of complete and consistent rule sets, and do not include obviously irrelevant features. The concept of feature relevancy is illustrated in the coverage spaces context as defined in Chap. 3. Furthermore, this chapter proposes a criterion for total and relative irrelevancy that has a quality-preserving property with respect to the resulting rules. We also discuss techniques for handling missing and numerical values.

After a brief introduction to the concept of features (Sect. 4.1), we start with an overview of different feature types in Sect. 4.2. In Sect. 4.3, we describe a simple algorithm for transforming attribute-value data into a feature-based representation in the form of a covering table. In Sect. 4.4, we introduce and discuss our notion of feature relevancy. This section includes a theoretical analysis of the concept of features, feature relevancy, and discriminating pairs of examples, as well as theoretical results that describe relations among features, relations between features and complete and consistent rule sets, and properties of the presented algorithm for simple propositional feature construction. Finally, Sects. 4.5 and 4.6 discuss different approaches for handling unknown and imprecise attribute values during feature construction.

[†]Parts of this chapter are based on Lavrač, Fürnkranz, and Gamberger (2010).

J. Fürnkranz et al., *Foundations of Rule Learning*, Cognitive Technologies,
DOI 10.1007/978-3-540-75197-7_4,

4.1 Introduction

Recall from Sect. 2.3 that rule learning algorithms construct rules of the general form

$$c_i \leftarrow \mathbf{f}_1 \wedge \mathbf{f}_2 \wedge \ldots \wedge \mathbf{f}_L$$

The body of the rule is a logical conjunction of features, where a *feature* is a test that checks whether the object to classify has the specified property or not. Because a complete hypothesis is built as a set of rules, and each rule is built as a conjunction of features, features are indeed the basic elements used in the rule construction process. In this way, features and their properties determine the form and quality of knowledge that can be induced by a rule learning algorithm. Formally speaking, features, their properties, and the way they are defined are an essential part of the hypothesis description language.

Features describe the presence or absence of certain properties of an instance. Thus, features are always Boolean-valued, i.e., either true or false. As illustrated in Table 4.1, features can test a value of a single attribute, like $\mathbf{A}_i > 3$, or they can represent complex logical and numerical relations, integrating properties of multiple attributes, like $\mathbf{A}_k < 2 \cdot (\mathbf{A}_j - \mathbf{A}_i)$.

It is important to realize that features differ from the attributes that describe instances in the input data. Attributes can be numerical variables (with values like 7 or 1.5) or nominal/discrete variables (with values like red or female). Also, in contrast to attributes, a feature cannot have a missing or unknown value. As a result, features are different from binary attributes even for binary-valued attributes that have values true and false. For example, in Table 4.2 for a binary attribute $\mathbf{A}_m$, the corresponding features are $\mathbf{A}_m = \text{true}$ and $\mathbf{A}_m = \text{false}$. Of course, more complex features like $\mathbf{A}_m \neq \mathbf{A}_n$ or $(\mathbf{A}_m = \text{true} \wedge \mathbf{A}_n = \text{true})$ are also possible.

Note that in this book, the use of the term *feature* is not fully aligned with the practice in the machine learning community, where terms like *feature extraction*, *feature construction*, or *feature selection* are commonly used to denote approaches that aim at finding a suitable set of descriptors for the training examples by including expert knowledge, increasing the quality of learning, or switching to a more expressive hypothesis language. As most learning algorithms, such as decision tree learners, focus on attributes, the term *feature* is frequently used as a synonym for *attribute*. In this book, we clearly distinguish between these two terms. We believe that the advantage of providing a clear semantics to the term *feature* outweighs the incompatibility with the fuzzy use of the term in some previous works.

Rule learning algorithms are *feature-based*, because rule learning algorithms employ features as their basic building blocks, whereas, for example, decision tree learning algorithms are attribute-based, because the nodes of decision trees are constructed from attributes.[1] For many rule learning systems this is not obvious because

[1]This can, for example, also be seen from the different definitions of information gain for decision tree learning (Quinlan, 1986) and rule learning (Quinlan, 1990).

Table 4.1 Illustration of simple and complex features in a domain with three instances described by three continuous attributes

Attributes			Features	
$\mathbf{A}_i$	$\mathbf{A}_j$	$\mathbf{A}_k$	$\mathbf{A}_i > 3$	$\mathbf{A}_k < 2 \cdot (\mathbf{A}_j - \mathbf{A}_i)$
7	1.5	2	True	False
4	3	−4	True	True
1.07	2	0	False	True

Table 4.2 Examples of features defined on a set of three instances with two binary attributes

Attributes		Features			
$\mathbf{A}_m$	$\mathbf{A}_n$	$\mathbf{A}_m$ = true	$\mathbf{A}_m$ = false	$\mathbf{A}_m \neq \mathbf{A}_n$	$\mathbf{A}_m$ = true $\wedge$ $\mathbf{A}_n$ = true
True	True	True	False	False	True
True	False	True	False	True	False
False	False	False	True	False	False

the process of transformation of attributes into features is implicit and is tightly integrated into the learning algorithm (Clark & Niblett, 1989; Cohen, 1995). In these systems, feature construction is part of the rule building process. The main reason for this strategy is simplicity and especially memory usage efficiency. Explicit usage of features requires that so-called *covering tables* are constructed, such as those shown in Tables 4.1 or 4.2. For a large number of attributes, these tables may be quite large. Nevertheless, there are rule learning algorithms that explicitly construct covering tables before starting the rule construction process. A classical example is the LINUS system for converting relational learning problems into a propositional form (Lavrač & Džeroski, 1994a; Lavrač, Džeroski, & Grobelnik, 1991; Lavrač & Flach, 2001).

As features may be considered as simple logical propositions with truth values true and false, the term *propositional learning*, which is frequently used to refer to any kind of learning that uses an attribute-value representation of the data, actually refers to an implicit interpretation of this table as a set of simple features. However, the term propositional learning has the intention to stress the difference to the use of first-order logic in relational rule learning. Thus, we suggest the term *feature-based learning* to emphasize that we intend to unify both scenarios in a single framework based on features and covering tables, as we will see later in this chapter.

In this chapter, we focus on propositional domains, but also briefly show that the concepts are applicable to relational domains as well; the latter will be discussed in more detail in Chap. 5. In the following, we describe procedures for the transformation of a set of attributes into a set of features, prove relevant relations that hold among features in the concept learning framework, and point out practical advantages of an explicit representation of the covering table. We will formally define the concept of feature relevancy, which enables a significant reduction of the number of features that effectively enter into the rule construction process. Finally, we also briefly discuss techniques for handling unknown or missing attribute values, and for soft thresholding of numerical attributes.

4.2 Feature Types

Features as the central component of rule learning algorithms have been defined and used in all previous works on rule learning algorithms. However, the terminology used differs from paper to paper. Michalski (1973) introduced the term *selector*. Some rule learning systems, such as CN2 (Clark & Niblett, 1989), use the term *attribute test*. Others, like RIPPER (Cohen, 1995), prefer the term *condition*. In the descriptive induction framework of association rule learning, features correspond to *items*, and conjunctions of features correspond to *itemsets*. Applications related to inductive logic programming and relational rule learning typically borrow the term *literal* from the logic programming terminology. Predefined conjunctions of literals are also known as *templates* or *schemata*. In this book, we use the term *feature* as a generalization of all these terms in the particular context of rule learning.

4.2.1 Propositional Features

In many cases, features are derived from an attribute-value description of the data, where each attribute gives rise to several features, typically corresponding to the different values of an attribute. Each feature refers only to information from a single attribute; relations between multiple attributes are not considered. We refer to such features as *propositional features*.

Selectors. A *selector* is a descriptor of the form $(\mathbf{A}_i \sim v_{i,j})$, where an example's value for a certain attribute $\mathbf{A}_i$ is related to a constant value $v_{i,j}$ of its domain via relations like $=$, $>$, or $<$. Equality relations are typically used for symbolic attributes, while inequalities are used for numeric, in particular for continuous attributes. Often the negations of these relations (i.e., $\neq$, $\leq$, $\geq$) are also available.

Algorithms of the AQ family (Bergadano, Matwin, Michalski, & Zhang, 1992; Michalski, 1980; Michalski, Mozetič, Hong, & Lavrač, 1986) are able to extend these elementary features to attribute sets (*internal conjunctions*), value sets (*internal disjunctions*), and intervals (*range operators*). A description of these extensions can be found in Michalski. Internal disjunctions are the most popular among these extensions (they are, e.g., also used in the BEXA rule learner (Theron & Cloete, 1996). They allow the learner to specify a subset of possible attribute values:

$$\mathbf{A}_i = \{v_1, v_2, v_3\}$$

which means that the attribute $\mathbf{A}_i$ can take any of the specified values. Note that such internal disjunctions can be transformed into conjunctions by using inequalities. If attribute $\mathbf{A}_i$ has five different values $v_1 \dots v_5$, the above internal disjunction is equivalent to the conjunction

$$(\mathbf{A}_i \neq v_4) \wedge (\mathbf{A}_i \neq v_5)$$

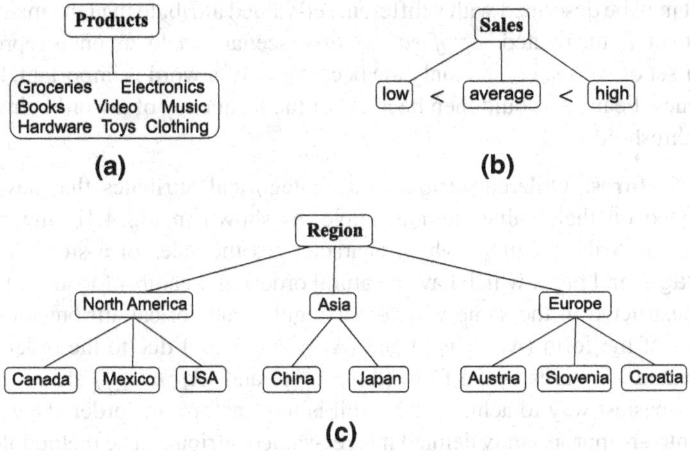

Fig. 4.1 Different structured attribute types. **(a)** Set-valued attribute. **(b)** Ordered attribute. **(c)** Hierarchical attribute

Many rule learners, such as RIPPER, do not explicitly deal with internal disjunctions, but can realize them indirectly via inequalities.

Set-valued features. Some attributes are set-valued in the sense that an object cannot only be described by a single attribute value, but by a set of possible values. Features are derived from set-valued attributes by testing whether one of the possible values is present in the object's description or not, i.e., the features have the form $v_{i,j} \in \mathbf{A}_i$ or $v_{i,j} \notin \mathbf{A}_i$. Essentially, a set-valued attribute is only a shorthand notation for a set of binary attributes which indicate for each possible value of the set whether the value is present or not. Following Cohen (1996), we call such features *set-valued features*.

For example, Fig. 4.1a shows an attribute **Products**, which encodes which groups of products a store sells. Obviously, stores can sell more than one of these product groups, and each store can be characterized with the set of products it sells. The features derived from this set-valued attribute are, e.g., Books ∈ **Products** or Clothing ∈ **Products**.

Association rule discovery algorithms (see Sect. 1.6.1) operate on a single set-valued attribute, a set of possible *items*. The standard application for such tasks is *market basket analysis*, where the items are the products that can be bought by store's customers, and the learning task is to discover which *itemsets* are frequently bought together. As there is typically only one set-valued attribute for all products, items are typically denoted without reference to an attribute name. For example, items diapers and beer correspond to the features diapers ∈ **Basket** and beer ∈ **Basket**.

Another scenario where set-valued features occur naturally is text classification, where a document is often described as a *set of words*. Note that different set-valued attributes may be used to describe the same object. In text classification tasks, these attributes may correspond to different parts of the text. For example, the title of a document may be described with a different set-valued attribute than the main text of the document. In the related *bag of words (BoW)* scenario, a document is represented as a multi-set of words, i.e., not only the occurrence of a word is important, but also its frequency. Features would then be a test of the frequency of a word being above a certain threshold.

Ordered features. Ordered attributes are categorical attributes that have some order defined on their values. For example, as shown in Fig. 4.1b, the possible values for an attribute Sales, which characterizes the sales of a store, might be low, average, and high, which have a natural ordering. Features for such attributes can be constructed in the same way as for regular categorical attributes resulting by features of the form $(\mathbf{A}_i = v_{i,j})$ and $(\mathbf{A}_i \neq v_{i,j})$. But due to the order among attribute values, features of the form $(\mathbf{A}_i > v_{i,j})$ and $(\mathbf{A}_i < v_{i,j})$ are also appropriate. The easiest way to achieve the result is to transform the ordered categorical attribute into an appropriately defined integer-valued attribute. The methodology for handling integer-valued attributes described at the end of Sect. 4.3.2 will result in a set of propositional features corresponding to the expected set of ordered features.

It must be noted that some categorical attributes may be ordered although they do not seem to be. An example is a 'color' attribute with values yellow, red, and blue. Depending on the goals of induction, an appropriate ordering may be based on the wavelength of these colors, resulting in blue < yellow < red.

A different situation is with a categorical attribute having days of the week as its values. Here we have clear ordering Sunday—Monday—Tuesday—...—Saturday, but there is a closed loop problem that after Saturday follows Sunday. In such cases there does not exist a straightforward transformation into an integer-valued attribute that leads to the construction of all potentially relevant ordered features.

Hierarchical features. In some application domains values of an underlying categorical attribute are organized in a tree structure. Strictly speaking, features based on such attributes are still propositional in the sense that they only involve a single attribute. However, in many cases, the mutual relations between the resulting features may be useful for the induction process.

Figure 4.1c shows an example which describes the geographic location of a store with two levels of granularity. The top level specifies the continent (North America, Asia, or Europe), each of which can be refined into different countries (e.g., USA, China, or Austria). Of course, the hierarchy could be deeper, encoding states (Pennsylvania) and cities (Pittsburgh) in lower levels. The features that are defined upon these values will cover monotonically increasing sets of examples. Such attributes are frequently supported by data warehouses, where operations like *roll-up* and *drill-down* allow one to navigate through these hierarchies (Han & Kamber, 2001). These operations essentially correspond to generalization and specialization operators, and can also be directly exploited for induction, as, e.g., in Cai, Cercone, and Han (1991) and Han, Cai, and Cercone (1992). In the propositional framework the easiest way to handle hierarchical feature construction is to substitute such an attribute by a set of attributes so that each of them describes one level of the hierarchy.

4.2.2 Relational Features

While all previous feature types only involve a single attribute and can thus be directly mapped to propositional features, relational features differ because they relate the values of two different attributes to each other. In the simplest case, one could, for example, test for the equality or inequality of the values of two attributes of the same type, such as Length > Height. Note that here the equality (and inequality) relations are used in a different way: while propositional features can only compare the value of one attribute with a certain domain constant, general literals are able to compare the values of two attributes with each other. The two right-most features of Table 4.2 are relational.

In many cases these background relations cannot only make use of the available attributes, but also introduce new variables that can be used for subsequent tests in this rule. For example, if relations are used to encode a graph, a relation like link(OldNode,NewNode) can be used to subsequently test properties of a new node in the graph. A problem often caused by conditions introducing new variables is that they have no discriminatory power when used for distinguishing between the positive and negative examples of a concept. Top-down hill-climbing algorithms that learn rules by adding one condition at a time will thus attribute low importance to these conditions, because they appear to be unable to discriminate between positive and negative examples.

Because of this added complexity, we will deal with relational features separately in Sect. 5.3. However, note that relational feature construction can be made explicit in the same way as propositional feature construction, as we will see in the next section. In both cases, the subsequent rule learning phase can operate on a covering table and need not be concerned how this table was generated.

4.3 Feature Construction

The construction of features for the given set of training examples is the first step in the rule learning process. It can also be viewed as the transformation from the attribute space into the space of features. Although the construction of simple features is a straightforward and a rather simple task, construction of an appropriate set of sophisticated features is a hard task. The first part of the problem lies in the fact that, for a given set of attributes, the set of possible features may be very large. This is particularly true in situations when features include numerical relations based on more than one continuous attribute; if in such a situation the complexity of relations is not strictly limited, the number of potentially interesting features can explode. The second part of the problem is the fact that the appropriateness of the set of features is domain-dependent.

Generating interesting features can be viewed as a separate knowledge discovery task. Its goal is to construct a so-called *covering table*, which represents the input

Table 4.3 A covering table

Examples		Features				
ID	Class	$\mathbf{f}_1$	$\mathbf{f}_2$	$\mathbf{f}_3$	...	$\mathbf{f}_F$
p_1	$\oplus$	True	False	False	...	True
p_2	$\oplus$	False	True	False	...	True
n_1	$\ominus$	False	True	True	...	True
n_2	$\ominus$	True	False	True	...	False
n_3	$\ominus$	False	True	False	...	False

data for rule learning through the discovered data properties. Covering tables are introduced in Sect. 4.3.1. Construction of set-valued features is straightforward while a simple algorithm for generating a set of propositional features from an attribute-value training set is presented in Sect. 4.3.2. Special cases of ordered and hierarchical attributes are handled via a transformation into one or more attributes that may enter the regular propositional feature construction process, as described in Sect. 4.2.1. The construction of relational features is briefly discussed in Sect. 4.3.3 but will be further elaborated in Chap. 5.

4.3.1 Covering Tables

Consider a two-class problem (a concept learning setting) where the positive class is denoted as $\oplus$, and the negative class as $\ominus$. A *covering table* is a table which has examples as its rows and features as its columns. A covering table has only values true and false as its elements. These truth values represent the covering properties of features on the given set of examples. Together with example classes, the covering table is the basic information necessary for the rule induction process. Rule construction from the available set of features can be done by any supervised rule induction algorithm. The actual attribute values are no longer needed for rule construction. The significant difference of explicit feature construction compared to the implicit feature construction that is performed by most standard rule learning algorithms is only that we do not generate features from attribute values during rule construction. Instead, the possible features with corresponding covering properties are obtained from the precomputed covering tables (Table 4.3).

The price of explicit feature construction is the increase of the space complexity because the covering table will have considerably more columns than the corresponding table presenting the examples by their attributes. However, explicit feature construction provides a much clearer view on the rule induction process. Some of the key advantages are: a possibility to use the same rule building blocks both for propositional and relational rule learning, a possibility to introduce feature relevancy and ensure that only the most relevant features really enter the rule learning process (Sect. 4.4), and the possibility to handle unknown attribute values (Sect. 4.5) and imprecision of continuous attributes (Sect. 4.6) during covering table construction.

```
function GENERATEFEATURES(ℰ,𝒜)
  Input:
     ℰ = 𝒫 ∪ 𝒩: a set of positive and negative training examples
     𝒜: a set of attributes

  Algorithm:
     ℱ := ∅
     for each attribute A ∈ 𝒜 do
          𝒱_𝒫 := all different values of A appearing in the positive examples
          𝒱_𝒩 := all different values of A appearing in the negative examples
          if A is a discrete attribute
          then for each v ∈ 𝒱_𝒫 do ℱ := ℱ ∪ {A = v}
                for each v ∈ 𝒱_𝒩 do ℱ := ℱ ∪ {A ≠ v}
          if A_i is a continuous attribute
          then 𝒱 := {v_1, ... v_|𝒱|} a sorted set of the attribute values in 𝒱_𝒫 ∪ 𝒱_𝒩
                for i ∈ {1...|𝒱| − 1} do
                      v := (v_i + v_{i+1})/2
                      if v_i ∈ 𝒱_𝒫 and v_{i+1} ∈ 𝒱_𝒩
                      then ℱ := ℱ ∪ {A < v}
                      if v_i ∈ 𝒱_𝒩 and v_{i+1} ∈ 𝒱_𝒫
                      then ℱ := ℱ ∪ {A ≥ v}
                endfor
          endif
     endfor
  Output:
     ℱ a set of features
```

Fig. 4.2 An algorithm for the transformation of a set of attributes into a set of propositional features

4.3.2 Propositional Feature Construction

Gamberger and Lavrač (2002) proposed an algorithm enabling the construction of simple features from attributes for a concept learning problem, which is presented in Fig. 4.2. The algorithm generates features separately and independently for each attribute. Although the algorithm does not generate all the features, it will be shown in Sects. 4.4.2 and 4.4.3 that it generates feature sets which are sufficiently rich for building complete and consistent hypotheses.

If an attribute is discrete then all distinct values in the positive examples are detected and for them features of the form $\mathbf{A}_i = v_{i,j}$ are generated. Also all distinct values for negative examples are detected and from them features of the form $\mathbf{A}_i \neq v_{i,j}$ are generated. The number of generated features for a discrete attribute is equal to the sum of distinct attribute values occurring in positive and negative examples.

For a continuous attribute, the features are generated in the following way. We first identify pairs of neighboring values that originate from examples with different

Table 4.4 A dataset with two positive and three negative examples, described by three attributes

Examples		Attributes		
ID	Class	$\mathbf{A}_1$	$\mathbf{A}_2$	$\mathbf{A}_3$
p_1	⊕	Red	1.2	4
p_2	⊕	Blue	1.3	2
n_1	⊖	Green	2.1	3
n_2	⊖	Red	2.5	1
n_3	⊖	Green	0.6	1

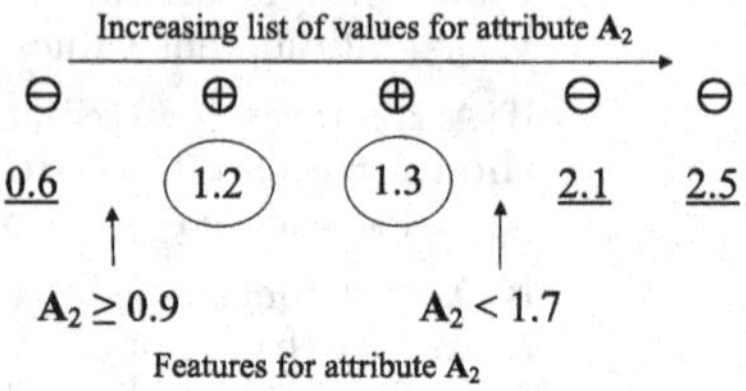

Fig. 4.3 Illustration of the process of feature construction for continuous attribute $\mathbf{A}_2$ from Table 4.4

class labels. From these pairs, we compute the mean value $\bar{v}$ of the two neighboring values. If the smaller of the two values is from the positive class we generate the feature $\mathbf{A}_i < v$, while if the smaller of the values is from the negative class we generate the feature $\mathbf{A}_i \geq v$. Note that the sorted set of values $\mathcal{V}$ does not contain duplicate values, but every value $v \in \mathcal{V}$ may appear in multiple positive and negative examples. Thus, it is also possible that both features are generated for the same pair of neighboring values. Thus, the maximum number of values is $2 \cdot (|\mathcal{V}| - 1)$, but the number that is actually generated is typically smaller, depending on the grouping of classes in the sorted value set $\mathcal{V}$.

For the domain presented in Table 4.4, consisting of two positive examples $P = \{p_1, p_2\}$ and three negative examples $N = \{n_1, n_2, n_3\}$, the feature construction algorithm generates the following features:

- $\mathbf{A}_1$: features $\mathbf{A}_1 =$ red and $\mathbf{A}_1 =$ blue are constructed for the values detected in the positive examples, and features $\mathbf{A}_1 \neq$ green and $\mathbf{A}_1 \neq$ red for the $\mathbf{A}_1$ values detected in negative examples.
- $\mathbf{A}_2$: features $\mathbf{A}_2 \geq 0.9$ and $\mathbf{A}_2 < 1.7$ are constructed as illustrated in Fig. 4.3.
- $\mathbf{A}_3$: the algorithm constructs features $\mathbf{A}_3 \geq 1.5$ for the value pair $(2, 1)$, $\mathbf{A}_3 < 2.5$ for pair $(2, 3)$, and $\mathbf{A}_3 \geq 3.5$ for pair $(4, 3)$.

Table 4.5 shows a part of the covering table constructed for the features generated for our example domain (Table 4.4).

It is important to note that for continuous attributes only features of the form $\mathbf{A}_i \geq v$ and $\mathbf{A}_i < v$ are generated. Numeric equalities like $\mathbf{A}_i = v_{i,j}$ are typically not implemented although practically they could be. The reason is that for most continuous attributes it does not make much sense to test the exact value because typically only one or a few objects will have exactly the same feature value. Besides, there is also the problem of imprecise measurements, which we will deal with later in this chapter (Sect. 4.6). However, there are also cases where it is completely justified to generate features of the form $\mathbf{A}_i = v_{i,j}$ for continuous attributes. This

Table 4.5 Part of the covering table generated for the domain of Table 4.4

Examples		Features					
ID	Class	$A_1 = \text{red}$	$A_1 \neq \text{red}$	$A_2 \geq 0.9$	$A_2 < 1.7$	$A_3 \geq 1.5$	...
p_1	⊕	True	False	True	False	True	...
p_2	⊕	False	True	True	False	True	...
n_1	⊖	False	True	True	True	True	...
n_2	⊖	True	False	True	True	False	...
n_3	⊖	False	True	False	False	False	...

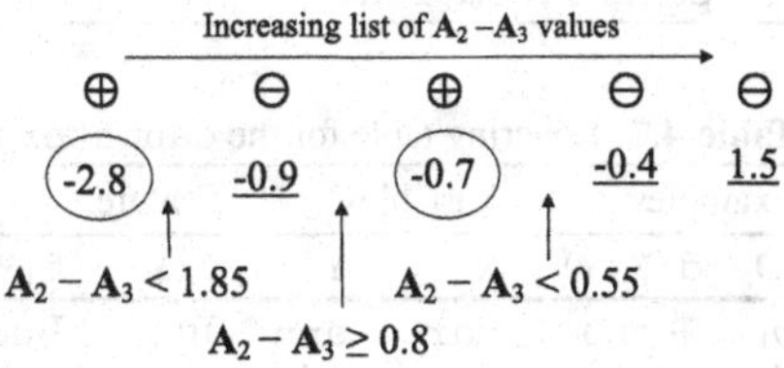

Fig. 4.4 Illustration of the construction process for relational features of the form $\mathbf{A}_2 - \mathbf{A}_3 \geq v$ or $\mathbf{A}_2 - \mathbf{A}_3 < v$ for the example data of Table 4.4

is the case in situations when attribute values are integers representing distinct, well specified, and disjunctive concepts, such as floors in a building, or school years as educational levels. In such cases, it is good practice to treat such attributes as being both continuous and discrete, resulting in both types of features. An example may be attribute $\mathbf{A}_3$ in Table 4.4 for which, in addition to the already mentioned features, one can also generate features $\mathbf{A}_3 = 4$, $\mathbf{A}_3 = 2$, $\mathbf{A}_3 \neq 3$, and $\mathbf{A}_3 \neq 1$.

Another way for dealing with numerical features is to *discretize* them into a set of ordered values. A variety of algorithms has been proposed for this problem; we refer to Yang, Webb, and Wu (2005) for a survey.

4.3.3 Relational Feature Construction

Similarly, the algorithm can also deal with arbitrary numerical functions and generate features of the form $f(\mathbf{A}_{i_1},\ldots\mathbf{A}_{i_m}) < v$ or $f(\mathbf{A}_{i_1},\ldots\mathbf{A}_{i_m}) \geq v$, where $f(\mathbf{A}_{i_1},\ldots\mathbf{A}_{i_m})$ is any fixed, parameter-free function of one or more continuous attributes. Again, we generate a temporary numerical attribute $\mathbf{A}_f$ which represents all possible values of $f(\mathbf{A}_{i_1},\ldots\mathbf{A}_{i_m})$, and apply GENERATEFEATURES to this attribute. Figure 4.4 illustrates this possibility for features $f(\mathbf{A}_2,\mathbf{A}_3)$ of the form $\mathbf{A}_2 - \mathbf{A}_3 < k$ or $\mathbf{A}_2 - \mathbf{A}3 \geq k$ for attributes $\mathbf{A}_2$ and $\mathbf{A}_3$ from the domain presented in Table 4.4. By applying the described algorithm to this attribute, we obtain features $f(\mathbf{A}_2,\mathbf{A}_3) = \{(\mathbf{A}_2 - \mathbf{A}_3 < -1.85), (\mathbf{A}_2 - \mathbf{A}_3 \geq -0.85), (\mathbf{A}_2 - \mathbf{A}_3 < -0.55)\}$.

Note that the described algorithm for the construction of propositional features may also be adapted to the construction of first-order relational features of the form $r(\mathbf{X}_1,\mathbf{X}_2,\ldots\mathbf{X}_r)$, where r is a relation that is defined over some variables $\mathbf{X}_i$. The $\mathbf{X}_i$ can be a subset of the attributes $\mathcal{A}$ as above, but may also consist of additional variables that are linked to the training examples via various relations in the background knowledge.

Table 4.6 Training examples and background knowledge for learning the `daughter` relation

Positive examples ⊕	Negative examples ⊖
p_1: `daughter(mary,ann)`	n_1: `daughter(tom,ann)`
p_2: `daughter(eve,tom)`	n_2: `daughter(eve,ann)`
Background knowledge	
`parent(ann,mary)`	`female(ann)`
`parent(ann,tom)`	`female(mary)`
`parent(tom,eve)`	`female(eve)`
`parent(tom,ian)`	

Table 4.7 Covering table for the `daughter` relation

Examples		Variables		Features					
ID	`d(X,Y)`	`X`	`Y`	`f(X)`	`f(Y)`	`p(X,X)`	`p(X,Y)`	`p(Y,X)`	`p(Y,Y)`
p_1	⊕ true	`mary`	`ann`	True	True	False	False	True	False
p_2	⊕ true	`eve`	`tom`	True	False	False	False	True	False
n_1	⊖ false	`tom`	`ann`	False	True	False	False	True	False
n_2	⊖ false	`eve`	`ann`	True	True	False	False	False	False

As an illustration, consider a simple relational problem of learning family relations. The task is to define target relation `daughter(X,Y)`, which states that person `X` is daughter of person `Y`, in terms of background knowledge relations `female` and `parent`. All the variables are of type `person`, defined as `person` = {`ann`, `eve`, `ian`, `mary`, `tom`}. There are two positive and two negative examples of the target relation and some background knowledge, as shown in Table 4.6.

Relational features are constructed by determining all possible applications of the background predicates on the arguments of the target relation, taking into account argument types. Each such application introduces a new feature. In our example, all variables are of the same type `person`. The corresponding covering table is shown in Table 4.7, where `d` stands for `daughter`, `f` stands for `female`, and `p` for `parent`. The tuples in the covering table are generalizations (relative to the given background knowledge) of the individual facts about the target relation. The columns entitled 'Variables' show the instantiations of the target relation, while the 'Features' columns denote the newly constructed relational features, which are used as the features for learning.

The illustrated approach to first-order relational feature construction has been introduced in LINUS, a system for converting relational learning problems into propositional form (Lavrač & Džeroski, 1994a; Lavrač et al., 1991). This approach is described in further detail in Sect. 5.3.

4.4 Feature Relevancy

This section defines conditions under which irrelevant features can be eliminated from the learning process, thus leading to more effective and efficient rule learning. We will start by defining sufficiently rich feature sets and proving a simple theorem that allows us to determine from the covering table whether a given feature set has this property (Sect. 4.4.1). In the following sections, we will establish different degrees of irrelevancy for features, such as total irrelevancy (Sect. 4.4.2), relative irrelevancy (Sect. 4.4.3), and constrained irrelevancy (Sect. 4.4.4). In these sections, we will also prove the correctness of the GENERATEFEATURES algorithm in the sense that we will show that it does not discard relevant features. We conclude with a discussion of the importance of relevancy filters for rule induction (Sect. 4.4.5).

4.4.1 Sufficiently Rich Feature Set

Although the construction of complete and consistent rule sets is seldom the ultimate learning goal, it is nevertheless of practical and theoretical interest whether such a rule set exists for a given feature set.

Definition 4.4.1 (Sufficiently rich feature set). A *sufficiently rich feature set* for a given training set of examples is a set of features from which a complete and consistent rule set can be found.

We will now work towards establishing a theorem that allows us to establish from its covering table whether a feature set is sufficiently rich. We follow Lavrač, Gamberger, and Jovanoski (1999) by defining the concept of pn-pairs and discriminating features.

Definition 4.4.2 (pn-pair). A *pn-pair* is a pair of training examples p/n where $p \in \mathcal{P}$ and $n \in \mathcal{N}$.

Definition 4.4.3 (Discriminating feature). A feature $\mathbf{f} \in F$ discriminates[2] a pair p/n iff $\mathbf{f}$ correctly classifies both examples, i.e., if it is *true* for p and *false* for n. We denote this as $\mathbf{f} \sqsupset p/n$.

From Definition 4.4.3 we can see that a good feature—one that is true for many positive examples and false for many negative examples—discriminates many pn-pairs.

Theorem 4.4.1. *$\mathcal{F}$ is sufficiently rich if and only if for every possible pn-pair p/n of the training set $\mathcal{E} = \mathcal{P} \cup \mathcal{N}$ there exists at least one feature $\mathbf{f} \in \mathcal{F}$ which discriminates the pn-pair, i.e., for which $\mathbf{f} \sqsupset p/n$.*

Proof. Necessity: Assume that a complete and consistent rule set $\mathcal{R}$ can be found using only features from $\mathcal{F}$. Suppose that a pn-pair exists that is not discriminated by any feature $\mathbf{f} \in \mathcal{F}$. Then no rule built of features from $\mathcal{F}$ will be able to discriminate between these two examples. Consequently, a description which is both complete and consistent cannot be found.

[2] Lavrač et al. (1999) used the term *coverage* of pn-pairs instead of pn-pair discrimination.

Sufficiency: Assume that for each possible pn-pair p/n from training set $\mathcal{E}$ there exists at least one feature $\mathbf{f} \in \mathcal{F}$ with $\mathbf{f} \sqsupset p/n$. Take a positive example p_i and select all features $\mathcal{F}_i$ which discriminate p_i from some negative example n_j, i.e.,

$$\mathcal{F}_i = \{\mathbf{f}_{i,j} \in \mathcal{F} | j \in \{1 \dots N\} \wedge (\mathbf{f}_{i,j} \sqsupset p_i/n_j)\}$$

From these features, we can now construct a conjunctive rule $\mathbf{r}_i$

$$\mathbf{r}_i = (\oplus \leftarrow \mathbf{f}_{i,1} \wedge \dots \wedge \mathbf{f}_{i,N})$$

for $\mathbf{f}_{i,1} \dots \mathbf{f}_{i,N} \in \mathcal{F}_i$. $\mathbf{r}_i$ will now cover p_i because each feature $\mathbf{f} \in \mathcal{F}_i$ covers p_i (by construction), and it will not cover any $n_j \in \mathcal{N}$, because at least feature $\mathbf{f}_{i,j}$ will not cover n_j (again by construction). Obviously, this construction can be repeated for each positive example in $\mathcal{P}$, resulting in a complete and consistent rule set $\mathcal{R} = \{\mathbf{r}_1, \dots, \mathbf{r}_P\}$. □

4.4.2 Total Irrelevancy

An immediate consequence of Theorem 4.4.1 is that all features that do not discriminate between *any* pn-pair are useless for rule construction. We call such features *totally irrelevant*.

Definition 4.4.4 (Total irrelevancy). Feature $\mathbf{f} \in \mathcal{F}$ is *totally irrelevant* iff $\nexists p/n$ s.t. $\mathbf{f} \sqsupset p/n$.

Totally irrelevant features can be removed by a so-called *relevancy filter* from any feature set because they cannot help in building rules that are true for positive examples and false for negative examples.

Note that for checking whether a feature is totally irrelevant, we do not need to construct all pn-pairs, as shown by the following theorem.

Theorem 4.4.2. *Feature* **f** *is totally irrelevant iff it is* false *for all positive examples or* true *for all negative examples.*

Proof. A feature is totally irrelevant if it does not discriminate any pn-pairs. This can be the case if it does not cover any positive examples. On the other hand, if it covers at least one positive example, it must also cover all negative examples because otherwise there would exist a pn-pair that is discriminated by **f**. □

Figure 4.5 shows the position of totally irrelevant features in *coverage space* (cf. Sect. 3.3). Such features either cover no positive examples ($\hat{P} = 0$) or cover all negative examples ($\hat{N} = N$), i.e., they can be found at the lower and right borders of the coverage space.

Note that one must not confuse feature irrelevancy with attribute irrelevancy. If we have a medical test T that perfectly recognizes cancer patients (i.e., T is

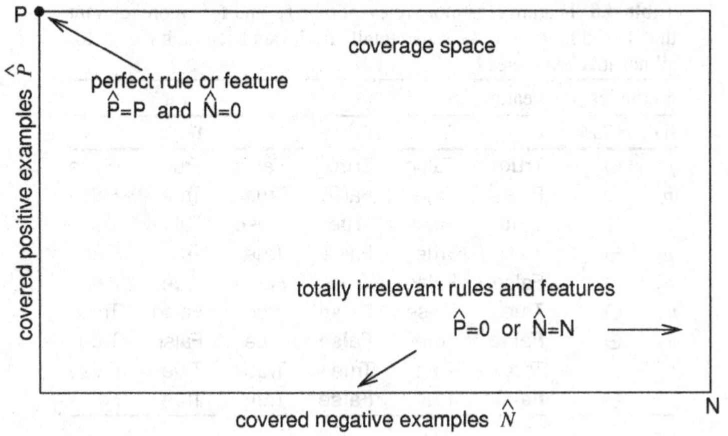

Fig. 4.5 The position of ideal rules or features and the position of totally irrelevant rules and features in the coverage space

positive if the patient has cancer, and is negative if the patient does not have cancer), one can be misled to believe that T is maximally relevant for the task of recognizing cancer, but totally irrelevant for the task of recognizing patients without cancer. However, the correct interpretation is that T actually corresponds to two features: $T =$ positive is a relevant feature for predicting cancer and is irrelevant for predicting non-cancer, whereas the opposite is true for $T =$ negative.

We can now prove that the `GenerateFeatures` algorithm does not discard relevant discrete attributes.

Theorem 4.4.3. *For a discrete attribute, all features that are not generated by the* `GenerateFeatures` *algorithm are totally irrelevant.*

Proof. Assume a discrete attribute **A**. The GENERATEFEATURES algorithm generates a feature $\mathbf{A} = v_i$ for all values v_i that occur in the positive examples. Conversely, feature $\mathbf{A} = v_j$ is not generated if v_j does not occur in the positive examples. Thus, v_j only occurs in the negative examples and is therefore false for all positive examples. By Theorem 4.4.2, feature $\mathbf{A} = v_j$ is totally irrelevant.

Analogously, feature $\mathbf{A} \neq v_j$ is not generated if v_j does not occur in the negative examples, i.e., $\mathbf{A} \neq v_j$ has the value true for all negative examples. By Theorem 4.4.2, feature $\mathbf{A} \neq v_j$ is totally irrelevant. □

A consequence of this theorem is that for discrete attributes, the algorithm `GenerateFeatures` generates sufficiently rich feature sets for building complete and consistent rule sets. The only exception is the case when the training set $\mathcal{E}$ contains contradictory information, which occurs when a positive and a negative example have the exact same representation, i.e., they have the same values for all attributes. But in the case of contradictions, no feature construction procedure, regardless of how complex it may be, can help in distinguishing these two examples.

Table 4.8 Feature $\mathbf{f}_3$ is more relevant than $\mathbf{f}_1$, and $\mathbf{f}_5$ is more relevant than $\mathbf{f}_2$ and $\mathbf{f}_4$. Feature $\mathbf{f}_4$ is also totally irrelevant because it is true for all negative examples

Examples		Features					
ID	Class	$\mathbf{f}_1$	$\mathbf{f}_2$	$\mathbf{f}_3$	$\mathbf{f}_4$	$\mathbf{f}_5$	$\mathbf{f}_6$
p_1	⊕	**True**	False	**True**	False	True	False
p_2	⊕	False	True	False	True	True	False
p_3	⊕	**True**	False	**True**	False	False	True
p_4	⊕	False	True	False	True	True	False
p_5	⊕	False	False	**True**	False	True	False
n_1	⊖	True	False	**False**	True	False	True
n_2	⊖	**False**	True	**False**	True	False	True
n_3	⊖	True	True	True	True	True	False
n_4	⊖	**False**	True	**False**	True	True	False

Consequently, the procedure will generate a sufficiently rich feature set so that complete and consistent rule sets can be built whenever it is possible to generate such sets. A similar theorem regarding continuous attributes will be proved in the next section.

4.4.3 Relative Irrelevancy

The previous section showed that the quality of features depends on their usefulness for building rules, which in turn depends on their ability to correctly classify pn-pairs of examples. Recall that totally irrelevant features do not discriminate any pn-pair and can consequently be immediately excluded from the rule construction process. But, besides total irrelevancy, weaker forms of feature irrelevancy exist.

In the following, we denote with $\mathcal{PN}_\mathbf{f}$ the set of pn-pairs discriminated (correctly classified) by feature $\mathbf{f}$, with $\hat{\mathcal{P}}_\mathbf{f}$ the set of covered correctly classified positive examples (true positives), and with $\bar{\mathcal{N}}_\mathbf{f}$ the set of not covered, thus correctly classified negative examples (true negatives). Analogously, $\bar{\mathcal{P}}_\mathbf{f}$ denotes the set of uncovered positives (false negatives) for feature $\mathbf{f}$, and $\hat{\mathcal{N}}_\mathbf{f}$ the set of covered negatives (false positives).

Definition 4.4.5 (Relative irrelevancy). If $\mathcal{PN}_\mathbf{f} \subseteq \mathcal{PN}_{\mathbf{f}_{rel}}$ for two features $\mathbf{f}, \mathbf{f}_{rel} \in \mathcal{F}$, we say that $\mathbf{f}_{rel}$ is *more relevant* than $\mathbf{f}$ and that feature $\mathbf{f}$ is *relatively irrelevant* because of the existence of feature $\mathbf{f}_{rel}$. We denote this by $\mathbf{f} \preceq \mathbf{f}_{rel}$.

Table 4.8 illustrates the concept of relative relevance. Features $\mathbf{f}_1$ and $\mathbf{f}_3$ discriminate (correctly classify) the following pn-pairs:

$$\mathcal{PN}_{\mathbf{f}_1} = \{p_1/n_2, p_1/n_4, p_3/n_2, p_3/n_4\}, \text{ and}$$

$$\mathcal{PN}_{\mathbf{f}_3} = \{p_1/n_1, p_1/n_2, p_1/n_4, p_3/n_1, p_3/n_2, p_3/n_4, p_5/n_1, p_5/n_2, p_5/n_4\}$$

$\mathbf{f}_3$ is more relevant than $\mathbf{f}_1$ because $\mathcal{PN}_{\mathbf{f}_1} \subset \mathcal{PN}_{\mathbf{f}_3}$.

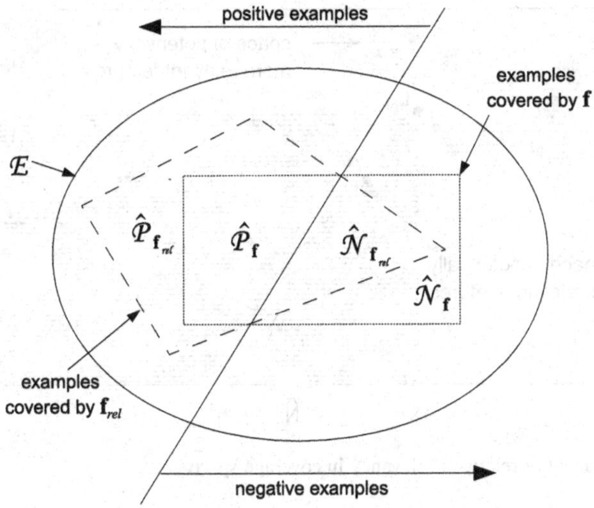

Fig. 4.6 Feature $\mathbf{f}_{rel}$ is more relevant than $\mathbf{f}$ iff $\hat{\mathcal{P}}_{\mathbf{f}} \subseteq \hat{\mathcal{P}}(\mathbf{f}_{rel})$ and $\hat{\mathcal{N}}(\mathbf{f}_{rel}) \subseteq \hat{\mathcal{N}}_{\mathbf{f}}$

Just as with total irrelevancy (Theorem 4.4.2), it is not necessary to actually construct pn-pairs for determining relative relevancy, even though the property of relative relevancy is based on the concept of pn-pairs.

Theorem 4.4.4. *For features* $\mathbf{f}, \mathbf{f}_{rel} \in \mathcal{F}$: $\mathbf{f} \preceq \mathbf{f}_{rel}$ *iff*

1. $\hat{\mathcal{P}}_{\mathbf{f}} \subseteq \hat{\mathcal{P}}_{\mathbf{f}_{rel}}$ *(or equivalently* $\bar{\mathcal{P}}_{\mathbf{f}_{rel}} \subseteq \bar{\mathcal{P}}_{\mathbf{f}}$*)* and
2. $\hat{\mathcal{N}}_{\mathbf{f}_{rel}} \subseteq \hat{\mathcal{N}}_{\mathbf{f}}$ *(or equivalently* $\bar{\mathcal{N}}_{\mathbf{f}} \subseteq \bar{\mathcal{N}}_{\mathbf{f}_{rel}}$*)*

Proof. Suppose that conditions 1 and 2 hold but $\mathbf{f} \not\preceq \mathbf{f}_{rel}$. Then there exists a pn-pair p_i/n_j that is discriminated by $\mathbf{f}$, but not discriminated by $\mathbf{f}_{rel}$. This implies that $p_i \in \hat{\mathcal{P}}_{\mathbf{f}}$ and that $n_j \in \bar{\mathcal{N}}_{\mathbf{f}}$, and that either $p_i \notin \hat{\mathcal{P}}(\mathbf{f}_{rel})$ or $n_j \notin \bar{\mathcal{N}}_{\mathbf{f}_{rel}}$. Both conclusions are contradictions to the assumptions that $\hat{\mathcal{P}}_{\mathbf{f}} \subseteq \hat{\mathcal{P}}_{\mathbf{f}_{rel}}$ and $\bar{\mathcal{N}}_{\mathbf{f}} \subseteq \bar{\mathcal{N}}_{\mathbf{f}_{rel}}$. Conversely, if $\mathbf{f} \preceq \mathbf{f}_{rel}$ then by definition $\mathcal{PN}_{\mathbf{f}} \subseteq \mathcal{PN}(\mathbf{f}_{rel})$, and therefore also conditions 1 and 2 must hold. □

Figure 4.6 illustrates the two conditions of the theorem for $\mathbf{f} \preceq \mathbf{f}_{rel}$ by schematically drawing the sets of positive and negative examples covered by $\mathbf{f}$ and $\mathbf{f}_{rel}$, so that $\hat{\mathcal{P}}_{\mathbf{f}} \subseteq \hat{\mathcal{P}}_{\mathbf{f}_{rel}}$ and $\hat{\mathcal{N}}_{\mathbf{f}_{rel}} \subseteq \hat{\mathcal{N}}_{\mathbf{f}}$.

A consequence of this theorem is that relative relevancy of features can also be directly detected from covering tables. Feature $\mathbf{f}_{rel}$ is more relevant than feature $\mathbf{f}$ if for all positives examples for which $\mathbf{f}$ is true $\mathbf{f}_{rel}$ is also true, and if for all negative examples for which $\mathbf{f}$ is false $\mathbf{f}_{rel}$ is also false. To see this, compare the values that are printed in bold font for features $\mathbf{f}_1$ and $\mathbf{f}_3$ of Table 4.8.

On the other hand, Fig. 4.7 illustrates the concept of relative relevancy of features in coverage space. If $\mathbf{f} \preceq \mathbf{f}_{rel}$, then $\mathbf{f}_{rel}$ must lie in the rectangle that is spanned

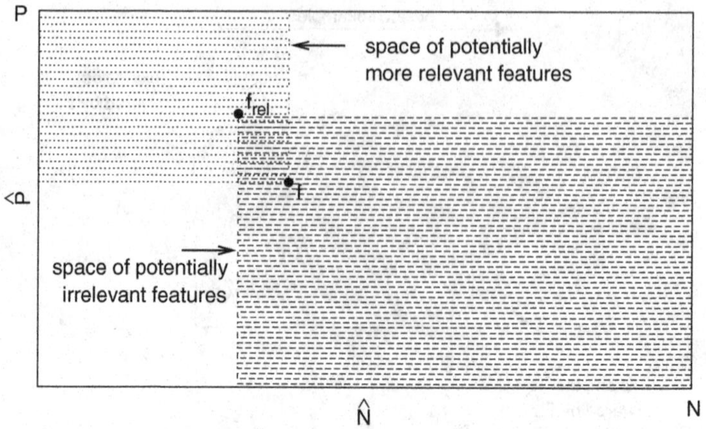

Fig. 4.7 The concept of relative irrelevancy in coverage space

between the perfect feature in the upper left corner and **f**. Note, however, that the coverage space of Fig. 4.7 does not actually show whether $\hat{\mathcal{P}}_{\mathbf{f}} \subseteq \hat{\mathcal{P}}_{\mathbf{f}_{rel}}$ and $\hat{\mathcal{N}}_{\mathbf{f}_{rel}} \subseteq \hat{\mathcal{N}}_{\mathbf{f}}$, it only shows the relative sizes of these sets: $\hat{P}_{\mathbf{f}} \leq \hat{P}(\mathbf{f}_{rel})$ and $\hat{N}(\mathbf{f}_{rel}) \leq \hat{N}_{\mathbf{f}}$. Thus, the fact that $\mathbf{f}_{rel}$ lies in the rectangle between **f** and the perfect feature does *not* imply that $\mathbf{f} \preceq \mathbf{f}_{rel}$.

Relative irrelevancy is also important for understanding the `Generate-Features` algorithm (Fig. 4.2).[3]

Theorem 4.4.5. *For a continuous attribute, all features that are not generated by the* `GenerateFeatures` *algorithm are totally or relatively irrelevant.*

Proof. Assume a continuous attribute **A**. Consider all features of type $\mathbf{A} < v$ and $\mathbf{A} \geq v$ for all values $v \in \mathbb{R}$. Note that all such values between a pair of neighboring values (v_i, v_{i+1}) are equivalent in the sense that for all $v : v_i \leq v < v_{i+1}$, features $\mathbf{A} < v$ and $\mathbf{A} \geq v$ cover the same positive and negative examples. Thus, in terms of example coverage, the means $\frac{v_i + v_{i+1}}{2}$ that are considered by the algorithm are equivalent to all other values in this range.[4]

GENERATEFEATURES only adds values that are between pairs (v_i, v_{i+1}) where v originates from a positive example ($v_i \in \mathcal{V}_{\mathcal{P}}$) and v_{i+1} from a negative example ($v_{i+1} \in \mathcal{V}_{\mathcal{N}}$), or vice versa. Thus, the values that are ignored are those where v_i and v_{i+1} both are (only) in $\mathcal{V}_{\mathcal{P}}$ or both in $\mathcal{V}_{\mathcal{N}}$.

---

[3] A similar result was first observed by Fayyad and Irani (1992) for discretization of numerical values.

[4] One reason for choosing the mean value between two neighboring points of opposite classes is that this point maximizes the *margin*, i.e., the buffer towards the decision boundary between the two classes.

Assume that $v_i, v_{i+1} \in \mathcal{V}_\mathcal{P}$: In this case, feature $\mathbf{f}' := \mathbf{A} \geq \frac{v_i + v_{i+1}}{2}$ is ignored by GENERATEFEATURES. Let $v_\mathcal{N} \in \mathcal{V}_\mathcal{N}$ be the largest value from a negative example with $v_\mathcal{N} < v_i < v_{i+1}$, and let $v_\mathcal{P} \in \mathcal{V}_\mathcal{P}$ be the smallest value with $v_\mathcal{N} < v_\mathcal{P} \leq v_i$:

1. If there is no such value $v_\mathcal{N}$, then $v > v_{i+1}$ for all $v \in \mathcal{V}_\mathcal{N}$. Therefore feature $\mathbf{f}$ covers all negative examples and is thus, by Theorem 4.4.2, totally irrelevant.
2. If such a value $v_\mathcal{N}$ exists, the feature $\mathbf{f}_{rel} := \mathbf{A} \geq \frac{v_\mathcal{N} + v_\mathcal{P}}{2}$ would be added by the algorithm. By construction, $\hat{\mathcal{P}}_{\mathbf{f}_{rel}} \supset \hat{\mathcal{P}}_{\mathbf{f}'}$ because (at least) the positive example corresponding to value $v_\mathcal{P}$ will be covered by $\mathbf{f}_{rel}$ but not by $\mathbf{f}'$. Furthermore, $\mathbf{f}'$ and $\mathbf{f}_{rel}$ cover the same set of negative examples, i.e., $\hat{\mathcal{N}}_{\mathbf{f}'} = \hat{\mathcal{N}}_{\mathbf{f}_{rel}}$. Thus, by Theorem 4.4.4, $\mathbf{f}' \preceq \mathbf{f}_{rel}$.

Analogous arguments can be made for feature $\mathbf{A} < \frac{v_i + v_{i+1}}{2}$, and for the case when both values occur only in negative examples ($v_i, v_{i+1} \in \mathcal{V}_\mathcal{N}$). □

The introduced concept of relative irrelevancy of features points out the fact that a feature is not necessarily irrelevant because of its low number of true positive examples $\hat{P} = |\hat{\mathcal{P}}_\mathbf{f}|$ or low number of true negative examples $\bar{N} = |\bar{\mathcal{N}}_\mathbf{f}|$, but because there exists some other, more relevant feature with better discrimination properties. Note that relative relevancy filtering also eliminates *redundant* features. Two features that cover exactly the same examples are mutually relatively irrelevant, and, for induction purposes, it will suffice to keep one of them. Conventional feature filtering approaches that are based on computing a quality measure for each individual feature (or attribute) are not able to cope with redundancy, because they do not put features into relation to each other.

Relevancy filtering using the concept of relative irrelevancy of features eliminates only features for which a more relevant feature exists. In this way it preserves the quality of rules that can be built from the reduced set of features. In addition, and even more important from the point of view of avoiding overfitting, it ensures that rule learners will use only the best discriminating features available.

In order to demonstrate the quality-preserving property of relative irrelevancy, let $\mathcal{F}$ be a sufficiently rich feature set, and let further $\mathcal{F}_{min}$ be the smallest among subsets of $\mathcal{F}$ with the same property, i.e., $\mathcal{F}_{min} \subset \mathcal{F}$ and there is no $\mathcal{F}' \subset \mathcal{F}_{min}$ which is sufficiently rich. According to Theorem 4.4.1, both $\mathcal{F}$ and $\mathcal{F}_{min}$ contain for each pn-pair at least one feature that correctly discriminates it.

Theorem 4.4.6. *Let $\mathcal{F}$ be a sufficiently rich feature set, and $\mathcal{F}'$ a subset of $\mathcal{F}$ obtained by eliminating all relatively irrelevant features from $\mathcal{F}$. Let further $\mathcal{F}'_{min}$ be a minimal sufficiently rich subset of $\mathcal{F}'$, and $\mathcal{F}_{min}$ be a minimal sufficiently rich subset of $\mathcal{F}$. Then $|\mathcal{F}'_{min}| = |F_{min}|$.*

Proof. If all features that are elements of $\mathcal{F}_{min}$ are also elements of $\mathcal{F}'$, then $\mathcal{F}_{min}$ is identical to $\mathcal{F}'_{min}$ and their cardinality is equal. If there is a feature $\mathbf{f} \in \mathcal{F}_{min}$ and $\mathbf{f} \notin \mathcal{F}'$ then there exists a feature $\mathbf{f}_{rel} \in \mathcal{F}'$ so that $\mathbf{f} \preceq \mathbf{f}_{rel}$. We can now transform $\mathcal{F}_{min}$ to $\mathcal{F}'_{min}$ by replacing each such feature $\mathbf{f}$ with its corresponding $\mathbf{f}_{rel}$. By construction, $|\mathcal{F}'_{min}| \leq |\mathcal{F}_{min}|$. As $\mathcal{F}_{min}$ is minimal, $|\mathcal{F}'_{min}| = |\mathcal{F}_{min}|$ must hold. □

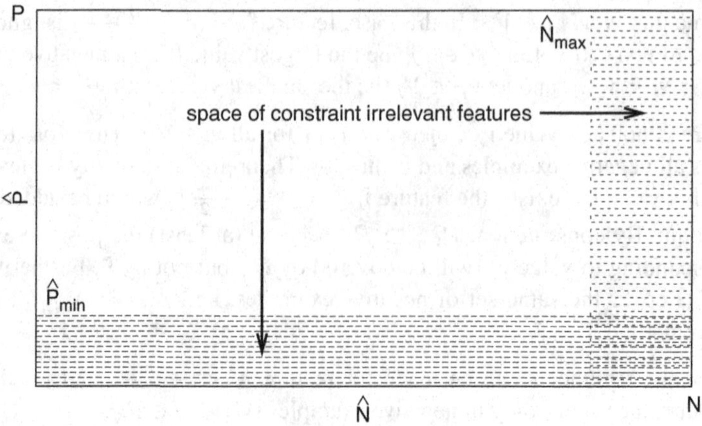

Fig. 4.8 The concept of constrained irrelevancy in the coverage space

In Theorem 4.4.6 we used the condition that the starting set of features $\mathcal{F}$ is sufficiently rich. Note that this is not a necessary condition for the applicability of relative irrelevancy. Also for feature sets not satisfying this property, filtering by relative relevancy will reduce the size of the feature set without reducing its discriminative power. If we substitute each feature **f** with the corresponding $\mathbf{f}_{rel}$, the obtained feature set cannot be larger and is still sufficiently rich because each $\mathbf{f}_{rel}$ discriminates a superset of pn-pairs that are discriminated by **f**.

4.4.4 Constrained Irrelevancy

In addition to total irrelevancy and relative irrelevancy, one can employ additional constraints to further filter the feature sets. For example, we can slightly generalize total irrelevancy by considering features irrelevant if they only cover a small number of positive examples (instead of none) or if they cover almost all negative examples (instead of all).

Definition 4.4.6 (Constrained irrelevancy). Feature **f** is irrelevant if either $\hat{P} < \hat{P}_{min}$ or $\hat{N} > \hat{N}_{max}$, where $\hat{P}_{min}$ and $\hat{N}_{max}$ are user-defined constraints on the numbers of covered positive and negative examples.

The part of the coverage space of features with constrained irrelevancy is represented by the shaded area of Fig. 4.8.

In principle, other constraints could be used as well for constraint-based filtering. For example, one could imagine to filter all features that are close to the diagonal of the ROC-space, which represents features whose discriminative power is close to random guessing. Although the significance of rules is proportional to their distance from the diagonal in the coverage space, this property is not appropriate as a quality criterion for feature filtering, because logical combinations of features lying on the diagonal or very near to it can result in very significant conjunctions of features (rules). Thus only relatively irrelevant features and constrained irrelevant features should be eliminated from the rule construction process. The main problem with constrained relevancy is that both $\hat{P}_{min}$ and $\hat{N}_{max}$ are user-defined values, and that no value can guarantee the actual relevancy of a feature when used as a building block for rule construction.

4.4.5 Discussion

The concept of relevancy and the possibility to detect and eliminate some features as irrelevant even before entering the rule construction process is important for several reasons. The first is that relevancy filtering reduces the complexity of the rule construction task. The second is that the elimination of irrelevant features is useful for overfitting avoidance by reducing the hypothesis search space through the elimination of features and their combinations with low covering properties. The third is that through *constructive induction* the learner can introduce many potentially interesting features while the relevancy approach can be used to eliminate—already in preprocessing—features that do not have a chance to be useful in the rule construction process.

The importance of criteria for feature relevancy also extends beyond preprocessing and reducing the set of features that enter the rule construction process. Logical combinations of features may again be features, and complete rule bodies can also be complex features. Thus, the relevancy of features is also applicable to rules and complex features. The consequence is that feature relevancy can be integrated into the covering algorithm with the intention to reduce the hypothesis search space and to speed up the algorithm execution.

4.5 Handling Missing Attribute Values

In practical applications, not all attribute values are known, and learning algorithms have to be able to deal with *unknown values* and missing information. Covering tables, however, are meant to be directly used by the rule learner and do not have missing values. Thus, we need to deal with missing attribute values in the feature construction phase, while the covering table is being constructed. In this section, we first briefly review the problem of missing values (Sect. 4.5.1) and common methods for addressing it (Sect. 4.5.2). We then continue by discussing a recent proposal, which takes a pessimistic stance, in a bit more depth (Sect. 4.5.3).

4.5.1 Types of Unknown Values

Note that there are several types of missing information, which may be dealt with in different ways. One can distinguish at least three different types of unknown attribute values (let **A** be an attribute, and *e* be an example for which the attribute value is missing):

- *missing value: e* should have a value for **A**, but it is not available (e.g., because it has not been measured)
- *not applicable value:* the value of **A** cannot be measured for *e* (e.g., the value of the attribute pregnant for male patients)
- *don't care value:* attribute **A** could assume any value for *e* without changing its classification

However, in practical applications one typically does not know which type of unknown value is the case. Most rule learning systems therefore do not discriminate between the different types and treat all unknown values uniformly.

4.5.2 Strategies for Handling Unknown Values

In this section, we briefly review several options for dealing with unknown values, before we discuss one strategy in more detail in the next section. Essentially, we can distinguish three principal approaches to handling unknown values. The first tries to circumvent the problem by removing all problematic cases:

Delete strategy. The simplest strategy is to completely ignore examples with unknown values. This does not require any changes in the learning algorithm, and the changes in the feature construction algorithm reduce to adding a simple filter that blocks examples with unknown values. The key disadvantage of this method is, of course, the waste of training data.

The second family of techniques tries to directly deal with the unknown value, without considering any context.

False value strategy. This strategy simply ignores attribute values with unknown values, i.e., they cannot be covered by any feature. Thus, every feature that tests for the unknown value will be assigned the truth value false. Like the *delete* strategy, this strategy can be easily realized in the feature construction phase, but is less wasteful with training data. It is, e.g., realized in RIPPER.

True value strategy. This strategy does the opposite of the *false value* strategy: while the latter adds value false for all entries in the covering table, this strategy adds value true for all entries that involve a test for the unknown attribute. This strategy corresponds to interpreting unknown values as *don't care* values.

Special value strategy. Another straightforward approach is to treat an unknown value as a separate value for this attribute. In the covering table, this results in a separate feature for each attribute with unknown values, which indicates whether the value of the attribute is unknown or known. This corresponds to interpreting unknown values as *not applicable* values.

Finally, the third group of techniques assumes that the unknown value is actually *missing*, and consequently tries to predict its real value. The methods differ in the amount of information they use for making their guess.

Common value strategy. The simplest approach is to replace unknown values of discrete attributes by the most common value (the *mode* value), and unknown values of continuous attributes by their average value (the *mean* value). This method is quite straightforward to realize, but it has its drawbacks, especially when the number of unknown attribute values is high.

Predicted value strategy. Always guessing the mean or mode values for the missing value has the obvious disadvantage that it is not sensitive to the example in question. One can, instead, train a separate classifier for each attribute, which can then be used for replacing the missing value with an educated guess. In principle, any classifier can be used for this task, but nearest-neighbor classifiers are particularly popular for this task, because they do not need to be trained unless a value is needed.

Distributed value strategy. Finally, one can predict a probability distribution over all possible values. The example with the missing value is then divided into fractions. Each of these fractions receives a different feature-based representation, and an example weight $w_e \in [0, 1]$, which indicates the degree to which this example can be covered (conventional, unweighted examples only have values 0 or 1). Implementing this approach requires a nontrivial modification of the rule learner that adapts all counting routines to weighted examples. A variant of this technique is, e.g., used in the CN2 rule learner which, in the learning phase, uses Laplace-corrected example weights, but for computing rule weights and for classifying new examples, uses equal fractions for all possible values of the missing attribute.

All these strategies have their respective advantages and disadvantages. For deeper discussions of their respective properties, we refer to Bruha and Franek (1996) and Wohlrab and Fürnkranz (2011). However, all strategies have in common that they can all be realized in the feature construction phase, with the possible exception of the *distributed value* strategy, which assumes a learner that can handle example weights, and needs to represent them in the covering table. Also note that the use of covering tables allows one to combine different strategies. For example, one can always combine the special value strategy with any other strategy, by having features that indicate whether the example has a missing or a known value.

4.5.3 Pessimistic Value Strategy

In this section we present a recent approach to solving the problem of handling unknown or missing values in the process of covering table construction (Gamberger, Lavrač, & Fürnkranz, 2008; Lavrač et al., 2010). Its key motivation is

Table 4.9 Illustration of the *pessimistic value* strategy for handling unknown attribute values: for an unknown attribute value, the feature truth value is false for positive examples and true for negative examples (shown in *bold font*)

Examples		Attributes			Features			
Ex.	Cl.	$\mathbf{A}_1$	$\mathbf{A}_2$	$\mathbf{A}_3$	$\mathbf{A}_1$ = red	$\mathbf{A}_2$ = green	$\mathbf{A}_3 < 1.6$	...
p_1	⊕	Red	?	1.2	True	**False**	true	
p_2	⊕	?	Green	?	**False**	True	**False**	
n_1	⊖	?	Blue	2.2	**True**	False	False	
n_2	⊖	Red	?	?	True	**True**	**True**	
n_3	⊖	Green	Blue	2.0	False	False	False	

that unknown attribute values should not affect the quality and potential usefulness of a feature in rule construction. Thus, feature values should be set in a way that prevents that unknown values are used for discriminating examples of different classes.

We have seen that a good feature discriminates many pn-pairs, i.e., it is true for many positive examples and false for many negative examples. The key idea of this strategy is that a feature should not be able to discriminate a pn-pair, unless the feature values of both involved examples are known. Thus, if we do not know how to determine a correct truth value for a feature, we set its value to false for positive examples and to true for negative examples. By doing so, we ensure that the resulting pn-pair is not discriminated by this feature. Thus, we effectively reduce the number of discriminated pn-pairs and penalize features based on unknown attribute values. For this reason we call the strategy *pessimistic value strategy*. Table 4.9 shows how it deals with unknown values.

The suggested approach is intuitively clear for positive examples as it implements the following intuition: if we do not know the attribute value, let us set all related feature values to false. In Table 4.9, by setting value false for example p_2 for feature $\mathbf{A}_1$ = red we prevent that p_2/n_3 is discriminated by this feature although n_3 has value false. However, for negative examples, setting also value false for unknown attribute values (as in the *false value* strategy) would, by the same principle, be incorrect, because such a negative example would form discriminating pn-pairs with all positive examples that have a true feature value. Therefore, as shown in the example of Table 4.9, we must set value true for feature $\mathbf{A}_1$ = red for negative example n_1. Otherwise we would enable that the feature discriminates the pair p_1/n_1, which we do not want because the attribute value for n_1 is unknown. Note that this strategy may be viewed as a mixture between the *false value* and the *true value* strategies, employing the former for positive and the latter for negative examples.

4.5.4 Discussion

The advantage of the *pessimistic value* approach is its simplicity; a possible disadvantage is its conservativeness: an example that has a missing value for an

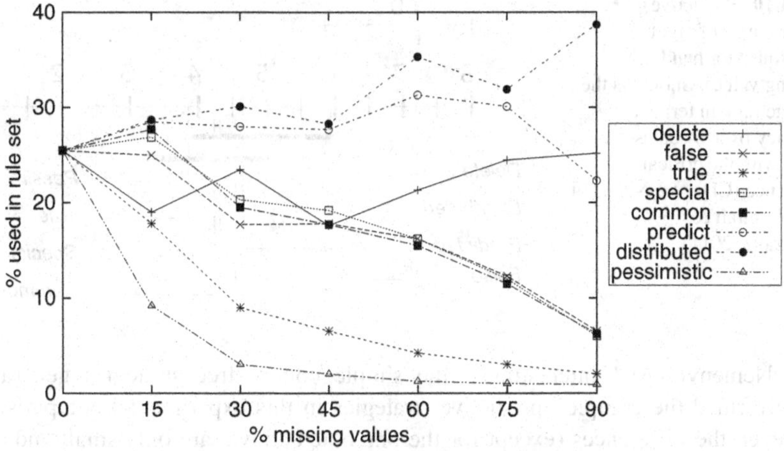

Fig. 4.9 Percentage of attributes with missing values that appear in the learned rules over varying percentages of missing values for eight techniques for handling missing values (Wohlrab & Fürnkranz, 2011)

attribute is incorrectly classified by every feature based on this attribute. This can be clearly seen in Fig. 4.9, which plots the percentage of features used in the final rule set of a separate-and-conquer learner, which are based on three key attributes. For these three attributes, the percentage of missing values was systematically varied from 0 to 90 % in steps of 15 %. The curve shown is the average over three datasets from the UCI collection of machine learning databases (*segment*, *krkp*, *credit-g*).

The results show that attributes that have been handled with the *pessimistic value* strategy are considerably less frequently used for rule construction than those for which other strategies have been used. On the other hand, the *delete* strategy is not much affected by varying percentages of missing values, while the strategies that learn a predictor for the missing values (*predict* and *distributed*) seem to implement a bias towards rules with such features.

In terms of predictive accuracy, the results were quite diverse. On the three above-mentioned datasets, the *pessimistic value* was twice among the worst, and once among the best strategies. However, it is questionable whether the simulated missing values are representative for real-word applications. Figure 4.10 shows the results of an experimental comparison of the 8 techniques on 24 databases with missing values from the UCI collection of machine learning databases. On each dataset, we ranked the algorithms according to their accuracy when used with a simple separate-and-conquer rule learner. The figure shows the average rank over the 24 datasets.

On these datasets, the *pessimistic value* strategy performed best, and was the only approach that significantly outperformed the *delete* strategy according

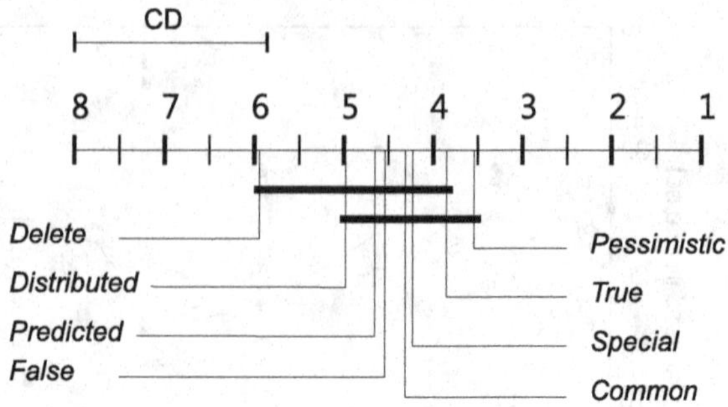

Fig. 4.10 Predictive performance of eight techniques for handling missing values. Shown is the average rank in terms of accuracy over 24 datasets, along with the critical difference (CD) of a Nemenyi test (Wohlrab & Fürnkranz, 2011)

to a Nemenyi test.[5] Interestingly, the simple context-free strategies generally outperformed the complex predictive strategies in this experimental comparison. However, the differences (except for the one noted above) are only small and are not statistically significant.

4.6 Handling Imprecise Attribute Values

Another problem encountered in practical rule learning applications is that of imprecise values. Imprecision is inherent to most noninteger continuous attributes. There are two main reasons for the necessity of imprecision handling. The first is that some continuous attributes are not or cannot be measured with high precision (like some biological properties) or their values are significantly fluctuating (e.g., human blood pressure). In such cases, the feature truth values are unreliable if the underlying example values are very near to the thresholds used in the features. The second reason is that although some attributes like income or human age can be known very precisely, building rules from features that have supporting examples very near to the thresholds used in the features may be a bad practice. The reasoning is that such features may lead to rules with substantial overfitting, or, in the case of descriptive induction, may result in nonintuitive rules like that people of age 20 may have different properties from those of age 19 years and 11 months. In this section, we briefly discuss two principal approaches to handling numerical imprecision: one is the reduction of the problem to the problem of missing values (Sect. 4.6.1), the other is the induction of fuzzy rules (Sect. 4.6.2).

[5]The *Nemenyi test* (Nemenyi, 1963) is a post hoc test to the Friedman test for rank differences (Friedman, 1937). It computes the length of the critical distance, which is the minimum difference in average rank from which one can conclude statistical significance of the observed average ranks. This is typically visualized in a graph that shows the average ranks of various methods connecting those that are within the same critical distance range. For its use in machine learning we refer to Demšar (2006).

4.6.1 Reduction to Missing Values

An approach to dealing with inherent attribute imprecision, regardless of which type it is, is to treat attribute values near to the feature thresholds as unknown values, in order to prevent such close values from affecting feature quality. Such *soft* unknown values are handled as standard unknown attribute values as described in Sect. 4.5. It must be noted, however, that—as the actual attribute value is known—it may happen that for a feature with some threshold v_{t_1}, the value is treated as unknown while for some other feature with a different threshold v_{t_2} it may be treated as a regular known value. More precisely, we should adjust a selected threshold v_t such that all attribute values v in the range $v_t - \delta < v < v_t + \delta$ are treated as unknown attribute values, for δ being the user-defined attribute imprecision boundary.

For illustration, in Table 4.9 there is attribute $\mathbf{A}_3$ and feature $\mathbf{A}_3 < 1.6$ that tests its value. The feature values in Table 4.9 for the feature $\mathbf{A}_3 < 1.6$ are determined assuming no or very small imprecision for this attribute. But if we assume that—from the meaning of attribute $\mathbf{A}_3$ or the way the data are collected—we know that the expected imprecision of $\mathbf{A}_3$ values is ± 0.5, we can take this into account when determining the feature truth values. In this case it means that value 1.2 which is true for feature $\mathbf{A}_3 < 1.6$ may be false in the case of imprecision of $+0.5$. Similarly, the value 2.0 is false for this feature but it may be true in case of assumed imprecision -0.5. The appropriate approach would be to treat both values as if they were unknown, resulting in value false for example p_1 and value true for example n_3.

The example presented in Table 4.10 demonstrates an artificial situation with two very similar continuous attributes which can both be used to generate perfect features discriminating all 25 pn-pairs. However, when the notion of imprecision of continuous attributes is introduced, the situation may change significantly. The right part of the table presents covering properties of the same features but with imprecision of 0.17. Now none of the features is perfect. The first discriminates 9 and the second 16 pn-pairs. Although with assumed imprecision 0.0 both features seem equally good, with assumed imprecision 0.17 we have a clear preference for the second feature. By analyzing the attribute values it really seems that it is better to use attribute $\mathbf{A}_2$ because it may turn out to be a more reliable classifier in an imprecise environment.

The above example assumes that the *pessimistic value* strategy was used for handling missing values. However, other strategies are also possible. For example, the C4.5 decision tree learner employs a strategy that distributes example values near a numerical threshold into both outcomes: part of the example is considered to be larger than the threshold, and part of the example is considered to be smaller than the threshold. This is essentially identical to treating numerical values near the threshold as missing and employing a variant of the *distributed value* strategy. It is also related to the key ideas of learning fuzzy rules, which we discuss in the next section.

Table 4.10 Attribute values near the feature decision value are treated as unknown values that are handled with the *pessimistic value* strategy. A user-defined constant δ defines the imprecision range. The example demonstrates the difference in feature covering properties for imprecision values 0.0 and 0.17, respectively

Examples		Attributes		Features (with $\delta = 0.0$)		Features (with $\delta = 0.17$)	
ID	Class	$\mathbf{A}_1$	$\mathbf{A}_2$	$\mathbf{A}_1 < 1.95$	$\mathbf{A}_2 < 1.95$	$\mathbf{A}_1 < 1.95$	$\mathbf{A}_2 < 1.95$
p_1	$\oplus$	1.5	1.5	True	True	True	True
p_2	$\oplus$	1.6	1.6	True	True	True	True
p_3	$\oplus$	1.7	1.65	True	True	True	True
p_4	$\oplus$	1.8	1.7	True	True	**False**	True
p_5	$\oplus$	1.9	1.8	True	True	**False**	**False**
n_1	$\ominus$	2.0	2.1	False	False	**True**	**True**
n_2	$\ominus$	2.1	2.2	False	False	**True**	False
n_3	$\ominus$	2.2	2.25	False	False	False	False
n_4	$\ominus$	2.3	2.3	False	False	False	False
n_5	$\ominus$	2.4	2.4	False	False	False	False

4.6.2 *Learning of Fuzzy Rules*

The key idea of fuzzy sets (Dubois & Prade, 1980; Hüllermeier, 2011; Zadeh, 1965) is that the Boolean set membership is replaced with a fuzzy set membership. In our context, this means that a feature **f** will not return the values true and false but will instead return a numerical value $d \in [0, 1]$ that indicates the degree to which an example is covered, where $d = 0$ corresponds to false and $d = 1$ corresponds to true. This mechanism can be employed to make the thresholds of numerical values less crisp and to allow a smoother transition between coverage and noncoverage of a numerical feature.

A straightforward but very effective algorithm that follows this approach has been implemented in the FURIA (Hühn & Hüllermeier, 2009b) rule learner, which extends RIPPER through the use of fuzzy intervals. The key idea of its approach is to maintain the full coverage of a feature $\mathbf{A} < v_t$, but to extend it with fuzzy coverage for the region $v_t \leq \mathbf{A} < v_u$ with a linearly decreasing degree of coverage so that $\mathbf{A} = v_t$ means full coverage ($d = 1$), $\mathbf{A} = v_u$ means no coverage ($d = 0$), and values in-between are covered to the degree $d = \frac{\mathbf{A}-v_t}{v_u-v_t}$. The value v_u is selected by maximizing a fuzzified version of *precision*, i.e., the percentage of positive among all covered examples, where each example is counted by its coverage degree d.

Note that fuzzy sets can also be defined for categorical attributes, by replacing their values with so-called *linguistic variables*. Essentially, these mean that an example is covered by all values of the attribute, but, again, with different degrees. The FUZZY BEXA algorithm (Cloete & Van Zyl, 2006) is a nice example of how such linguistic variables can be incorporated into a separate-and-conquer rule learning algorithm.

4.7 Conclusion

We have defined propositional and relational features as the basic rule building blocks, and presented algorithms for propositional and relational feature construction. Explicit feature construction and presentation of their covering properties in the covering table has several important advantages. The basic idea is that the complete rule construction process can be done using only the information from the covering tables, which are an appropriate representation format for various learning algorithms.

The most relevant advantage is a possibility of systematic handling of relevancy of features with the possibility to detect and eliminate irrelevant features from the process of classification rule learning. We have shown that the notion of pn-pairs is substantial for detecting relevant feature covering properties. We have also presented a simple algorithm for the construction of propositional features that provably generates only the potentially relevant features both for discrete and continuous attributes, where these sets of features enable the construction of complete and consistent rules if there are no contradictions in the learning examples.

We have also demonstrated that the explicit definition of features is useful for systematic handling of unknown attribute values. In contrast to existing approaches, which try to substitute an unknown value with some good approximation, we have addressed the problem of handling unknown values by appropriately defining feature truth values for such attribute values. The approach is simple, applicable also to complex features based on many attributes, and well justified.

Finally, we proposed an approach to handle the expected imprecision of continuous attributes, which is seldom analyzed in the rule learning framework, but which can be very relevant in real-life domains. The problem is solved so that attribute values near to the feature decision value are treated as unknown attribute values. The approach relies on the user's estimation of the expected imprecision levels and the possibility to efficiently handle unknown attribute values through covering properties of generated features.

An application of the proposed approach to feature construction, feature filtering, and handling of missing values to a real-life problem of DNA gene expression analysis is briefly described in Sect. 12.3. In this scientific discovery task, described in more detail in Gamberger, Lavrač, Zelezny, and Tolar (2004), the effectiveness of the described methods is proved by good feature filtering results and good prediction results on an independent test set as well as by the expert interpretation of induced rules.

Chapter 5
Relational Features

While typical data mining approaches find patterns/models from data stored in a single data table, *relational data mining* and *inductive logic programming* approaches (Džeroski & Lavrač, 2001; Lavrač & Džeroski, 1994a) find patterns/models from data stored in more complex data structures, such as graphs, multiple tables, etc., involving multiple relations. This chapter shows how to construct relational features and how to derive a covering table from such complex data structures.

After an introduction in Sects. 5.1 and 5.2 gives two examples of simple relational learning tasks where the outputs are relational rules. Section 5.3 gives examples of relational features, and Sect. 5.4 demonstrates approaches to relational feature construction. Section 5.5 considers various relational knowledge representation formalisms, clarifies the logic programming terminology, and relates it to relational and deductive databases.

5.1 Introduction

In Chap. 4, we have only considered propositional features derived from attribute-value data. First-order logic provides a much richer representation formalism, which allows classification of objects whose *structure* is relevant to the classification task. Besides the ability of relational data mining systems to deal with structured data and to use the powerful language of logic programs for describing the induced patterns/hypotheses, learning can be performed by taking into account generally valid domain knowledge (usually in the form of logic programs). While propositional rule learning assumes that the learner has no prior knowledge about the problem and that it learns exclusively from examples, relational learning relies on *background knowledge*, which can be used by the learner to increase the expressiveness of the hypothesis space.

§This chapter is based on Flach and Lavrač (2003).

J. Fürnkranz et al., *Foundations of Rule Learning*, Cognitive Technologies,
DOI 10.1007/978-3-540-75197-7_5, © Springer-Verlag Berlin Heidelberg 2012

Consider a data mining scenario where there is an abundance of expert knowledge available, in addition to the experimental data. Incorporating expert knowledge in propositional machine learning is usually done by introducing new features, whose values are computed from existing attribute values. Many of the existing data mining tools provide simple means of defining new columns as functions of other data columns. This is not always sufficient since the available expert knowledge may be structural or relational in nature. Consider chemical molecules and their properties, involving single or double bonds between atoms constituting individual molecules, benzene rings, and numerous other structural features. It is hard to imagine how to express this type of knowledge in a functional form, given that it is structural, and can be most naturally expressed in a relational form.

The ability to take into account such background knowledge, and the expressive power of the language of discovered patterns are distinctive features of relational learning, which is also often referred to as *first-order learning*, *relational learning* (Quinlan, 1990) or *inductive logic programming* (ILP) (Lavrač & Džeroski, 1994a; Muggleton, 1992), as the patterns they find are expressed in the relational formalism of first-order predicate logic. Inductive logic programming is an area of rule learning employing clausal logic as the representation language (De Raedt, 1995; Džeroski & Lavrač, 2001; Lavrač & Džeroski; Muggleton, 1992), but other declarative languages can be used as well (Flach, Giraud-Carrier, & Lloyd, 1998; Hernández-Orallo & Ramírez-Quintana, 1999). Note that the main ingredients of predicate logic are *predicates*, referring to relations in the domain, and *terms*, referring to objects in the domain. The main difference with respect to propositional logic is the use of variables and quantification to refer to arbitrary objects. The term 'first-order' indicates that quantification is restricted to variables ranging over domain objects, while higher-order logic allows variables to range over relations.

Where does the term 'inductive logic programming (ILP)' come from? The term was coined in the early 1990s to represent a new research area (Lavrač & Džeroski, 1994a; Muggleton, 1992), at that time usually described as being at the 'intersection of inductive machine learning and logic programming'. In ILP, *inductive* denotes that algorithms follow the inductive machine learning paradigm of learning from examples, and *logic programming* denotes that first-order representational formalisms developed within the research area of logic programming (Prolog in particular) are used for describing data and induced hypotheses. The term was intended to emphasize that ILP has its roots in two research areas: its sound formal and theoretical basis from computational logic and logic programming is complemented with an experimental approach and orientation towards practical applications which it inherits from inductive machine learning.

The field of ILP resulted in many implemented systems and successful applications. Nowadays many people prefer to use terms such as *relational data mihing* or *multirelational data mining* instead of ILP, mainly because research directions and application interests have changed. Early ILP algorithms concentrated almost exclusively on logic program synthesis, while more recent systems are capable of solving many other types of data mining problems, such as learning constraints (De Raedt & Dehaspe, 1997), association rules (Dehaspe & Toivonen, 2001), first-order clustering (Kirsten, Wrobel, & Horvath, 2001), database restructuring (Flach, 1993), subgroup discovery (Flach & Lachiche, 2001; Wrobel, 2001), and other forms of descriptive rule learning.

To summarize, relational data mining and inductive logic programming are to be used for multirelational data mining tasks with data stored in relational databases and tasks with abundant expert knowledge (background knowledge) of a relational or structural nature. Compared to traditional inductive learning techniques, ILP is thus more powerful in several respects. First, it uses an expressive first-order rule formalism enabling the representation of concepts and hypotheses that cannot be represented in an attribute-value framework of traditional inductive machine learning. Next, it facilitates the representation and use of background knowledge which broadens the class of problems for which inductive learning techniques are applicable. Furthermore, many techniques and theoretical results from computational logic can be used and adapted for the needs of inductively generating theories from specific observations and background knowledge.

5.2 Examples of Relational Learning Tasks

The most commonly addressed task in ILP is the task of learning logical definitions of relations, where tuples that belong or do not belong to the target relation are given as examples. From training examples, an ILP system then induces a logic program (a predicate definition) corresponding to a database view that defines the target relation in terms of other relations given in the background knowledge. This classical ILP task is addressed, for instance, by the seminal MIS system (Shapiro, 1982, 1991), rightfully considered as one of the most influential ancestors of modern ILP, and FOIL (Quinlan, 1990), one of the best-known ILP systems.

As an illustration, consider again the task of defining the `daughter(X,Y)` relation, addressed in Sect. 4.3.3, which states that person `X` is a daughter of person `Y`, in terms of the background knowledge relations `female` and `parent`. Recall the following facts from Table 4.6 of Sect. 4.3.3, defining the target predicate `daughter` and the background knowledge predicates `parent` and `female`:

- Two facts constitute the positive examples of the target relation: `daughter(mary,ann)` and `daughter(eve,tom)`,
- Two facts constitute the negative examples of the target relation: `daughter(tom,ann)` and `daughter(eve,ann)`,
- Several facts defining background knowledge relations: `parent(ann,mary)`, `parent(ann,tom)`, `parent(tom,eve)`, `parent(tom,eve)`, `female(ann)`, `female(mary)`, `female(eve)`.

From these facts it is possible to formulate the following rule defining the target relation `daughter`:

```
daughter(X,Y):- female(X), parent(Y,X).
```

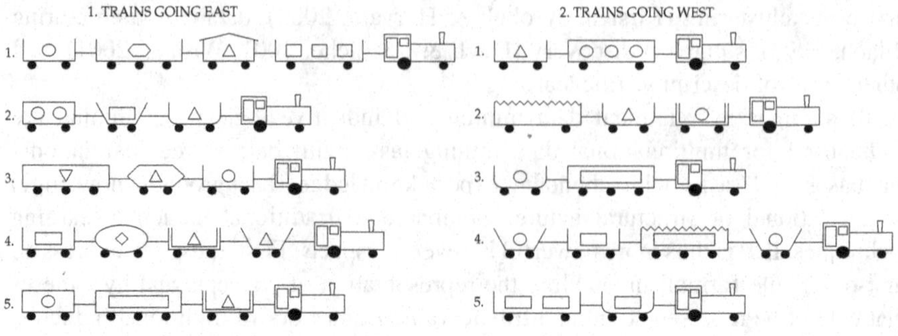

Fig. 5.1 The ten-train East–West challenge

This rule has the form of a Prolog *clause* with a conjunction of two conditions—two *relational features* `female(X)` and `parent(Y,X)`—in the clause *body* and the target relation in the clause *head* (see Sect. 2.3 in which the clause formalism was introduced). This rule covers all the positive examples and none of the negative examples of the target relation. This definition is therefore *complete* (as all the positive examples of the target concept are covered) and *consistent* (as no negative examples of the target concept are covered) in terms of the given training examples and background knowledge.

The target predicate definition may consist of a set of rules, such as

```
daughter(X,Y) :- female(X), mother(Y,X).
daughter(X,Y) :- female(X), father(Y,X).
```

if the relations `mother` and `father` were given in the background knowledge instead of the `parent` relation.

As another illustrative example used throughout this chapter, consider Michalski's classical trains example (Michie, Muggleton, Page, & Srinivasan, 1994), shown in Fig. 5.1. The learning task is to induce a Prolog program classifying trains as eastbound or westbound. The training examples consist of ten trains `t1`,…, `t10`, where the predicates `eastbound` and `westbound` indicate whether the train is eastbound or westbound: `eastbound(t1)`, `eastbound(t2)`,…, `eastbound(t5)`, and `westbound(t6)`, `westbound(t7)`, …, `west bound(t10)`.

Each train is described by a set of facts which describe the properties of the train. Each train consists of two to four cars; the cars have attributes like shape (`rectangular`, `oval`, `u-shaped`, …), length (`long`, `short`), number of wheels (`2`, `3`), type of roof (`no_roof`, `peaked`, `jagged`, …), shape of load (`circle`, `triangle`, `rectangle`, …) and number of loads (`1–3`). A set of Prolog facts describing the first train car is shown in Fig. 5.2.

For this small dataset consisting of ten instances, a rule distinguishing between eastbound and westbound trains could be

```
eastbound(T) :-
hasCar(T,C), clength(C,short), not croof(C,no_roof).
```

```
eastbound(t1).

cnumber(t1,4).

hasCar(t1,c11).              hasCar(t1,c12).
cshape(c11,rect).            cshape(c12,rect).
clength(c11,long).           clength(c12,short).
croof(c11,no_roof).          croof(c12,peak).
cwheels(c11,2).              cwheels(c12,2).
hasLoad(c11,l11).            hasLoad(c12,l12).
lshape(l11,rect).            lshape(l12,tria).
lnumber(l11,3).              lnumber(l12,1).

hasCar(t1,c13).              hasCar(t1,c14).
cshape(c13,rect).            cshape(c14,rect).
clength(c13,long).           clength(c14,short).
croof(c13,no_roof).          croof(c14,no_roof).
cwheels(c13,3).              cwheels(c14,2).
hasLoad(c13,l13).            hasLoad(c14,l14).
lshape(l13,hexa).            lshape(l14,circ).
lnumber(l13,1).              lnumber(l14,1).
```

Fig. 5.2 Prolog representation of the first train in the East–West challenge

which states that *'A train* T *is eastbound if it contains a short closed car'*.

In general, depending on the background knowledge, the hypothesis language, and the complexity of the target concept, the target predicate definition may consist of a set of rules. For example, the westbound trains cannot be described with a single rule, but could be described by the following set of two rules:

```
westbound(T)  :- cnumber(T,2).
westbound(T)  :- hasCar(T,C), croof(C,jagged).
```

which state that *'A train* T *is westbound if it contains exactly two cars or if it has a car with a jagged top.'*

The above Prolog notation for rules will be explained in more detail in Sect. 5.5, which aims at presenting alternative relational representations. Before doing so, we describe relational features and present a simple approach to relational feature construction in Sect. 5.3 below.

5.3 Examples of Relational Features

When learning about properties of objects in relational domains, feature construction can be guided by the structure of individual objects. Consider the East–West challenge problem. As it will be seen later in this chapter, the data can be stored in

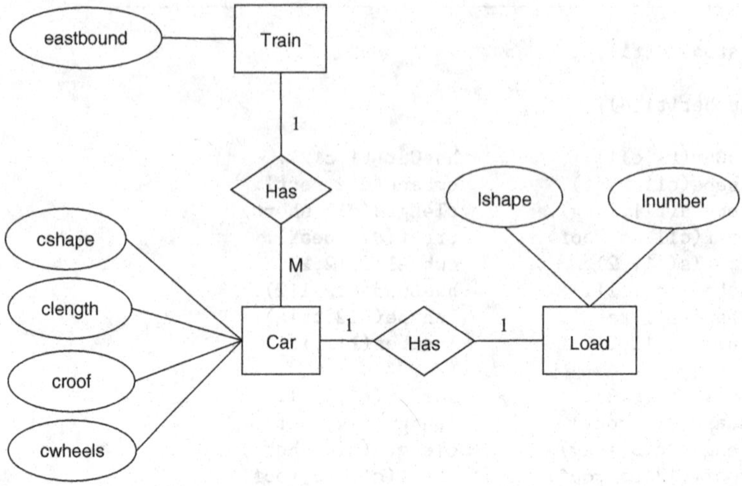

Fig. 5.3 Entity-relationship diagram for the East–West challenge

TRAIN

trainID	eastbound
t1	true
...	...
t5	true
t6	false
...	...

LOAD

loadID	lshape	lnumber	car
l11	rect	3	c14
l12	tria	1	c13
l13	hexa	1	c12
l14	circ	1	c11
...	...	...	...

CAR

carID	cshape	clength	croof	cwheels	train
c11	rect	long	no_roof	2	t1
c12	rect	short	peak	2	t1
c13	rect	long	no_roof	3	t1
c14	rect	short	no_roof	2	t1
...	...	...	...	...	...

Fig. 5.4 A relational database representation of the East–West challenge

three relational tables (shown in Fig. 5.4), where the structure of the problem can be represented by an entity-relationship diagram (shown in Fig. 5.3).

The following two Prolog rules, describing eastbound trains, were learned from the training examples and background knowledge.

```
eastbound(T) :-
   hasCar(T,C), clength(C,short), not croof(C,no_roof).
eastbound(T) :-
   hasCar(T,C1), clength(C1,short),
   hasCar(T,C2), not croof(C2,no_roof).
```

The first rule, which we have seen several times already, expresses that trains which have a short closed car are going east. The second rule states that trains which have a short car and a (possibly different) closed car are going east. This rule is clearly more general than the first, covering all instances that the first one covers, and in addition instances where the short car is different from the closed car.

We say that the body of the first rule consists of a single *relational feature*, while the body of the second rule contains two distinct relational features. Formally, a feature is defined as a minimal set of literals such that no local (i.e., existential) variable occurs both inside and outside that set. The main point of relational features is that they localize variable sharing: the only variable which is shared among features is the global variable occurring in the rule head. This can be made explicit by naming the features:

```
hasShortCar(T)   :-
    hasCar(T,C), clength(C,short).
hasClosedCar(T)  :-
    hasCar(T,C), not croof(C,no_roof).
```

Using these named features as background predicates, the second rule above can now be translated into a rule without local variables:

```
eastbound(T) :- hasShortCar(T), hasClosedCar(T).
```

This rule only refers to properties of trains, and hence could be captured extensionally by a single table describing trains in terms of those properties. In the following, we will see how we can automatically construct such relational features and the corresponding covering table.

5.4 Relational Feature Construction

In Sect. 4.3.3, we saw an example of a covering table for the daughter relation, which was automatically generated from relational background knowledge. The process of converting a relational knowledge representation into a propositional representation such as a covering table is also known as *propositionalization*. In this section, we will briefly discuss the key ideas; a more extensive survey of propositionalization approaches can be found in Kramer, Lavrač, and Flach (2001).

This approach of converting a relational representation into a propositional representation forms the basis of LINUS (Lavrač & Džeroski, 1994a; Lavrač, Džeroski, & Grobelnik, 1991), one of the early ILP algorithms. The key idea of this approach is to generate one feature for each possible variabilization of a relation in the background knowledge. Through the generation of simple relational features, the learning problem is transformed from a relational to a propositional form and is solved by an attribute-value learner.

Initially, this approach was feasible only for a restricted class of ILP problems where the hypothesis language was restricted to function-free program clauses which are *typed* (each variable is associated with a predetermined set of values), *constrained* (all variables in the body of a rule also appear in the head), and *nonrecursive* (the predicate symbol in the rule head does not appear in any of the literals in the body), i.e., to function-free constrained deductive hierarchical database clauses. In the following, we describe a more general procedure, inspired by the entity-relationship data model, that has been used by several ILP systems employing propositionalization such as ONEBC or LINUS (Flach & Lachiche, 1999; Lavrač & Flach, 2001).

For this purpose, we distinguish between two types of predicates:

- *structural predicates*, which introduce new variables, and
- *properties*, which are *predicates* that consume variables.

From the perspective of the entity-relationship data model, predicates defining properties correspond to attributes of entities, while structural predicates correspond to relationships between entities. A *relational feature* of an individual object is a conjunction of structural predicates and properties such that:

1. There is exactly one free variable which will play the role of the global variable in rules;
2. Each structural predicate introduces a new existentially quantified local variable, and uses either the global variable or one of the local variables introduced by other structural predicates;
3. Properties do not introduce new variables (this typically means that one of their arguments is required to be instantiated);
4. All variables are used either by a structural predicate or a property.

The actual relational feature generation can be restricted by parameters that define the maximum number of literals constituting a feature, maximum number of variables, and the number of occurrences of individual predicates. The following relational feature can be generated in the trains example, allowing for four literals and three variables:

```
trainFeature42(T) :-
    hasCar(T,C), hasLoad(C,L), lshape(L,triangle).
```

As can be seen here, a typical feature contains a chain of structural predicates, closed off by one or more properties. Properties can also establish relations between parts of the individual: e.g., the following feature expresses the property of *'having a car whose shape is the same as the shape of its load'*.

```
trainFeature978(T) :-
    hasCar(T,C),  cshape(C,CShape),
    hasLoad(C,L), lshape(L,LShape),
    CShape = LShape.
```

We demonstrate the extended propositionalization algorithm, implemented in SINUS (Krogel, Rawles, Železný, Flach, Lavrač, & Wrobel, 2003) and RSD (Zelezný & Lavrač, 2006), by applying it to the East–West trains example. In the

Table 5.1 Propositional form of the eastbound trains learning problem

Examples		Variable	Features					
ID	eb(T)	T	f1(T)	f2(T)	f3(T)	f4(T)	...	f190(T)
p_1	True	t1	True	False	False	True	...	False
p_2	True	t2	True	True	False	True	...	True
	...	...	...	...	...	...		
n_1	False	t6	False	False	True	True	...	True
n_2	False	t7	True	False	False	False	...	False
	...	...	...	...	...	...		

example below we simply provide the algorithm with all the features that use at most two structural predicates introducing local variables, and at most two predicates describing properties. There are 190 such features, whose truth values are recorded in a single covering table (Table 5.1).

Taking Table 5.1 as input (using column eb(T) as the class), an attribute-value rule learner such as CN2 can be used to induce propositional rules. For example, the following rule can be induced:

```
IF    f1(T) = true AND f4(T) = true
THEN  eb(T) = true
```

Translated back into Prolog, the above rule has the following form:

```
eastbound(T) :- f1(T), f4(T).
```

The actual result of learning eastbound trains by applying the propositional rule learner CN2 on the covering table with 10 rows and 190 columns describing the dataset from Fig. 5.1 are the rules shown below (using self-explanatory names for the generated features):

```
eastbound(T,true) :-
    hasCarHasLoadSingleTriangle(T),
    not hasCarLongJagged(T),
    not hasCarLongHasLoadCircle(T).
eastbound(T,false) :-
    not hasCarEllipse(T),
    not hasCarShortFlat(T),
    not hasCarPeakedTwo(T).
```

The key disadvantage of an explicit construction of relational features in covering tables is that the number of features that need to be generated grows exponentially with the variable depth that is used for generating the features. Several authors have addressed this problem by trying to reduce the number of generated features without sacrificing predictive accuracy. For example, Kramer, Pfahringer, and Helma (2000) propose to use a stochastic search for good rules and use the resulting rule bodies as features for a propositional learner. Other algorithms specialize on propositionalization in graph-like structures (Geibel & Wysotzki, 1996; Kramer &

Frank, 2000). In Sect. 5.5.5 we will discuss a few simple and general techniques that allow us to explicitly define and restrict the size of the search space.

5.5 Alternative Knowledge Representation Formalisms

Logic is a powerful reasoning mechanism as well as a versatile knowledge representation formalism. However, its versatility also means that there are usually many different ways of representing the same knowledge. What is the best representation depends on the task at hand. In this section, we discuss several ways of representing a particular first-order task in logic, pointing out the strengths and weaknesses of each.

5.5.1 Logic Programming and Prolog

Patterns discovered by ILP systems are typically expressed as logic programs, an important subset of first-order (predicate) logic. In this section, we briefly recapitulate basic logic programming concepts as can be found in most textbooks on this subject (Lloyd, 1987). Our notation will follow conventions used in Prolog, arguably the most popular logic programming language (Bratko, 1990; Flach, 1994; Sterling & Shapiro, 1994).

A *logic program*, also called a *clausal theory*, consists of a set of *clauses*. We can think of clauses as first-order rules, where the conclusion part is termed the *head* and the condition part the *body* of the clause. In a first-order context, we will thus sometimes use the terms 'clause' and 'rule' interchangeably. A set of rules with the same predicate in the head is called a *predicate definition*. Predicate definitions may consist entirely of ground facts, enumerating all the instances of the predicate; these are referred to as *extensional* predicate definitions. Predicate definitions involving variables are called *intensional*.

The definitions for `eastbound` and `westbound` in the previous sections can be used as examples for clauses. The predicate `eastbound` describes the property 'a train heading east'. This is the property of interest in our learning problem: we want to learn the definition of predicate `eastbound(T)` in terms of other predicates. Predicate `hasCar` is a binary predicate, describing a relation between trains and cars: `hasCar(T,C)` is true if and only if train `T` contains a car `C`. The symbol '`:-`' is the Prolog notation for the backward implication ←, the comma in the body of the rule denotes conjunction, and the period marks the end of the rule. White space in the rule definition is ignored by the interpreter.

Head and body are made up of *literals*, which are applications of predicates to terms. We restrict ourselves to *function-free* Prolog, where a *term* can only be a *variable* or a *constant*.[1] For instance, in the literal `clength(C,short)` the predicate `clength` is applied to the terms `C` (a variable) and `short` (a constant). Note the Prolog convention of starting variables with an uppercase letter and everything else with a lowercase letter.

In general, both head and body of a clause may contain an arbitrary number of literals. However, we will only consider *definite clauses*, which contain exactly one literal in the head. The last literal in the above rule for `eastbound` is a *negative* literal, in which the predicate is preceded by the negation symbol `not`. Un-negated literals are also referred to as *atoms*. Definite clauses that allow negative literals in the body are also known as *program clauses*.

All variables that occur in the head of a clause are *universally quantified*,[2] whereas variables that are introduced in the body of a rule are *existential* variables.[3] Typically, these refer to parts of individuals (cars, loads). A rule like *'for all trains* `T` *and cars* `C`*, if* `C` *occurs in* `T` *and* `C` *is short and closed then* `T` *is eastbound'* can be reformulated as *'for all trains* `T`*, if there exists a car* `C` *in* `T` *which is short and closed, then* `T` *is eastbound'*. That is, for a variable only occurring in the body of a rule, quantifying it universally over the whole rule is equivalent to quantifying it existentially over the body of the rule.

If variables `T` and `C` are replaced by constants denoting a particular train and a particular car, we obtain the *fact* `hasCar(t1,c11)` which tells us that train `t1` contains car `c11`. In logic programming terminology, `hasCar(t1,c11)` is called a *ground fact* since it doesn't contain any variables. A fact can be seen as a rule with an empty body.

Each predicate has an associated *arity* which specifies the number of terms the predicate is applied to: e.g., the predicate `clength` has arity 2, which is denoted as `clength/2`. We also say that it is a binary predicate. Conversely, `eastbound/1` is a *unary* predicate.

Note that the hypothesis language is typically a subset of the language of program clauses. As the complexity of learning grows with the expressiveness of the hypothesis language, restrictions have to be imposed on hypothesized clauses. For instance, even though

```
eastbound(T) :- hasCar(T,C), eastbound(C).
```

is a syntactically well-formed Prolog rule, `eastbound/1` should only apply to trains, and not to cars. ILP systems therefore often use some form of *typing*, assigning a type to each argument of a predicate. Other restrictions are the exclusion

---

[1]Full Prolog allows one to construct aggregate terms by means of functors. While this facility is crucial to Prolog as a programming language (essentially this is the mechanism for building up data structures), we ignore this possibility because only a few ILP systems are actually able to use functions, and structured terms can be converted into sequences of predicates via a process called *flattening* (Rouveirol, 1994).

[2]Variable Y is *universally quantified* if $\forall y \in Y$.

[3]Variable X is *existentially quantified* if $\exists x \in X$.

of recursion and restrictions on variables that appear in the body of the clause but not in its head (so-called new variables).

We can now represent the East–West challenge problem in Prolog. The training examples consist of ten trains `t1`, ..., `t10`, where the predicates `eastbound` and `westbound` indicate the sought target information, i.e., `eastbound(t1)`, `eastbound(t2)`, ..., `eastbound(t5)`, and `westbound(t6)`, `west bound(t7)`,..., `westbound(t10)`. Each train is described by a set of facts which provide an extensional definition of the properties of this particular train. For example, Fig. 5.2 presented in Sect. 5.2 provided the representation of the first train.

In addition, some of the background knowledge could also be represented in the form of intensional clauses; for instance, if long cars always have three wheels and short cars have two, this might be expressed by means of the following clauses:

```
cwheels(C,3) :- clength(C,long)
cwheels(C,2) :- clength(C,short)
```

Testing whether instance `t1` is *covered* by the above clause for eastbound trains amounts to deducing `eastbound(t1)` from the hypothesis and the description of the example (i.e., all ground facts without the fact providing its classification) and possibly other background knowledge. In order to deduce this we can make use of the standard Prolog resolution mechanism (Lloyd, 1987). One of the great advantages of using Prolog for rule learning is that the rule interpreter comes for free.

In addition to this purely logical knowledge, an ILP system needs further knowledge about predicates such as their types, in order to generate meaningful hypotheses. We will now show that such knowledge can be derived from the structure of the domain in the form of a data model.

5.5.2 *Data Models*

A data model describes the structure of the data, and can, for instance, be expressed as an entity-relationship diagram (see Fig. 5.3). The boxes in this diagram indicate *entities*, which are individuals or parts of individuals. Here, the Train entity is the individual, each Car is part of a train, and each Load is part of a car. The ovals denote attributes of entities. The diamonds indicate *relationships* between entities. There is a *one-to-many relationship* from Train to Car, indicating that each train can have an arbitrary number of cars but each car is contained in exactly one train; and a *one-to-one relationship* between Car and Load, indicating that each car has exactly one load and each load is part of exactly one car.

The main point about entity-relationship diagrams is that they can be used to choose a proper logical representation for our data. For instance, if we store the data in a relational database the most obvious representation is to have a separate table for each entity in the domain, with relationships being expressed by *foreign keys*.[4] This

[4]In the context of relational databases, a foreign key is a field in a relational table that matches a candidate key of another table. The foreign key can be used to cross-reference tables.

is not the only possibility: for instance, since the relationship between Car and Load is one-to-one, both entities can be combined in a single table. Entities linked by a one-to-many relationship cannot be combined without either introducing significant redundancy (the size of the combined table is determined by the longest train) or loss of information. One-to-many relationships distinguish ILP from propositional learning. We will take a closer look at constructing a single aggregated table in Sect. 5.4.

If we use Prolog as our representation language, we can use the entity-relationship diagram to define types of objects in our domain. Basically, each entity will correspond to a distinct type. In both the database and Prolog representations, the data model constitutes a language bias that can be used to restrict the hypothesis space and guide the search. We will elaborate these representation issues below. Before we move on, however, we would like to point out that ILP classification problems rarely use the full power of entity-relationship modeling. In most problems, only individuals and their parts exist as entities, which means that the entity-relationship model has a tree-structure with the individual entity at the root and only one-to-one or one-to-many relations in the downward direction. Representations with this restriction are called *individual-centered representations*. We will use this restriction in Sect. 5.3 to define an appropriate language bias.

5.5.3 Relational Database Representation

A *relational database* (RDB) is a set of relations (Elmasri & Navathe, 2006; Ullman, 1988). A n-ary relation p is formally defined as a set of *tuples*: a subset of the Cartesian product of n domains $D_1 \times D_2 \times \cdots \times D_n$, where a *domain* (or a *type*) is a specification of the valid set of values for the corresponding argument.

A relational database representation of the train dataset is given in Fig. 5.4. The `train` attribute in the CAR relation is a foreign key referring to `trainID` in TRAIN, and the `car` attribute in the LOAD relation is a foreign key to `carID` in CAR. These foreign keys correspond to the relationships in the entity-relationship diagram. As expected, the first foreign key is many-to-one (from CAR to TRAIN), and the second is one-to-one.

If our data is stored in a relational database, we can employ the database query language SQL for multirelational data mining (Groff & Weinberg, 2002). For instance, the hypothesis 'trains with a short closed car are eastbound' can be expressed in SQL as:

```
SELECT DISTINCT TRAIN.trainID FROM TRAIN, CAR WHERE
    TRAIN.trainID = CAR.train AND
    CAR.clength = 'short' AND
    CAR.croof != 'no_roof'
```

Table 5.2 Database and logic programming terminology.

DB terminology	LP terminology
Relation name p	Predicate symbol p
Attribute of relation p	Argument of predicate p
Tuple $\langle a_1, \ldots, a_n \rangle$	Ground fact $p(a_1, \ldots, a_n)$
Relation p—defined by a set of tuples	Predicate p—defined extensionally by a set of ground facts
Relation q—defined by a view	Predicate q—defined intensionally by a set of rules (clauses)

This query performs a join of the TRAIN and CAR tables over `trainID`, selecting only short, closed cars.[5] To prevent trains that have more than one such car from being included several times, the DISTINCT construct is used. The query represents only the body of a classification rule, and so the head (the assigned class) must be fixed in advance—one could say that the goal of multirelational data mining is to construct the correct query for each class.

5.5.4 Datalog Representation

Logic has also been used as a representation language for databases, in particular in deductive databases. A *deductive database* is a set of database clauses (Ullman, 1988). *Database clauses* are *typed program clauses* of the form $T \leftarrow L_1, \ldots, L_m$, where T is an atom and $L_1, \ldots, L_m$ are literals. A popular logic-based data representation language is Datalog (Ceri, Gottlob, & Tanca, 1989, 1990), which may be viewed as a typed, function-free Prolog.

Mapping a database representation to first-order logic is fairly straightforward. A predicate in logic programming corresponds to a relation in a relational database, and the arguments of a predicate correspond to the attributes of a relation. Table 5.2 summarizes the relations between basic database and logic programming terminology.

Thus, a straightforward mapping of the relational representation of Fig. 5.4 to a Datalog representation is as follows: each table is mapped to a predicate, with as many arguments as the table has columns or attributes:

```
train(TrainID,Eastbound)
car(CarID,Cshape,Clength,Croof,Cwheels,Train)
load(LoadID,Lshape,Lnumber,Car)
```

[5] Put differently, SQL takes the Cartesian product of the tables in the FROM clause, selects the tuples that meet the conditions in the WHERE clause, and projects on the attributes in the SELECT clause.

Then, each row or tuple in each table corresponds to a ground fact:

```
train(t1,true)
car(c11,rect,long,no_roof,2,t1)
load(l11,rect,3,c11)
```

This representation is equivalent to the representation that we have chosen in Sect. 5.5.1, in which each predicate corresponds to an attribute. This form is usually preferred in ILP. The mapping from the database to this alternative Datalog representation is as follows: each non-key attribute in each table is mapped to a binary predicate involving that attribute and the key, i.e.,

```
eastbound(TrainID,Eastbound)
cshape(CarID,Cshape)
clength(CarID,Clength)
...
lshape(LoadID,Lshape)
lnumber(LoadID,Lnumber)
```

For the foreign keys it is more natural to use the direction from the entity-relationship diagram, in our example `hasCar(TrainID,CarID)` and `hasLoad`
`(CarID,LoadID)`, although this is not essential. A small simplification can be furthermore obtained by representing Boolean attributes as unary rather than binary predicates, since the predicate itself can be seen as a function mapping into Booleans. For instance, instead of `eastbound(t1,true)` we can simply write `eastbound(t1)`.

If we restrict database clauses to be nonrecursive, we obtain the formalism of deductive hierarchical databases. A *deductive hierarchical database* (DHDB) is a deductive database restricted to nonrecursive predicate definitions and nonrecursive types. Nonrecursive types determine finite sets of values which are constants or structured terms with constant arguments. While the expressive power of DHDB is the same as that of RDB, DHDB allows intensional relations which can be much more compact than a RDB representation. If recursive definitions are allowed, the language is substantially more expressive than the language of relational databases.

For a full treatment of the relations between logic programming, relational databases and deductive databases, we refer the reader to Ullman (1988), and Ceri, Gottlob, and Tanca (1990); good surveys can be found in Ceri, Gottlob, and Tanca (1989) and Grant and Minker (1992).

5.5.5 Declarative Bias

Considering all first-order literals as conditions in the body of a rule may lead to huge, even infinite search spaces of possible rules. For this reason, many relational data mining and inductive logic programming systems admit only more restrictive variants of relational features, so that not all possible combinations of structural predicates and properties have to be generated. Such approaches allow one to explicitly model the hypothesis space with a so-called *declarative bias*. Many different ways for specifying such a declarative bias have been proposed in the literature. In this section, we review just the most common and important ones, where the user defines a static set of possible features, which are then used by the learner either directly or by computing a covering table.

5.5.5.1 Determinate Literals

Several inductive logic programming algorithms, like GOLEM (Muggleton & Feng, 1990), restrict the set of conditions that may be used in the body of a rule to *determinate literals*, i.e., to literals that have at most one new value for each combination of values for the old variables. In FOIL, such determinate literals are added to the body of the rule when no other condition is given a high evaluation by the search heuristic (Quinlan, 1991). This is necessary, as determinate literals usually have a low heuristic evaluation, because they will typically have a valid ground instantiation for all positive and negative training examples. Thus they will not exclude any negative examples, and a rule that is refined by adding such a literal consequently receives a low heuristic value by most common heuristics.

DINUS (Lavrač & Džeroski, 1994b) uses determinate literals for constructing a covering table, where the entries in the column of a determinate literal do not correspond to its truth value, but to the values to which the newly introduced variables are bound. These values can then be used by property literals in subsequent columns of the covering table.

5.5.5.2 Syntactic Restrictions

We have already observed that in most applications the available variables have different types, so that many of the available background literals only make sense when their arguments are instantiated with variables of certain types. For example, list-processing predicates need list-valued variables, arithmetic operations need numbers, and so forth. These *argument types* can be specified in advance so that attributes are only used in appropriate places. The places where new variables can appear can be specified by so-called *mode declarations*. Arguments where only bound variables can appear are called *input variables*, whereas arguments where new variables can be used as well are called *output variables*. Type and mode declarations have already been used in early ILP systems such as MIS (Shapiro, 1981). Many predicates are also symmetric in some of their input arguments, i.e., they will produce the same result no matter in which order these arguments are given. These *symmetries* have been exploited by various programs.

Restrictions of this type can significantly reduce the hypothesis space. Similar effects can also be achieved by restricting the domains of constant arguments in the predicates. Several systems allow one to declare certain values of an attribute as constants that can be used in specified places of a relation. For example, it may be useful to specify that only the empty list may be used as a constant in list-valued attributes (Quinlan & Cameron-Jones, 1995a).

Many rule learning systems also place upper bounds on the complexity of the learned rules in order to restrict the search space. For instance, PROGOL has a parameter that can be used to specify a *maximum rule length*. FOIL allows the specification of a *maximum variable number* and a *maximum variable depth*.[6] It also makes sense to allow only *linked* literals, i.e., literals that share variables with the head of the clause or with another linked literal (Helft, 1989). Severe restrictions have to be used when learning recursive programs in order to avoid infinite recursions. This problem has been discussed at length in (Cameron-Jones & Quinlan, 1993).

5.6 Conclusion

In this chapter we have given an introduction to relational rule learning and inductive logic programming, trying to provide a link with propositional rule learning introduced in Chap. 2, and upgrading the propositional feature construction techniques from Chap. 4. While propositional and relational rule learning are usually treated as rather distinct, we have tried to present them in such a way that the connections become apparent. We have shown various ways in which ILP problems can be represented, and concentrated on individual-centered domains, and used the notion of first-order features to present ILP in such a way that the relations and differences with propositional rule learning is most easily seen.

As a result, some aspects of first-order rule learning may seem simpler than they actually are. Most notably, the set of first-order features is potentially infinite. For instance, as soon as an individual can have an arbitrary number of parts of a particular kind (e.g., trains can have an arbitrary number of cars) there is an existential feature of the kind we discussed above ('train having a short car') for every *value* of every attribute applicable to that part and to its subparts. To that end, we reviewed a variety of methods that use a declarative bias to explicitly specify the space of possible hypotheses.

Alternatively to their use for constructing covering tables, which is proposed here, many of these techniques can also be directly used in the refinement operator of the search algorithm. For example, the generation of features based on structural predicates and properties, which we described in Sect. 5.3, has been proposed in a very similar form for the use in a refinement operator in a first-order rule learning algorithm (Peña Castillo & Wrobel, 2004). Such an approach may have advantages.

---

[6]The original attributes have depth 0. A new variable has depth $i + 1$, where i is the maximum depth of all old variables of the literal where the new variable is introduced.

In particular, if it is not known in advance how deep the structural predicates can be chained, it may be advantageous to use an approach that can dynamically adjust the number of generated features.

We believe that the presentation of propositional and relational rules in a unifying view based on covering tables leads to a better understanding of the nature of ILP, and how work in propositional rule learning can be carried over to the ILP domain. However, our perspective has been biased towards approaches that are more easily connected with propositional learning, and many exciting topics such as learning recursive rules and logic program synthesis had to be left out.

Chapter 6
Learning Single Rules

This chapter describes algorithms and strategies for constructing single rules in the concept learning framework. The starting point in Sect. 6.2 is a generic algorithm for finding the best rule which searches the hypothesis space for a rule that optimizes some quality criterion. Section 6.3 presents alternative search algorithms (heuristic search algorithms including hill-climbing and beam search, as well as exhaustive search algorithms). Section 6.4 describes different strategies for traversing the generality lattice using any of the previously described search algorithms. The chapter concludes with Sect. 6.5, which presents fundamental approaches to relational rule construction.

6.1 Introduction

Let us assume that we have a sufficiently rich set of features from which we will construct a rule for the target class. The learning task addressed in this chapter is to construct a rule $\mathbf{r}$ (a partial theory) that captures the logical relationship between the features $\mathbf{f}_i$ and the class value c (all interpreted as propositional truth values). In the simplest case, such a relation can be formulated as a rule $\mathbf{r}$ in the form of a logical implication:

$$c \leftarrow \mathbf{f}_1 \wedge \mathbf{f}_2 \wedge \cdots \wedge \mathbf{f}_L$$

The goal of learning is to select a set of features $\mathbf{f}_i$ whose conjunction will strongly correlate with the class.

Note that we restrict ourselves to conjunctive combinations of features in the rule body, which severely constrains the expressiveness of the hypothesis language. However, recall that more expressive Boolean combinations can already be handled during feature construction, where arbitrary logical combinations of features could form a new feature. Moreover, disjunctions can also be dealt with by learning a set of rules for a given class. In brief, a rule set is a logical formula in disjunctive normal form, i.e., represented as a disjunction of conjuncts (cf. Sect. 8.1 for an example).

J. Fürnkranz et al., *Foundations of Rule Learning*, Cognitive Technologies,
DOI 10.1007/978-3-540-75197-7_6, © Springer-Verlag Berlin Heidelberg 2012

Each conjunct in this representation may be regarded as a single rule. Typical rule learning algorithms learn such expressions by learning one rule at a time, following the covering or separate-and-conquer strategy. Thus, as every Boolean formula may be reduced to an expression in disjunctive normal form, there is no principal loss of expressiveness by the limitation to sets of conjunctive rules.

In this chapter, we take a closer look at different ways of learning a single rule, while combining single rules into a multirule theory is the topic of Chap. 8. We define the rule construction task as a search problem for the rule with highest quality based on an unspecified *rule quality measure*, where for different quality measures the result of the search may be different (we have already seen some examples of such criteria in Sect. 2.5.3, while a more systematic treatment of this subject will follow in Chap. 8). Similarly, at this point we also mostly ignore the problem of *overfitting*, i.e., the problem that the rule that optimizes the given measure on the training data may not be optimal on an independent test set. Overfitting is typically avoided by including constraints—so-called *stopping criteria*—that must be satisfied by every potentially acceptable rule. Additional constraints may be defined with the intention to make the search easier or faster. A frequently used constraint is that the support of the rule should be higher than some user-defined threshold. Typically it is not difficult to also include other types of constraints in the search procedures, regardless which search strategy is used (Chap. 9 gives an overview and analysis of different overfitting avoidance techniques).

6.2 Single Rule Construction

As first noted by Mitchell (1982), learning may be considered as a complex *search problem*: The *search space* is the space of all possible rules (conjunctions of features), and the *search goal* is to find the rule that optimizes a given quality criterion. Recall the search space of propositional rules discussed in Sect. 2.5.1, and the simple LEARNONERULE algorithm from Fig. 2.8. It constructs a single rule by consecutively adding conditions to the rule body so that a given quality criterion is greedily optimized. Thus it follows a simple greedy hill-climbing strategy for finding a local optimum in the hypothesis space defined by the feature set. However, this is only one of several options for searching a given hypothesis space for acceptable rules.

While most rule learning algorithms can be formulated in a similar fashion, different problems may require different modifications of this simple algorithm. For example, if the data are noisy, it may be necessary to implement additional constraints, stopping criteria, or post-processing methods to be able to learn simpler theories, which are often more accurate on unseen data. Other algorithms replace the top-down general-to-specific search with a bottom-up specific-to-general search, where a starting rule (which, e.g., corresponds to a single positive example) is successively generalized. Yet other algorithms do not use hill-climbing, but employ less myopic search algorithms like beam search or best-first search.

```
function FindBestRule(ℰ)
  Input:
    ℰ = 𝒫 ∪ 𝒩: a set of positive and negative examples for a class c,
               represented with a set of features ℱ

  Algorithm:
    r* := InitializeRule(ℰ)
    h* := EvaluateRule(r*)
    ℛ := {r*}
    while ℛ ≠ ∅ do
      ℛ_cand := SelectCandidates(ℛ, ℰ)
      ℛ := ℛ \ ℛ_cand
      for r ∈ ℛ_cand do
          ρ(r) = RefineRule(r, ℰ)
          for r' ∈ ρ(r) do
              if RuleStoppingCriterion(r', ℰ)
              then  next r'
              h' := EvaluateRule(r', ℰ)
              ℛ := InsertSort(r', ℛ)
              if h' > h*
              then  h* := h', r* := r'
          endfor
      endfor
      ℛ := FilterRules(ℛ, ℰ)
    endwhile
  Output:
    r*: the best rule for the given set of examples
```

Fig. 6.1 A generic search algorithm for finding the best rule

Figure 6.1 presents a generic algorithm FINDBESTRULE, which searches the hypothesis space for a rule that optimizes a given quality criterion defined in EVALUATERULE. Typically, the value of this heuristic function increases with the number of positive examples covered by the candidate rule, and decreases with the number of covered negatives. FINDBESTRULE maintains $\mathcal{R}$, a sorted list of candidate rules, which is initialized by the function INITIALIZERULE. New rules will be inserted in appropriate places (INSERTSORT), so that $\mathcal{R}$ will always be sorted in decreasing order of the heuristic evaluation of the rules. At each iteration of the **while**-loop, SELECTCANDIDATES selects a subset $\mathcal{R}_{cand} \subset \mathcal{R}$ of candidate rules, which are then refined using REFINERULE. Refining a rule simply means that the rule is replaced with multiple successor rules. Thus, REFINERULE could implement specializations, generalizations, random mutations, etc. Each refinement $\mathbf{r}'$ is evaluated and inserted into the sorted list $\mathcal{R}$ unless the RULESTOPPINGCRITERION prevents this. If $\mathbf{r}'$ is better than the best rule found previously, $\mathbf{r}'$ becomes the new best rule $\mathbf{r}^*$. FILTERRULES can be used to trim the list of candidate rules for the next iteration. When all candidate rules have been processed, the best rule $\mathbf{r}^*$ will be returned as the result of the search.

Different choices of the various subroutines allow the definition of different biases for the separate-and-conquer learner. The search bias is defined by the choice of a search strategy (INITIALIZERULE and REFINERULE), a search algorithm (SELECTCANDIDATES and FILTERRULES), and a search heuristic (EVALUATERULE). The refinement operator REFINERULE constitutes the language bias of the algorithm. An overfitting avoidance bias can be implemented via an appropriate RULESTOPPINGCRITERION and/or in a post-processing phase. Several common options for these choices will be discussed in subsequent sections and chapters.

To obtain the simple, greedy top-down search that is implemented in the LEARNONERULE algorithm of Fig. 2.8, we have to make the following choices: in order to implement a top-down search, INITIALIZERULE has to return the rule $(c \leftarrow \emptyset)$ which always predicts the desired class c. REFINERULE will always extend the current rule with all rules that result from adding a feature to the rule body:

$$\rho(\mathbf{r}) = \{\mathbf{r}' | \mathbf{r}' := \mathbf{r} \cup \mathbf{f}, \mathbf{f} \in \mathcal{F}\}$$

This can be optimized if dependencies between features are considered. For example, it is not necessary to add another feature of the form $\mathbf{A}_i = v_{i,j}$ if the rule body already contains one feature that tests for a specific value of the attribute $\mathbf{A}_i$. Among all candidates, only the best one is used as a candidate rule in the next iteration, i.e., FILTERRULES($\mathcal{R}$) will return only the first element of the sorted list $\mathcal{R}$. Thus, as there is only one candidate rule, SELECTCANDIDATES($\mathcal{R}$) will not modify the list $\mathcal{R}$. Similarly, there is no stopping criterion, and therefore RULESTOPPINGCRITERION will always return `true`.

6.3 Search Algorithms

This section discusses various options for searching the space of possible rules that can be implemented into the FINDBESTRULE procedure of Fig. 6.1. The simplest method is to systematically generate all possible rules and check each of them for consistency. Rivest (1987) discusses such an algorithm that examines all rules up to a maximum rule length L. Whenever it finds a rule that covers only positive examples, it adds this rule to the current rule set, removes the covered examples, and continues the search until all examples are covered. This simple algorithm will find a complete and consistent concept description if there is one in the search space. For a fixed length L, the time complexity of the algorithm is polynomial in the number of tests to choose from. However, this algorithm was only developed for theoretical purposes.

It is worth describing more clearly how brute-force exhaustive search methods operate. Exhaustive search basically searches all possible rules to find the one that has the highest value for the given quality criterion. However, in practice, large parts of this search space can be pruned because of the ordering of the search space, which is due to properties of the quality criterion.

A severe drawback of exhaustive search methods is that they do not use any heuristics for pruning the search space, which makes these algorithms very inefficient. Rivest (1987) notes that it would be advisable for practical purposes to generate the rules ordered by simplicity. Thus simple consistent rules will be found first and will therefore be preferred. For similar reasons most practical algorithms have access to a preference criterion that estimates the quality of a found rule as a trade-off between simplicity and accuracy (see also Chap. 7) and use this heuristic to guide their search through the hypothesis space. Heuristic search algorithms are described in Sect. 6.3.1 below.

6.3.1 Heuristic Search

6.3.1.1 Hill-Climbing

The most commonly used search algorithm in separate-and-conquer learning systems is *hill-climbing*, which tries to find a rule with an optimal evaluation by continuously choosing the refinement operator that yields the best refined rule and halting when no further improvement is possible. Hill-climbing tries to discover a *global* optimum by performing a series of *locally* optimal refinements.

Hill-climbing can be trivially implemented in the procedure FINDBESTRULE of Fig. 6.1. We have already discussed above how we can instantiate the generic algorithm into the simple LEARNONERULE algorithm of Fig. 2.8. The essential part of this transformation is that the FILTERRULES procedure will only let the best refinement pass for the next iteration, so that SELECTCANDIDATES will always have only one choice. Together these two procedures implement the hill-climbing search.

The basic problem of this method is its *myopia* due to the elimination of all refinements but one. If a refined rule is not locally optimal, but one of its refinements is the global optimum, hill-climbing will not find it (unless it happens to be a refinement of the local optimum as well). Nevertheless, many top-down, separate-and-conquer algorithms use hill-climbing because of its efficiency.

A simple technique for decreasing search myopia in hill-climbing is to look further ahead. This can be done by choosing the best rule resulting from performing n refinement steps at once instead of only one. This approach has been implemented in the ATRIS rule learning algorithm (Mladenić, 1993). Its major deficiency is its inefficiency, as the search space for each refinement step grows exponentially with n. A comparison of various techniques for fighting myopia can be found in Peña Castillo and Wrobel (2004).

6.3.1.2 Beam Search

Another strategy to alleviate the myopic behavior of hill-climbing is to employ a *beam search*. In addition to remembering the best rule found so far, beam search

Table 6.1 A parity problem, which is hard for heuristic algorithms because all features cover the same number of positive and negative examples. The class value corresponds to $\mathbf{A}_1 = \mathbf{A}_2$, which can be encoded with two rules, but in isolation, features based on $\mathbf{A}_1$ and $\mathbf{A}_2$ do not appear to be any different than features based on $\mathbf{A}_3$ and $\mathbf{A}_4$

$\mathbf{A}_1$	$\mathbf{A}_2$	$\mathbf{A}_3$	$\mathbf{A}_4$	c
0	0	1	1	$\oplus$
0	1	1	0	$\ominus$
1	0	0	1	$\ominus$
1	1	0	1	$\oplus$
0	0	0	0	$\oplus$
0	1	1	1	$\ominus$
1	0	0	0	$\ominus$
1	1	1	0	$\oplus$

also keeps track of a fixed number of alternatives, the so-called *beam*. While hill-climbing has to decide upon a single refinement at each step, beam search can defer some of the choices until later by keeping the b best rules in its beam. It can be implemented by modifying the FILTERRULES procedure of a hill-climbing algorithm so that it will return the b best elements of the refinements of the previous beam. Setting $b = 1$ results in hill-climbing. The SELECTCANDIDATES function will return all elements of $\mathcal{R}$.

Beam search effectively maintains the efficiency of hill-climbing (reduced by a constant factor), but can yield better results because it explores a larger portion of the hypothesis space. Thus many separate-and-conquer algorithms use beam search in their FINDBESTRULE procedures. Among them are AQ (Michalski, Mozetič, Hong, & Lavrač, 1986), CN2 (Clark & Niblett, 1989), *m*FOIL (Džeroski & Bratko, 1992), and BEXA (Theron & Cloete, 1996).

While it is able to alleviate the problem, beam search may still suffer from myopia. A classical example is the XOR problem or, more generally, parity problems. Such problems consist of R relevant and I irrelevant attributes, and the target class is whether the number of attributes with value 1 is even or odd. The irrelevant attributes do not influence the class value. The problem for hill-climbing and the beam search algorithm is that they cannot discriminate between relevant and irrelevant features because they all will cover about 50 % of the positive and 50 % of the negative examples. Table 6.1 illustrates this situation for a dataset with eight examples encoded by $R = 2$ relevant and $I = 2$ irrelevant attributes. The target rules learned from this dataset could be

$$\oplus \leftarrow (\mathbf{A}_1 = 1) \wedge (\mathbf{A}_2 = 1)$$
$$\oplus \leftarrow (\mathbf{A}_1 = 0) \wedge (\mathbf{A}_2 = 0)$$

However, neither rule can be found reliably because the relevant features $\mathbf{A}_1 = 1$, $\mathbf{A}_1 = 0$, $\mathbf{A}_2 = 1$, and $\mathbf{A}_2 = 0$ all cover half of the positive and half of the negative examples, and are thus indiscriminable from the irrelevant features $\mathbf{A}_3 = 1$, $\mathbf{A}_3 = 0$, $\mathbf{A}_4 = 1$, and $\mathbf{A}_4 = 0$. Hill-climbing search has to make an arbitrary choice between these eight features, and has thus a chance of $4/8$ to find the correct solution. Beam search with a beam of 2 can improve this chance to $6/28$ because there are 6 out of 28 possible subsets of size 2 which do not contain a relevant feature. Finally, beam search with $b \geq 5$ is guaranteed to find the correct solution in this case. Nevertheless, it is easy to see that for any beam size, one can construct a parity problem where the probability of finding the relevant features is arbitrarily small.

6.3.2 *Exhaustive Search*

While heuristic search may miss the rule with the highest quality, approaches based on exhaustive search are guaranteed to find it. However, there is considerable evidence that exhausting a search space can lead to worse results because the chances that rules are encountered that fit the training data by chance are increased. For example, Webb (1993) used the efficient best-first search algorithm OPUS (see below) for inducing decision lists in a covering framework and surprisingly found that the generalizations discovered by beam search with CN2 are often superior to those found by an exhaustive best-first search. Quinlan and Cameron-Jones (1995b) later investigated this problem of *oversearching* by studying the behavior of a CN2-like algorithm at different beam widths. They empirically verified in a variety of domains that too large a beam width may lead to worse results. Recent results have shown that the oversearching phenomenon also depends on the employed search heuristic (Janssen & Fürnkranz, 2009). The conclusion is that although exhaustive search can be favorable for effective induction of rules, special attention should be devoted to overfitting prevention when it is used.

In practice, we can find two main approaches to exhaustive search: best-first search, which always continues with the best possible refinement, and level-wise search, which employs a guided breadth-first search. What algorithm is preferable depends on the application. With respect to space complexity, best-first search is much better than the level-wise search. It is also much easier to incorporate arbitrary constraints into the search space. Level-wise search, on the other hand, is optimized for minimizing the number of runs through the dataset, which is an advantage if the dataset cannot be held in main memory. However, it may take a large amount of space to hold all current candidate rules.

6.3.2.1 Best-First Search

Hill-climbing and beam search algorithms are both limited by myopia that results from their restriction to storing only a fixed number of candidate rules and immediately pruning the others. Best-first search, on the other hand, selects the best candidate rule (SELECTCANDIDATES) and inserts *all* its refinements into the sorted list $\mathcal{R}$. FILTERRULES will not remove any rules. Thus best-first search does not restrict the number of candidate rules and may be viewed as a beam search with an infinite beam size $b = \infty$.

Nevertheless, the RULESTOPPINGCRITERION can be used to prune portions of the search space. ML-SMART (Bergadano, Giordana, & Saitta, 1988) implements such a strategy with several coverage-based pruning approaches that discard unpromising rules. In (Botta, Giordana, & Saitta, 1992) this approach has been shown to compare favorably to hill-climbing in an artificial domain. When no pruning heuristics are used (i.e., STOPPINGCRITERION always returns false), the search space will be completely exhausted and it is guaranteed that an optimal solution will be found. Nevertheless, the A* algorithm (Hart, Nilsson, & Raphael, 1968) allows one to prune large portions of the search space without losing an optimal solution. This optimality can be guaranteed if the used evaluation measure is *admissible*. An admissible evaluation measure usually consists of two components, one for evaluating the quality of a rule and one for computing an estimate for the quality of the rule's best refinement. It has to be ensured that the latter estimate will always return an optimistic value, i.e., it has to overestimate the quality of the best refinement of the rule. If this optimistic estimate is already worse than the evaluation of the best rule found so far, the refinements of the current rule need not be further investigated, because their true evaluation can only be worse than the optimistic estimate, which in turn is already worse than the best rule.

The first-order relational rule learner PROGOL (Muggleton, 1995) implements an A* best-first search. It generates the most specific rule in the hypothesis space that covers a randomly chosen example and searches the space of its generalizations in a top-down fashion. It guides this search by a rule evaluation measure that computes the number of covered positive examples minus the covered negative examples minus the length of the rule. Incomplete rules are evaluated by subtracting an estimate for the number of literals that are needed to complete the rule and adding the number of covered negative examples to this heuristic value. By doing so it is assumed that the completed rule will cover all positive examples the incomplete rules covers, but none of the negative examples, which is an optimistic assumption. With this admissible search heuristic PROGOL performs an exhaustive search through the hypothesis space. However, for longer rules (>4 conditions) this exhaustive search is too inefficient for practical problems (Džeroski, Schulze-Kremer, Heidtke, Siems, Wettschereck, & Blockeel, 1998). A similar heuristic is used in FOIL's hill-climbing algorithm for safely pruning certain branches of the search space (Quinlan, 1990). It might be worthwhile to try a best-first search with this heuristic.

6.3.2.2 Ordered Search

OPUS (Optimized Pruning for Unordered Search; Webb, 1995) is a particularly elegant framework for efficient exhaustive search in a propositional rule learner.

Its key idea is to ensure that each possible rule is only visited once, which not only avoids duplicate search efforts but also efficiently prunes large parts of the search space. The strategy is best combined with best-first search, but can also be used with breadth-first search (for quickly finding the shortest solutions) or with depth-first search (for memory efficiency).

OPUS orders the search space by assuming a (natural or artificial) order of the features, such as $\mathbf{f}_1 > \mathbf{f}_2 > \mathbf{f}_3 > \cdots > \mathbf{f}_F$. In the beginning, we test all rules $\oplus \leftarrow \mathbf{f}_i, i = 1 \ldots f$. In the next iteration, only rules of the form $\oplus \leftarrow \mathbf{f}_i \wedge \mathbf{f}_j$ with $i < j$ are considered, so that no rule is considered twice. In general, to each rule we will only consider the addition of all features that have a lower index than any in the current list of features. For example, if the current rule is $\oplus \leftarrow \mathbf{f}_3 \wedge \mathbf{f}_4 \wedge \mathbf{f}_6$, we will try to add $\mathbf{f}_1$ and $\mathbf{f}_2$. There is no reason to test features $\mathbf{f}_5$, $\mathbf{f}_7$ or higher as these will be found as a specialization of a different parent (e.g., as a specialization of $\oplus \leftarrow \mathbf{f}_5 \wedge \mathbf{f}_6$ or $\oplus \leftarrow \mathbf{f}_7$).

To further increase efficiency, OPUS also employs a few pruning criteria. For example, it will stop refining a rule if an optimistic estimate of the rule quality says that additional features cannot lead to a rule that has a better quality then the currently best rule. This optimistic estimate is computed by assuming that further refinement of the rule will lead to the best possible result, i.e., that all remaining negative examples will no longer be covered, while still maintaining the coverage of all positive examples that are covered by the current rule. When considering specializations of $\oplus \leftarrow \mathbf{f}_i \ldots \mathbf{f}_j$, once the algorithm determines that no specialization of $\oplus \leftarrow \mathbf{f}_i \ldots \mathbf{f}_j \wedge \mathbf{f}_k$ can lead to a rule of better quality than the best rule discovered so far, then $\mathbf{f}_k$ is removed from consideration in combination with any other feature and $\mathbf{f}_i \ldots \mathbf{f}_j$. This means that every pruning action when considering the addition of a feature to a conjunction of features $\mathbf{f}_i \ldots \mathbf{f}_j$ approximately halves the search space that needs to be explored in combination with those features.

This pruning strategy is very important for the algorithm effectiveness. The algorithm execution speed can significantly increase if we manage to detect a very good solution right at the beginning. Thus, ordering the features by estimated relevance may further improve efficiency, for example, by ordering of features by the decreasing number of covered p/n pairs, as motivated by the discussion in Sect. 4.4. It can also be useful to employ a fast heuristic search as a preprocessing step of the exhaustive search in order to find a bound on the quality of the final solution, which can then be used to cut off unpromising combinations.

6.3.2.3 Level-Wise Search

The most popular exhaustive search algorithm is the APRIORI algorithm (Agrawal, Mannila, Srikant, Toivonen, & Verkamo, 1995) for induction of association rules. Its basic idea is to find so-called *frequent itemsets*, which in our terminology correspond to frequently co-occurring feature combinations. The key idea of this algorithm is to generate in parallel all rules with a certain minimum quality in parallel with a so-called *level-wise search*. The level-wise search first generates all feature sets with one feature, then all feature sets with two features, and so on, thereby performing a breadth-first search. However, from each generation to the next, a large number of feature sets can be pruned if the given quality criterion satisfies the so-called *anti-monotonicity* property, which basically states that a superset $\mathcal{F}' \supset \mathcal{F}$ of a feature set $\mathcal{F}$ cannot have a better quality than $\mathcal{F}$ itself. A simple quality criterion that satisfies this constraint is the support of a feature set, i.e., the number of examples that are covered by a conjunction of features: if a feature set is extended with a new feature, the resulting feature set $\mathcal{F}' = \mathcal{F} \cup \mathbf{f}$ will only cover a subset of the examples covered by $\mathcal{F}$. These frequent feature sets are then used for constructing rules in a post-processing phase. For detailed accounts of algorithms for frequent pattern mining and association rule discovery we refer to (Goethals, 2005; Hipp, Güntzer, & Nakhaeizadeh, 2000; Zhang and Zhang, 2002).

Variants of the algorithm can be easily adapted to classification problems, e.g., by mining frequent sets for each class separately or by including the class value as a feature and removing all frequent feature sets that do not include a class feature. The prototypical algorithm in this class is CBA (Liu, Hsu, & Ma, 1998) and its many successors, which we briefly discussed in Sect. 2.10.9. Several strategies for pruning the search space can make the search for the special problem of discovering classification rules more efficient (Bayardo, 1997). If there is a specified target class, the feature construction algorithm presented in Sect. 4.3.2 can construct features for this class, and APRIORI can be used to induce rules only for this selected class. This is implemented in APRIORI-C (Jovanoski & Lavrač, 2001) and APRIORI-SD (Kavšek & Lavrač, 2006) algorithms.

The main problem of all APRIORI variants is that they are tailored to minimize the number of database accesses, but may have also enormous space complexity if input features cover many examples. For this reason, alternative algorithms have been proposed that rely on variants of best-first search (Han, Pei, Yin, & Mao, 2004; Webb, 2000; Zaki, Parthasarathy, Ogihara, & Li, 1997). These algorithms are typically more efficient, but assume that the entire database can be held in main memory so that the quality of arbitrary feature sets can be computed efficiently.

6.3.3 Stochastic Search

Another approach to escape the danger of getting stuck in local optima is the use of *stochastic search*, i.e., by allowing randomness in the refinement operator REFINERULE. In that case, a given rule will not be refined step by step, but the refinement operator may (with a certain probability) also perform bigger leaps, so that the learner has the chance to focus on entirely new regions of the hypothesis space. In the simplest case, each call to REFINERULE will return a random rule of the search space. More elaborate methods employ randomized generalization and specialization operators. The probability for selecting an operator is often correlated with the quality of the resulting rule, so that better rules are selected with a higher chance, but seemingly bad candidates also get a fair chance to be improved with further refinement steps (*stochastic hill-climbing*). The probability for selecting a suboptimal rule may also decrease over time so that the algorithm will eventually stabilize (*simulated annealing* (Kirkpatrick, Gelatt, & Vecchi, 1983)).

Kononenko and Kovačič (1992) and Mladenić (1993) present and compare a variety of such algorithms, ranging from an entirely random search to an approach based on Markovian neural networks (Kovačič, 1991). The latter algorithm has later been generalized into a first-order framework (Kovačič, 1994b). The resulting system, MILP, performs a stochastic hill-climbing search with simulated annealing. Whenever it reaches a local optimum, it backtracks to a previous rule, whose successors have not yet been examined, in order to get a new starting point. A similar approach is implemented in the SFOIL algorithm (Pompe, Kovačič, & Kononenko, 1993).

Another family of stochastic separate-and-conquer rule learning algorithms choose a *genetic algorithm* (Goldberg, 1989) for finding good rules. One such system, SIA (Venturini, 1993), selects a random starting example and searches for a suitable generalization in a bottom-up fashion. It maintains a set of candidate rules—a *generation*—which is initialized with random generalizations of the selected examples. The next generation is obtained by randomly generating new rules, randomly generalizing old rules, or randomly exchanging conditions between rules. The resulting rules are evaluated using a weighted average of their accuracy and complexity, and a fixed number of them are retained to form the next generation. This process is repeated until the best rule remains stable for a certain number of generations. A similar approach was used in GA-SMART for learning first-order rules in a top-down fashion (Giordana & Sale, 1992).

6.4 Search Strategies

An important decision that has to be made is in which direction the hypothesis space will be searched. The search space can be structured into a *generality lattice*, where a rule $\mathbf{r}_A$ is considered to be more general than rule $\mathbf{r}_B$ iff $\mathbf{r}_A$ covers all examples that are covered by $\mathbf{r}_B$. $\mathbf{r}_B$ is then said to be more specific than $\mathbf{r}_A$ (see Definition 2.5.1). *Refinement operators* allow the search algorithm to navigate through this space, typically (but not necessarily) by moving along the defined generality relation.

Definition 6.4.1 (Refinement operator). A *refinement operator* ρ for a language bias $\mathcal{L}$ is a mapping from $\mathcal{L}$ to $2^{\mathcal{L}}$. The rules $\mathbf{r}' \in \rho(\mathbf{r})$ are called *refinements* of rule $\mathbf{r}$.

Typically, refinement operators are defined in a minimal way, so that a systematic navigation between different nodes in the structured search space is possible.

Definition 6.4.2 (Minimal refinement operator). A *minimal refinement operator* is a refinement operator for which every element in $\rho(\mathbf{r})$ is a neighbor of $\mathbf{r}$ in the generality lattice of rules.

A rulelearning algorithm can employ different strategies for searching this lattice: *top-down* (general-to-specific), *bottom-up* (specific-to-general), and *bidirectional*. These options can be implemented using suitable rule initializers (INITIALIZERULE) and refinement operators (REFINERULE).

6.4.1 Top-Down Search

Top-down search is most commonly used in rule learning algorithms. The hypothesis space of possible rules is searched by repeatedly specializing candidate rules. Typically, the list of candidate rules is initialized with the most general rule in the hypothesis space. This rule, with the sole condition true, can be viewed as the rule with an empty body: $\oplus \leftarrow \emptyset$.

Candidate rules are then refined using *specialization operators*: refinement operators which only return the specializations of the current rule. In order to be able to reach every rule with repeated application of a specialization operator, one is typically only interested in maximally general specializations (i.e., in a minimal specialization operator).

Definition 6.4.3 (Specialization operator). A *specialization operator* σ maps a rule $\mathbf{r}$ to a set of rules $\sigma(\mathbf{r})$, such that all rules in $\sigma(\mathbf{r})$ are less general than $\mathbf{r}$, i.e., $\forall \mathbf{r}' \in \sigma(\mathbf{r}) : \mathbf{r}' \subset \mathbf{r}$.
The specialization operator is *minimal* if $\sigma(\mathbf{r})$ is the set of maximally general specializations of rule $\mathbf{r}$, i.e.,

$$\sigma(\mathbf{r}) = \{\mathbf{r}' | (\mathbf{r}' \subset \mathbf{r}) \wedge (\nexists \bar{\mathbf{r}} : \mathbf{r}' \subset \bar{\mathbf{r}} \subset \mathbf{r})\}$$

The most typical specialization operator simply adds a single feature to the body of the rule. Thus, the refinements $\mathbf{r}'$ of a rule

$$\mathbf{r} = (\oplus \leftarrow \mathbf{f}_1 \wedge \ldots \wedge \mathbf{f}_L)$$

are obtained by adding a feature $\mathbf{f}_{L+1}$ to the conjunction of conditions in the rule body, so that we get

$$\mathbf{r}' = (\oplus \leftarrow \mathbf{f}_1 \wedge \ldots \wedge \mathbf{f}_L \wedge \mathbf{f}_{L+1}).$$

In practice, it is often not necessary to extend the current rule with all remaining features. For example, we have already mentioned above that it is not necessary to add another feature of the form $\mathbf{A}_i = v_{i,j}$ if the rule body already contains one feature that tests for a specific value of the attribute $\mathbf{A}_i$. However, there may be more complex strategies for cutting down the search space. AQ, for example, selects a random *seed example* and repeatedly specializes the most general rule until it still covers the selected example, but none of the negative examples. Thus, the refinement operator only admits features that cover this seed example. A similar strategy is followed by the ILP system PROGOL.

6.4.2 Bottom-Up Search

In bottom-up search, the hypothesis space is examined by repeatedly generalizing a most specific rule. This can be achieved in INITIALIZERULE by randomly selecting a positive *seed example* and finding the most specific rule that covers only this example. Usually this rule is constructed by adding all features that cover the seed example to the body of the rule. This starting rule is then successively generalized via generalization operators to increase the number of covered examples.

A straightforward generalization operator is analogous to the specialization operator of Definition 6.4.3.

Definition 6.4.4 (Minimal generalization operator). A *generalization operator* γ maps a rule $\mathbf{r}$ to a set of rules $\gamma(\mathbf{r})$, such that all rules in $\gamma(\mathbf{r})$ are more general than $\mathbf{r}$, i.e., $\forall \mathbf{r}' \in \gamma(\mathbf{r}) : \mathbf{r}' \supset \mathbf{r}$.
The generalization operator is *minimal* if $\gamma(\mathbf{r})$ is the set of least general specializations of rule $\mathbf{r}$, i.e.,

$$\gamma(\mathbf{r}) = \{\mathbf{r}' | (\mathbf{r}' \supset \mathbf{r}) \wedge (\nexists \bar{\mathbf{r}} : \mathbf{r}' \supset \bar{\mathbf{r}} \supset \mathbf{r})\}$$

A straightforward implementation of this generalization operator would be to simply delete an existing feature from the body of the rule. Thus, the rule $\mathbf{r} = (\oplus \leftarrow \mathbf{f}_1 \wedge \ldots \wedge \mathbf{f}_L)$ has L refinements, one for each possible deletion of a feature $\mathbf{f}_i$ from the body of the rule.

However, this simple strategy may fail when combined with a greedy hill-climbing or beam search (Fürnkranz, 2002a). Consider the sample database shown in Table 6.2. The target concept $\oplus \leftarrow \mathbf{A}_1 = 1$ is quite obvious and can be found without problems by top-down hill-climbing algorithms. However, one can easily verify that there is a high chance that it cannot be found by a bottom-up hill-climbing algorithm that uses the minimal generalization operator of the type described above. Assume that the first example is picked as the seed example. The corresponding rule would be $\oplus \leftarrow (\mathbf{A}_1 = 1) \wedge (\mathbf{A}_2 = 1) \wedge (\mathbf{A}_3 = 1)$.

If the learner now tries to generalize this rule by deleting the first feature, the resulting rule $\oplus \leftarrow (\mathbf{A}_2 = 1) \wedge (\mathbf{A}_3 = 1)$ does not cover any additional examples. This is good because we obviously do not want to drop the feature $\mathbf{A}_1 = 1$ in this case. However, the situation is analogous for the other two conditions: no matter what condition is deleted from the rule, the resulting generalization does not cover any additional examples, positive or negative. Also note that this situation does not depend on the choice of the seed example: one can construct an equivalent example where the same problem occurs for picking any of the other three features.

Table 6.2 An example where bottom-up hill-climbing cannot find the target concept $\oplus \leftarrow \mathbf{A}_1 = 1$

$\mathbf{A}_1$	$\mathbf{A}_2$	$\mathbf{A}_3$	c
1	1	1	$\oplus$
1	0	0	$\oplus$
0	1	0	$\ominus$
0	0	1	$\ominus$

The following simple observation shows that the above-mentioned case occurs whenever the seed example differs from every other example in at least two of its covering features.

Theorem 6.4.1. *Let $\hat{\mathcal{F}}_e$ be all features that cover a given example e. Bottom-up hill-climbing with n-step look-ahead is unable to discriminate between all immediate generalizations of the bottom rule $\mathbf{r}_e$ of e if no more than $|\hat{\mathcal{F}}_e| - n - 1$ features of $\hat{\mathcal{F}}_e$ are shared with any other example $e_i \neq e$, i.e.,*

$$|\hat{\mathcal{F}}_e \setminus \hat{\mathcal{F}}_{e_i}| \geq n + 1.$$

Proof. Let us first consider a one-step look-ahead ($n = 1$). Then, all generalizations of the bottom rule $\mathbf{r}_e$ are obtained by deleting single feature $\mathbf{f}_j$ from the conjunction of all features in $\hat{\mathcal{F}}_e$. Thus, the generalization shares all but one feature with the original rule. If such a generalization covers a second example $e_i \neq e$, then $\mathbf{f}_j$ is the only feature that covers e but does not cover e_i, i.e., $\hat{\mathcal{F}}_e \setminus \hat{\mathcal{F}}_{e_i} = \{\mathbf{f}_j\}$. Thus, if we assume that $|\hat{\mathcal{F}}_e \setminus \hat{\mathcal{F}}_{e_i}| \geq 2$, no generalization can cover a second example, and thus all generalizations obtained by one-step look-ahead are indiscriminable.

The generalization of this argument to general $n > 1$ is quite straightforward. In this case, the candidate generalizations of the conjunction are obtained by deleting (up to) n conditions from the rule. Thus, no generalization of e can cover additional examples *iff* all other examples differ from e in at least $n + 1$ features of $\hat{\mathcal{F}}_e$. □

In particular, this theorem states that simple bottom-up hill-climbing with a minimal generalization operator will not work well whenever the chances are high to pick a seed example e which differs in at least two features from all other examples. Fürnkranz (2002a) has shown that this situation occurs in many commonly used benchmark datasets.

Nevertheless, variants of bottom-up hill-climbing search are still used to prune rule sets. In particular, the reduced-error pruning algorithm shown in Fig. 9.8 is essentially the algorithm discussed above applied to starting theory that does not necessarily contain one rule per example, but has been generated by an initial, possibly overfitting learner (see Chap. 9).

For this reason, many bottom-up learning algorithms employ different generalization operators. Typically, they do not generalize a single rule, but generalize two rules (or, equivalently, a rule and an example). Well-known examples include the NEAR and RISE algorithms, which combine instance-based and rule-based learning. We will return to these algorithms in Sect. 8.4.1. Below, we briefly discuss two types of formal refinement operators for bottom-up induction, namely inverse resolution and least general generalizations.

6.4.2.1 Inverse Resolution

Early work in this area is based on inverting the resolution principle, which forms the basis of deductive inference systems such as Prolog. DUCE (Muggleton, 1987), the first system in this family, introduced six operators (*interconstruction*, *intraconstruction*, *absorption*, *identification*, *dichotomization*, and *truncation*), which can be used to progressively generalize a propositional rule base. Suggestions for rule transformations are proposed to the user, who has the possibility to confirm or reject the new rule. The system is also capable of suggesting new predicates. The space of all possible operator applications is searched in a best-first fashion, operators producing the largest symbol reduction being examined first. Later, Muggleton (1991) showed that intraconstruction, absorption, and identification may be viewed as operators that invert the resolution principle.

CIGOL[1] (Muggleton, 1988; Muggleton & Buntine, 1988) lifts this work to inverting resolution in first-order logic. Absorption and identification form the so-called *V-operators*, intra- and interconstruction the *W-operators*. The search through the rule space is guided by a measure of the textual complexity of the theory, similar to that used in DUCE. CIGOL only uses positive examples to generalize and is not capable of specializing when it encounters negative examples which contradict its theory. Because of its elegance, research on inverse resolution had many successors (Rouveirol, 1992; Rouveirol & Puget, 1990; Wirth, 1988). However, the complexity of the search space remained a problem.

6.4.2.2 Least General Generalization

The simplest and best-known bottom-up refinement operator is the *least general generalization* of two rules.

Definition 6.4.5 (Least general generalization (lgg)). The *least general generalization (lgg)* operator takes two rules $\mathbf{r}_1$ and $\mathbf{r}_2$ and returns a rule $\mathbf{r}' = lgg(\mathbf{r}_1, \mathbf{r}_2)$, such that

(a) $\mathbf{r}'$ is more general than both $\mathbf{r}_1$ and $\mathbf{r}_2$ ($\mathbf{r}' \supseteq \mathbf{r}_1 \wedge \mathbf{r}' \supseteq \mathbf{r}_2$)
(b) There is no more specific rule for which the above holds, i.e.,
$$\nexists \bar{\mathbf{r}} \subset \mathbf{r}' : \bar{\mathbf{r}} \supseteq \mathbf{r}_1 \wedge \bar{\mathbf{r}} \supseteq \mathbf{r}_2$$

There are only a few propositional bottom-up rule learning algorithms. One such example is SIA (Venturini, 1993) which uses a genetic algorithm to search the

[1]The name CIGOL is 'logic' spelled backwards.

space of generalizations of a randomly selected example. However, the stochastic search does not progress in a strictly bottom-up fashion. Another propositional rule learning system, DLG (Webb, 1992), successively generalizes a starting example by constructing a propositional least general generalization of the current rule and the next positive example. If the resulting rule covers more positive examples without covering any negative examples it is retained.

Bottom-up learning algorithms are more popular in relational learning and inductive logic programming. We will briefly return to them in Sect. 6.5.2, where we discuss the GOLEM algorithm, which is based on a version of *lggs* that generalizes relative to existing background knowledge in first-order logic.

6.4.3 Bidirectional Search

The third option for searching the hypothesis space is to combine the previous approaches into a *bidirectional* search algorithm, which can employ both specialization and generalization operators, and, in some cases, also arbitrary jumps in the search space during the search for good rules. For example, the basic induction algorithm of the SWAP-1 rule learning system (Weiss & Indurkhya, 1991) checks whether dropping or replacing a previously learned condition can improve the rule's purity before it tries to improve it by adding a new condition. Similarly, IBL-SMART (Widmer, 1993) can perform a generalization step by dropping a condition whenever its top-down search leads to a rule that covers too few positive examples (according to some predefined threshold). However, both algorithms preserve an overall top-down tendency in their search.

The JOJO algorithm (Fensel & Wiese, 1993), on the other hand, starts the search at an arbitrary point in the hypothesis space (e.g., a randomly generated rule) and improves it by applying generalization and specialization operators, i.e., by adding or dropping conditions. Recent additions allow the system to directly replace conditions in rules (Fensel & Wiese, 1994) and to use general first-order literals (Wiese, 1996). The ATRIS rule learning shell (Mladenić, 1993) allows one to perform a similar bidirectional search, but replaces JOJO's hill-climbing search with less myopic stochastic search procedures as in (Kononenko & Kovačič, 1992) or a generalized hill-climbing technique that performs a fixed number of refinement operations at a time.

6.5 Relational Rule Construction

In Chap. 5 we introduced relational background knowledge and showed how it can be mapped onto covering tables, upon which standard rule learning algorithms can operate. However, it is also possible to directly integrate first-order refinement operators into the learning process. This has the advantage that the entire covering table does not have to be calculated in advance, so that the algorithm may focus only on a few important refinements. For propositional learning algorithms, this difference is not so significant, but for relational learners it may reduce the complexity of the task, at an increased danger of running astray.

Let us recall the steps of rule construction from Sect. 2.4: feature construction, rule construction, and hypothesis construction. While the last hypothesis construction step (in which a set of rules is constructed) is beyond the scope of this section, we indicate how—in the case of first-order learning—the first two steps differ from propositional rule learning.

Feature construction. While propositional features usually take the form of attribute-value pairs, first-order features consist of literals. Individual first-order features are constructed using a feature refinement operator, as will be described below.

Rule construction. An individual first-order rule of the form H ← B is constructed by fixing the head H to a given predicate to be learned, and heuristically searching for the best rule body, where the rule body B is a conjunction of features. As in the FINDBESTRULE algorithm of Sect. 6.2, body construction is performed by body refinement, through adding features to the initially empty rule body.

We should add that, in practice, body construction is often done in a single step by adding literals one-by-one. However, this can be problematic because a single literal may not be added for the purpose of discriminating between positive and negative examples, but because it adds a new variable that may be used by subsequent literals. For example, in the East–West train challenge introduced in Sect. 5.5, literals such as `hasCar(T,C)` are not really meaningful unless such a literal is followed by another literal 'consuming' the local variable `C`. Furthermore, such a literal is unlikely to improve the quality of the rule (unless it is explicitly meant to distinguish between trains with and trains without cars). In practice, therefore, some sort of look-ahead is applied in order to add several literals at a time. The notion of first-order feature formalizes the need for look-ahead in a systematic way.

Refinement operators were defined in Sect. 2.5.2. If we distinguish between body construction and feature construction, we need a refinement operator for both. Because on the level of features each rule is propositional, the body refinement operator is fairly simple as it considers all possible ways to add a single feature to the body of the rule. A feature refinement operator, on the other hand, constructs features consisting of several literals sharing local variables and hence needs to assume some kind of language bias.

A thorough discussion of research in inductive logic programming and relational learning is beyond the scope of this section; we refer the reader to (De Raedt, 2008; Džeroski & Lavrač, 2001; Lavrač & Džeroski, 1994a). In this section we briefly show the integration of first-order refinement operators into heuristic rule learning algorithms using two exemplary systems, FOIL and GOLEM.

6.5.1 *Top-Down First-Order Rule Learning with* FOIL

Using refinement operators, it is easy to define a simple general-to-specific search algorithm for finding hypotheses. FOIL (Quinlan, 1990) is such a top-down ILP learner, employing the Datalog representation explained in the previous section, and using the covering algorithm explained in Sect. 2.6. Just as in propositional learning, a rule is constructed by repeatedly applying a refinement operator, until the rule is consistent with all negative examples for the predicate (no negative example is covered) or the encoding length restriction is violated. In FOIL, the search for the predicate definition of target predicate `p` starts with the maximally general rule `p(X1,...,Xn) :- true`, where `true` stands for the empty body.

In each refinement step a literal is added to the body of the rule, having one of the following four forms: (i) `q(Y1,...,Yk)` or (ii) `not q(Y1,...,Yk)`, where `q` is a background predicate or possibly the target predicate; (iii) `Yi` $=$ `Yj` or (iv) `Yi` $\neq$ `Yj`. It is required that at least one of the variables in the new literal has appeared before in the rule. This ensures that the rule is 'linked', i.e., there is a connection between each body literal and the head of the rule. The remaining variables in the new literal are new local variables. Some further restrictions are applied if the new literal contains the target predicate, in order to avoid possible nontermination problems.

The main difference between FOIL and CN2 is the way the current sets of positive and negative examples are extended if new local variables are added to the rule. For instance, in the East–West trains problem the initial hypothesis is `eastbound(T):- true`. This rule covers all positive examples `t1`, ..., `t5` as well as all negative examples `t6`, ..., `t10`. Suppose that we add the literal `hasCar(T,C)` to the rule, then the covered instances are extended to pairs of constants, one for each of the variables in the current rule. So the covered examples of the form $< \mathsf{Train}_i, \mathsf{Car}_{ij} >$ are listed below.

```
⊕  <t1,c11>    <t1,c12>    <t1,c13>    <t1,c14>
   <t2,c21>    <t2,c22>    <t2,c23>
   <t3,c31>    <t3,c32>    <t3,c33>
   <t4,c41>    <t4,c42>    <t4,c43>    <t4,c44>
   <t5,c51>    <t5,c52>    <t5,c53>

⊖  <t6,c61>    <t6,c62>
   <t7,c71>    <t7,c72>    <t7,c73>
   <t8,c81>    <t8,c82>
   <t9,c91>    <t9,c92>    <t9,c93>    <t9,c94>
   <t10,c101>  <t10,c102>
```

Extending the rule to `eastbound(T) :- hasCar(T,C),clength(C,short)` removes four positive and five negative examples, resulting in the following covered examples:

```
⊕  <t1,c12>    <t1,c14>
   <t2,c21>    <t2,c22>    <t2,c23>
   <t3,c31>    <t3,c32>
   <t4,c41>    <t4,c42>    <t4,c43>    <t4,c44>
   <t5,c51>    <t5,c53>

⊖  <t6,c62>
   <t7,c71>    <t7,c72>
   <t8,c82>
   <t9,c91>    <t9,c93>    <t9,c94>
   <t10,c101>
```

Finally, adding the literal `not croof(C,no_roof)` removes all the remaining negative examples, and the rule is added to the hypothesis.

A few points are worth noting here. First, structural predicates like `hasCar/2` can be easily overlooked by a greedy system such as FOIL, because they do not help to distinguish between eastbound and westbound trains, but facilitate the inclusion of properties which do discriminate. This is why FOIL gives special consideration to determinate literals in its search process (cf. Sect. 5.5.5.1). Secondly, extending the example in the above way (i.e., as a tuple of constants for *all* variables in the rule) can lead to counterintuitive results. In effect, this method of counting covered tuples amounts to counting the number of proofs that can be found for a covered instance. As a result, in the above example, if westbound trains tend to have more cars than eastbound trains, the rule

```
eastbound(T) :- hasCar(T,C).
```

will get a much worse evaluation than

```
eastbound(T) :- true.
```

whereas if eastbound trains tend to be longer the situation is the opposite. Adapting FOIL to work with first-order features would solve both these problems.

6.5.2 *Bottom-Up First-Order Rule Learning with* GOLEM

Until now we have concentrated on top-down approaches to rule learning. The final example we give in this chapter concerns the bottom-up ILP system GOLEM (Muggleton & Feng, 1990). Bottom-up systems start directly from the examples and construct specific rules which are subsequently generalized.

We illustrate GOLEM on a simplified version of the trains example, in which cars are only described by their length and roof-type. Suppose the first eastbound train `t1` has two cars, `c11` which is long with a flat roof, and `c12` short and open. A rule covering exactly this positive example is the following:

```
eastbound(t1) :-
  hasCar(t1,c11), clength(c11,long), croof(c11,flat),
  hasCar(t1,c12), clength(c12,short), croof(c12,no_roof).
```

Similarly, let the second eastbound train `t2` have two cars, `c21` which is short and open, and `c22` short with a flat roof. A rule covering only `t2` is the following:

```
eastbound(t2) :-
  hasCar(t2,c21), clength(c21,short), croof(c21,no_roof),
  hasCar(t2,c22), clength(c22,short), croof(c22,flat).
```

These rules are obviously too specific to be useful in learning. However, we can construct a single rule covering both positive examples, and potentially others as well, by an algorithm for computing the *lgg* (cf. Definition 6.4.5) for first-order rules (Plotkin, 1970). Essentially, this involves comparing the two rules, and in particular each pair of literals, one from each rule, with the same predicate: for each argument of the predicate, if they are the same constant then we copy that constant to the *lgg*, else the *lgg* will have a variable for that argument, such that the same variable will be used whenever that pair of constants is encountered.

The *lgg* of the two rules above is the following rule:

```
eastbound(T) :-
  hasCar(T,CA), clength(CA,L1), croof(CA,R1),
  hasCar(T,CB), clength(CB,L1), croof(CB,flat),
  hasCar(T,CC), clength(CC,short), croof(CC,no_roof),
  hasCar(T,CD), clength(CD,short), croof(CD,R2).
```

The variable `CA` results from comparing `c11` and `c21`, `CB` stands for `c11` and `c22`, and so on. The variable `L1` results from comparing a short car in the first train with a long car in the second, and thus occurs twice. However, these are variables in the Prolog sense, and thus match any constant, not just the ones from which they were constructed.

It may appear that the generalized rule requires four cars in an eastbound train, but since some of the features in the rule may coincide upon instantiation, this is not really the case. For instance, note that if `L1` is instantiated to `short` and `R1` to `no_roof`, the first and the third features become equivalent (and essentially describe car `c21`). By construction, the *lgg* covers `t1` and `t2`, which are trains with only two cars. Also, note that a rule can be simplified if it has a feature which is a more specific version of another feature: in that case, the latter is redundant. This results in the following simplified rule:

```
eastbound(T) :-
  hasCar(T,CB), clength(CB,L1), croof(CB,flat),
  hasCar(T,CC), clength(CC,short), croof(CC,no_roof).
```

That is, any train with a short open car and a car with a flat roof is eastbound, which is a very sensible generalization.

Notice that the size of the *lgg* grows quadratically with the size of the examples. In more realistic cases the *lgg* will still contain a lot of redundant information.

In this case, GOLEM uses the negative examples to search for literals that can be removed without covering negatives.

GOLEM (Muggleton & Feng, 1990) forms a starting rule by computing the *lgg* of a set of randomly chosen pairs of positive examples. Among these, it selects the best one, and then reduces it in the way just described. The resulting starting rule is then successively generalized by greedily selecting additional positive examples which will further generalize the rule by computing the *lgg* of the rule and the example. If no further generalizations improve the rule, GOLEM removes all the positives covered by the rule as in the covering algorithm, and iterates to find the next rule.

GOLEM can take background knowledge into account, but only in the form of ground facts. Before constructing the *lgg*, it adds all background facts to the bodies of both ground rules: the resulting *lgg* is called a *relative* least general generalization or *rlgg* (Plotkin, 1971).

Several ILP systems implement quite similar techniques. ITOU (Rouveirol, 1992) constructs a starting rule by adding all conditions than can be proved from the background knowledge (*saturation*). A representation change called *flattening* that removes all function symbols from the examples and the background knowledge allows one to implement generalization with a single operator that drops literals from rules (*truncation*) (Rouveirol, 1994). NINA (Adé, De Raedt, & Bruynooghe, 1995) is a bottom-up first-order rule learning algorithm that unifies several bottom-up ILP algorithms such as GOLEM (Muggleton & Feng, 1990), ITOU (Rouveirol, 1992), and CLINT (De Raedt, 1992). As the most specific rules can be exponentially large, even infinite in the case of general first-order Horn-rule logic (Plotkin, 1971), the hypothesis space has to be restricted to a subset of first-order logic using syntactic or semantic restrictions, as discussed in Sect. 5.3. PROGOLEM (Muggleton, Santos, & Tamaddoni-Nezhad, 2009) is a recent system that weakens this restriction by using an asymmetric generalization rule that is based upon the computation of a bottom rule as in PROGOL.

6.6 Conclusion

This chapter presented approaches for finding single rules. As laid out in Chap. 2, we will use these algorithms primarily as a building block for discovering a set of rules forming the entire rule-based theory, which we will discuss in more detail in Chap. 8. However, individual rules are already an interesting data mining result by themselves, and are often directly reported and put to use. In Chap. 11, we will put various such approaches into a common framework named *supervised descriptive rule learning*.

A key ingredient to both predictive and descriptive rule learning is the choice of an appropriate heuristic function that assesses the quality of candidate rules. This will eventually determine the result which is returned by the search algorithm. All algorithms discussed in this chapter work with essentially arbitrary rule quality measures. In the next chapter, we discuss various choices for this parameter.

Chapter 7
Rule Evaluation Measures

This chapter gives an overview of rule evaluation measures that are commonly used as search heuristics in rule learning. The behavior of various heuristics is analyzed by visualizing their dynamics in the coverage space (cf. Chap. 3). The chapter starts with some basic definitions, and introduces isometrics and equivalence of heuristics in Sect. 7.2. Section 7.3 reviews many commonly used heuristics, including accuracy, weighted relative accuracy, entropy, cross-entropy, and the Gini index. We show that they can be reduced to two fundamental prototypes: precision and a cost-weighted difference between covered positive and negative examples. A straightforward generalization of the m-estimate may be viewed as a generalization of both prototypes. Section 7.4 then discusses a few heuristics that do not conform to this model because they have nonlinear isometrics in the coverage space. Among them are correlation, FOIL's information gain, and Klösgen's measures, the last two being discussed in Sect. 7.5. Finally, we discuss a few complexity-based heuristics (Sect. 7.6) and general models for composing complex heuristics out of simpler ones (Sect. 7.7). After this survey, we attempt to give a few recommendations for the practical use of these heuristics, particularly focusing on the differences of their use in descriptive and predictive rule induction (Sect. 7.8).

7.1 Introduction

While the outer loop of the covering strategy is commonly shared by many rule learning algorithms, individual approaches differ in the way single rules are learned. In the previous chapter, in particular in Sect. 6.2, we saw that finding a single rule may be considered as a search problem through the hypothesis space of all possible rules. Typically, this search is guided by a heuristic function for evaluating

---

†Parts of this chapter are based on (Fürnkranz & Flach, 2005).

J. Fürnkranz et al., *Foundations of Rule Learning*, Cognitive Technologies,
DOI 10.1007/978-3-540-75197-7_7, © Springer-Verlag Berlin Heidelberg 2012

the quality of a candidate rule. In the rule learning literature one can find many rule quality measures that have been proposed for this purpose. While most of them have justifications based in statistics, information theory, or related fields, their relations to each other are not well understood.

In general, the search heuristics used in separate-and-conquer rule learning are similar to the heuristics used in other inductive learning algorithms such as decision-tree learning heuristics discussed in (Buntine & Niblett, 1992; Mingers, 1989b). The major difference between heuristics for rule learning and heuristics for decision tree learning is that the latter evaluate the average quality of a number of disjoint sets (one for each value of the attribute that is tested), while rule learning approaches only evaluate the quality of the set of examples that is covered by the candidate rule.

All common heuristics are based on determining several basic properties of a candidate rule, like the number of positive and negative examples that it covers. Minor variations in counting are possible, as, for example, in the first-order rule learning algorithm FOIL, which does not count covered instances but instead counts the number of covered tuples (Sect. 6.5.1).

The basic goal of search heuristics is to evaluate a rule by its ability to discriminate between positive and negative training examples. Thus, the main challenge is to simultaneously maximize $\hat{P}$, the number of positive examples that are covered by a rule (the true positives), and $\bar{N}$, the number of negative examples that are not covered by the rule (the true negatives). Maximizing the latter is equivalent to minimizing $\hat{N}$, the number of false positives. Most rule learning heuristics define a trade-off between $\hat{P}$ and $\hat{N}$, such as the ratio or the difference of true and false positives. This ensures that both completeness and consistency of rules are enforced.

However, in many cases we are not only interested in a rule's level of precision or accuracy, but also how this level is achieved. For example, in propositional rule learning it is always possible to learn a rule that only covers positive examples and does not cover any negative examples: one only has to use a conjunction of all attribute values of given positive examples.[1] However, learning such a rule is undesirable, because the rule does not generalize beyond the training examples. For this reason, many heuristics try to penalize the complexity of a rule (e.g., by considering the rule length: long rules will be penalized), or include a bias towards more general rules (e.g., by penalizing rules that cover few examples). As adding conditions to a rule can only specialize and never generalize a rule, rule complexity and rule generality may be considered mutually dependent concepts.

In summary, almost all rule learning heuristics may be interpreted as consisting of partial measures for the consistency, the completeness, or the complexity of the rule. In this chapter, we attempt to shed some light upon this issue by investigating commonly used evaluation measures. With few exceptions, we will focus on heuristics for rule learning that can be found in the literature on covering algorithms.

---

[1] We assume that the dataset is not contradictory, i.e., there are no examples with identical attribute-value representations that are assigned different class values.

In particular, the chapter focuses on heuristics used in classification rule learning; a reader with a strong interest in association rule learning can find excellent surveys of rule evaluation measures proposed in the data mining literature in Geng & Hamilton (2006), Tan, Kumar, & Srivastava (2002, 2004).

Note the terminology used in this chapter: we use interchangeably the terms *heuristics*, *rule evaluation metrics* and *rule quality measures*. We mostly use the term *heuristics* for pragmatic reasons: the term is shorter and reflects the fact that rule quality measures are most crucially needed to heuristically guide the search for rules when refining a rule by adding a feature to the rule body. We also use the term *metrics* in an informal way, without requiring that the quality measures necessarily conform to the strict mathematical properties that metrics have to satisfy. Some rule quality measures outlined in this chapter were already briefly introduced in Chap. 2. In this chapter we shed new light on these measures by providing their analysis and visualization in the coverage space.

7.2 Basic Properties of Heuristics

The goal of a rule learning algorithm is to find a simple set of rules that explains the training data and generalizes well to unseen data. This means that individual rules have to simultaneously optimize two criteria:

- *Completeness:* the number of positive examples that are covered by the rule should be maximized, and
- *Consistency:* the number of negative examples that are covered by the rule should be minimized.

Thus, each rule can be characterized by

- $\hat{P}$ and $\hat{N}$ (the positive/negative examples covered by the rule)
- P and N (the total number of positive/negative training examples)

Consequently, most rule learning heuristics depend on $\hat{P}$, $\hat{N}$, P, and N, but combine these values in different ways. Heuristics that use other parameters, such as the length of the rule (L), will be covered in Sect. 7.7. Given that P and N are constant for a given dataset, heuristics effectively only differ in the way they trade off completeness (maximizing $\hat{P}$) and consistency (minimizing $\hat{N}$). Thus a heuristic may be viewed as a function $H(\hat{P}, \hat{N})$. However, as both parameters depend on the rule, we will typically denote a heuristic with $H(\mathbf{r})$, or omit the argument entirely. All parameters used for defining the heuristics are then implicitly assumed to refer to rule $\mathbf{r}$.

Clearly, heuristics that effectively only depend on $\hat{P}$ and $\hat{N}$ are unable to discriminate between rules that cover the same number of positive and negative examples, i.e., two different rules that cover the same number of positive and negative examples will have the same heuristic evaluation. In other words, two different rules need not have different heuristic evaluation, i.e., $\mathbf{r}_1 \neq \mathbf{r}_2 \nRightarrow H(\mathbf{r}_1) \neq H(\mathbf{r}_2)$.

Let us now define some basic properties of rule evaluation metrics. The prime task of heuristics is to order all possible candidates according to some quality measure. Intuitively, if two heuristics order all rules in the same way, they can be considered to be equivalent. In the following, we will provide a more formal definition of this notion.

We first define that two heuristics are *compatible* if they rank the rules in exactly the same order.

Definition 7.2.1 (compatible). Two heuristic functions H and G are *compatible* iff for all rules $\mathbf{r}_i$:

$$H(\mathbf{r}_1) > H(\mathbf{r}_2) \Leftrightarrow G(\mathbf{r}_1) > G(\mathbf{r}_2).$$

Sometimes, two heuristics order the rules in exactly the opposite way, e.g., because one is a criterion that needs to be minimized, and the other is a criterion that needs to be maximized. We call such heuristics *antagonistic*.

Definition 7.2.2 (antagonistic). Two heuristic functions H and G are *antagonistic* iff for all rules $\mathbf{r}_1, \mathbf{r}_2$:

$$H(\mathbf{r}_1) > H(\mathbf{r}_2) \Leftrightarrow G(\mathbf{r}_1) < G(\mathbf{r}_2).$$

Obviously, compatibility and antagonicity are mutually dependent concepts: maximizing one is equivalent to minimizing the other.

Lemma 7.2.1. *H and G are antagonistic iff H and $-G$ are compatible.*

Proof. Assume H and G are antagonistic:

$$H(\mathbf{r}_1) > H(\mathbf{r}_2) \Leftrightarrow G(\mathbf{r}_1) < G(\mathbf{r}_2) \Leftrightarrow -G(\mathbf{r}_1) > -G(\mathbf{r}_2)$$

Thus, H and $-G$ are compatible. □

Next, we define two heuristics as *equality-preserving* if they have identical regions of equality:

Definition 7.2.3 (equality-preserving). Two heuristic functions H and G are *equality-preserving* iff for all rules $\mathbf{r}_1, \mathbf{r}_2$:

$$H(\mathbf{r}_1) = H(\mathbf{r}_2) \Leftrightarrow G(\mathbf{r}_1) = G(\mathbf{r}_2).$$

Now it is easy to show the following:

Theorem 7.2.1. *Compatible or antagonistic heuristics are equality-preserving.*

Proof. Assume the heuristics were not equality-preserving. This means that there exist rules $\mathbf{r}_1$ and $\mathbf{r}_2$ with $H(\mathbf{r}_1) = H(\mathbf{r}_2)$ but $G(\mathbf{r}_1) \neq G(\mathbf{r}_2)$. Without loss of generality assume $G(\mathbf{r}_1) > G(\mathbf{r}_2)$. This implies that $H(\mathbf{r}_1) > H(\mathbf{r}_2)$ (for compatibility) or $H(\mathbf{r}_1) < H(\mathbf{r}_2)$ (for antagonicity). Both cases contradict the assumption. □

Note that the opposite direction does not necessarily hold, i.e., equality-preserving heuristics are only compatible or antagonistic if we make some continuity assumptions. Although all subsequently discussed functions are continuous, we note that this does not have to be the case. For all practical purposes, a heuristic $H(\hat{P}, \hat{N})$ may be regarded as a look-up table that defines a value H for each integer-valued pair $(\hat{N}, \hat{P})$.

Based on the above, we can define the equivalence of heuristics. The basic idea is that heuristic functions are equivalent if they order a set of candidate rules in an identical way.[2]

Definition 7.2.4 (equivalence). Two heuristic functions H and G are *equivalent* $(H \sim G)$ if they are either compatible or antagonistic.

This definition also underlines the importance of *isometrics* for the analysis of heuristics. Informally, an isometric is a line which connects points in coverage space that have the same heuristic evaluation, very much like a contour line connects points on a map that have the same altitude. A formal definition is given below:

Definition 7.2.5 (isometric). An *isometric* of a heuristic H is a line (or curve) in the coverage space that connects all points $(\hat{N}, \hat{P})$ for which $H(\hat{P}, \hat{N}) = c$ for some value c.

Equality-preserving heuristics can be recognized by examining their isometrics and establishing that for each isometric line for H there is a corresponding isometric for G. Compatible (and antagonistic) heuristics can be recognized by investigating corresponding isometrics and establishing that their associated heuristic values are in the same (the opposite) order.

In the following sections, we will give a systematic overview of commonly used rule learning heuristics, starting with heuristics that have linear isometrics in coverage space (Sect. 7.3), heuristics with nonlinear isometrics (Sect. 7.4), as well as gain-based (Sect. 7.5), complexity-based (Sect. 7.6), and composite heuristics (7.7).

7.3 Heuristics with Linear Isometrics

The ultimate goal of learning is to reach point $(0, P)$ in the coverage space, i.e., to learn a correct theory that covers all positive examples, but none of the negative examples. This will rarely be achieved in a single step, but a *set* of rules will be needed for meeting this objective. The purpose of a rule evaluation metric is to estimate how close a rule takes you to this ideal point.

[2]The notion of equivalence should also capture the fact that some learning systems search for rules that minimize the heuristic function, while others search for a maximum (e.g., maximizing information gain is equivalent to minimizing entropy).

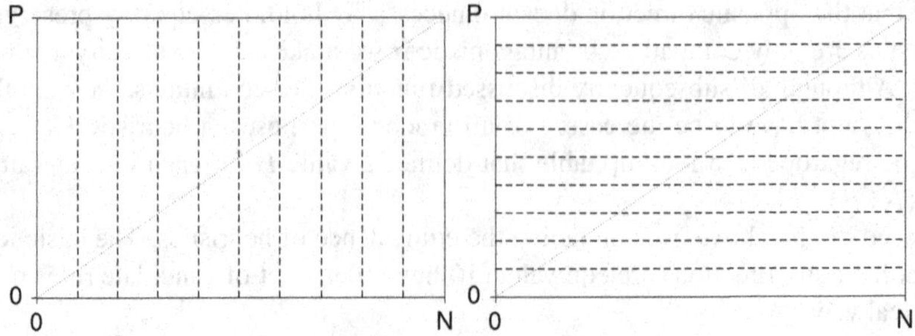

Fig. 7.1 Isometrics for minimizing false positives (*left*) and for maximizing true positives (*right*)

7.3.1 Elementary Heuristics

A straightforward strategy for finding a rule that covers some positive and as few negative examples as possible is to minimize the number of covered negative examples for each individual rule. This is equivalent to maximizing the *number of not-covered negative examples.*

$$\boxed{UncovNeg(\mathbf{r}) = \bar{N} = N - \hat{N} \sim -\hat{N}}$$

The left graph of Fig. 7.1 shows the isometrics for *UncovNeg*, which are parallel vertical lines. All rules that cover the same number of negative examples are evaluated equally, irrespective of the number of positive examples they cover.

In medical data analysis, the fraction of uncovered negative examples is known as the *specificity*; in data mining this is also known as the *true negative rate*.

$$\boxed{Specificity(\mathbf{r}) = \frac{\bar{N}}{N} = \bar{\nu}}$$

Obviously, *UncovNeg* and *Specificity* do not capture the intuition that we want to cover as many positive examples as possible. For this purpose, the *positive coverage*, the number of covered positive examples, could be used instead.

$$\boxed{CovPos(\mathbf{r}) = \hat{P}}$$

The isometrics of *CovPos* are parallel horizontal lines (right graph of Fig. 7.1) because all rules that cover the same number of positive examples are evaluated the same. Two rules, one covering no negative and one covering all negative examples, are considered to be equal as long as they cover the same number of positive examples. Despite this deficiency, *CovPos* has been used as a search heuristic (e.g., in the DLG rule learner; Webb, 1992), but it is more often used as a weighting function, such as, e.g., in FOIL's information gain (cf. Sect. 7.7).

More widely used, however, are *support* and *recall*. Recall is the percentage of covered positive examples among all positive examples. This is also known as the *true positive rate* and as *sensitivity*, the counterpart to specificity.

$$\mathit{Recall}(\mathbf{r}) = \mathit{Sensitivity}(\mathbf{r}) = \frac{\hat{P}}{P} = \hat{\pi}$$

As P is constant, *Recall* is obviously equivalent to *CovPos*.

Support is usually defined as the fraction of examples that satisfy both the head and the body of the rule, i.e., the fraction of positive examples among all examples.

$$\mathit{Support}(\mathbf{r}) = \frac{\hat{P}}{P+N}$$

Obviously, *Support* only has a different normalizing constant and is thus equivalent to both *Recall* and *CovDiff*. All of these measures may be considered to be simple measure for the *generality* of a rule (the more general the rule is, the more examples it covers).

It is trivial to find theories that maximize either of the above two types of heuristics: *UncovNeg* is maximal for the empty theory $\mathbf{r}_\perp$, which does not cover any negative examples, but also no positive examples, and *CovPos* is maximal for the universal theory $\mathbf{r}^\top$, which covers all positive and negative examples. Ideally, one would like to achieve both goals simultaneously.

7.3.2 *Classification Accuracy, Weighted Relative Accuracy, General Linear Costs*

A straightforward way for trading off covering many positives and excluding many negatives is to simply add up *UncovNeg* and *CovPos*, i.e., to compute the *difference between true and false positives*, in short the *coverage difference*.

$$\mathit{CovDiff}(\mathbf{r}) = \mathit{CovPos}(\mathbf{r}) + \mathit{UncovNeg}(\mathbf{r}) = \hat{P} - \hat{N}$$

The isometrics for this function are shown in the left graph of Fig. 7.2. Note that the isometrics all have a 45° angle. *CovDiff* has been used in this form as part of the admissible search heuristic used in PROGOL (Muggleton, 1995).

One can show that this heuristic is equivalent to *classification accuracy*:

$$\mathit{Accuracy}(\mathbf{r}) = \frac{\hat{P}+\bar{N}}{P+N} = \frac{\hat{P}+(N-\hat{N})}{P+N}$$

Classification accuracy evaluates the accuracy of the rule set containing only a single rule $\mathbf{r}$. It computes the percentage of correctly classified examples, i.e., the positive examples covered by the rule (*true positives*) plus the negative examples not covered

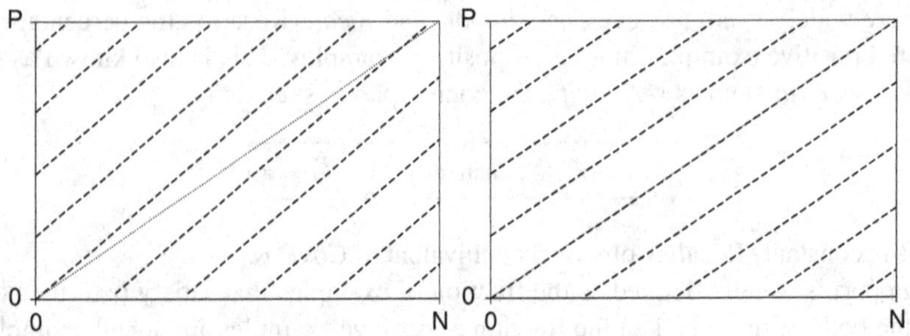

Fig. 7.2 Isometrics for coverage difference and classification accuracy (*left*) and weighted relative accuracy (*right*)

by the rule (*true negatives*). This idea, has, e.g., been used in the rule pruning algorithm I-REP (Fürnkranz & Widmer, 1994).

Theorem 7.3.1. *Coverage difference (CovDiff) is equivalent to classification accuracy (Accuracy).*

Proof. The isometrics of *Accuracy* are of the form $\frac{\hat{P}+(N-\hat{N})}{P+N} = c_{acc}$, and the isometrics of *CovDiff* are of the form $\hat{P} - \hat{N} = c_{cd}$. The following relation holds:

$$c_{cd} = c_{acc} \cdot (P + N) - N$$

Thus, because P and N are constant, the isometrics for *Accuracy* are identical to the isometrics of *CovDiff*. □

Accuracy is sometimes also used together with *rule coverage*, which is defined as the proportion of all examples that are covered by a rule.

$$\boxed{Coverage(\mathbf{r}) = \frac{\hat{P}+\hat{N}}{P+N}}$$

The isometrics of rule coverage are orthogonal to those for accuracy, i.e., they move in parallel lines from the empty rule (no coverage) to the universal rule (covering all examples). The lines have an angle of 45° with the N- and P-axes because for coverage there is no difference in importance for covering a positive or covering a negative example.

Optimizing accuracy gives equal weight to covering a single positive example and excluding a single negative example. There are cases where this choice is arbitrary, for example, when misclassification costs are not known in advance or when the samples of the two classes are not representative. In such cases, it may be advisable to normalize with the sample sizes, i.e., to compute the *difference between the true positive and false positive rates*, in short the *rate difference*

$$\boxed{RateDiff(\mathbf{r}) = \frac{\hat{P}}{P} - \frac{\hat{N}}{N} = \hat{\pi} - \hat{\nu}}$$

The isometrics of this heuristic are shown in the right part of Fig. 7.2. The main difference to accuracy is that the isometrics of *RateDiff* are parallel to the diagonal, which reflects that we now give equal weight to increasing the *true positive rate* ($\hat{\pi}$) or to decreasing the *false positive rate* ($\hat{\nu}$). Note that the diagonal encompasses all classifiers that make random predictions. Thus, optimizing *RateDiff* may be viewed as a preference for rules that are as far from random as possible.

RateDiff may also be viewed as a simplification of a heuristic known as *weighted relative accuracy* (Lavrač, Flach, & Zupan, 1999);

$$\boxed{WRA(\mathbf{r}) = \frac{\hat{P}+\hat{N}}{P+N} \cdot \left(\frac{\hat{P}}{\hat{P}+\hat{N}} - \frac{P}{P+N}\right)}$$

Variations of this rule evaluation metric are well-known in subgroup discovery (Klösgen, 1996; Lavrač, Kavšek, Flach, & Todorovski, 2004; Piatetsky-Shapiro, 1991), and its use as a heuristic for predictive rule learning has been explored in Todorovski, Flach, and Lavrač (2000).

Theorem 7.3.2. *Rate difference (RateDiff) is equivalent to weighted relative accuracy (WRA).*

Proof. Using equivalence-preserving transformations such as multiplications with constant values like $P + N$, we can transform *WRA* into

$$WRA = \frac{1}{P+N} \cdot \left(\hat{P} - \hat{P} \cdot \frac{P}{P+N} - \hat{N} \cdot \frac{P}{P+N}\right) \sim$$

$$\sim \hat{P} \cdot \frac{N}{P+N} - \hat{N} \cdot \frac{P}{P+N} \sim \hat{P} \cdot N - \hat{N} \cdot P \sim RateDiff.$$

□

CovDiff, which gives equal weights to positive and negative examples, and *RateDiff*, which weighs them according to their proportions in the training set, are special cases of a general function that allows us to incorporate arbitrary cost ratios between false negatives and false positives. The general form of this *linear cost metric* is

$$\boxed{LinCost(\mathbf{r}) = a \cdot \hat{P} - b \cdot \hat{N} \sim (1-c) \cdot \hat{P} - c \cdot \hat{N} \sim \hat{P} - d \cdot \hat{N}}$$

The parameters a, b, c, and d are different ways of specifying the cost trade-off: $a \in [0, \infty]$ denotes the benefit (negative costs) of a true positive, whereas $b \in [0, \infty]$ specifies the costs of a false negative. However, as we are only interested in comparing *LinCost* for different rules, the absolute size of a and b is irrelevant;

only their ratio is important for determining the order of the rules. Thus, the benefit of a true positive can be normalized to 1, which results in costs $d = \frac{b}{a}$ for a false positive; $d = 0$ means no costs for false positives (resulting in *CovPos*), whereas $d = \infty$ denotes infinite costs for false positives (i.e., *UncovNeg*). It is often more convenient to normalize this cost parameter to the range [0, 1], which results in $c = b = \frac{d}{d+1}$, and $1 - c = a = \frac{1}{d+1}$.

In the coverage graph, the cost trade-off is characterized by the slope $d = \frac{b}{a} = \frac{c}{1-c}$ of the parallel isometrics. As a consequence, arbitrary cost ratios can be modeled by *LinCost*. For example, accuracy is characterized by equal costs for false positives and false negatives. Therefore, its isometrics can be obtained with $a = b = d = 1$ or $c = \frac{1}{2}$. The isometrics of weighted relative accuracy can be obtained by setting $a = \frac{1}{P}$ and $b = \frac{1}{N}$, or $c = \frac{P}{P+N}$, or $d = \frac{P}{N}$.

A straightforward variant is the *relative linear cost metric*, which does not trade off the number of true and false positives, but instead their ratios.

$$\mathit{RelLinCost}(\mathbf{r}) = a \cdot \frac{\hat{P}}{P} - b \cdot \frac{\hat{N}}{N} = a \cdot \hat{\pi} - b \cdot \hat{\nu}$$

While *RelLinCost* and *LinCost* are equivalent for a particular dataset (one just needs to divide the parameters a and b of *LinCost* by P and N, respectively), it may make a practical difference if a uniform choice of parameters is used across multiple datasets. For example, Janssen and Fürnkranz (2010) have observed that in a CN2-like learner, the optimal parameter setting for *RelLinCost* significantly outperforms the optimal parameter setting for *LinCost* across 57 datasets.

7.3.3 Precision

The most commonly used heuristic for evaluating single rules is to look at the proportion of positive examples in all examples covered by the rule. This metric is known under many different names, e.g., *rule accuracy*, *confidence* in association rule discovery (with support as its counterpart), or *precision* in information retrieval (with recall as its counterpart). It may be regarded as the conditional probability that the head of a rule is true given that its body is true, or equivalently (for concept learning) the probability that the example is positive given that it is covered by the rule.

$$\mathit{Precision}(\mathbf{r}) = \frac{\hat{P}}{\hat{P}+\hat{N}}$$

Obviously, this measure will attain its optimal value when no negative examples are covered. However, it does not bias the learner towards covering many positive examples. Figure 7.3 shows the isometrics for this heuristic. Like *CovPos*, precision considers all rules that cover only positive examples to be of equal quality

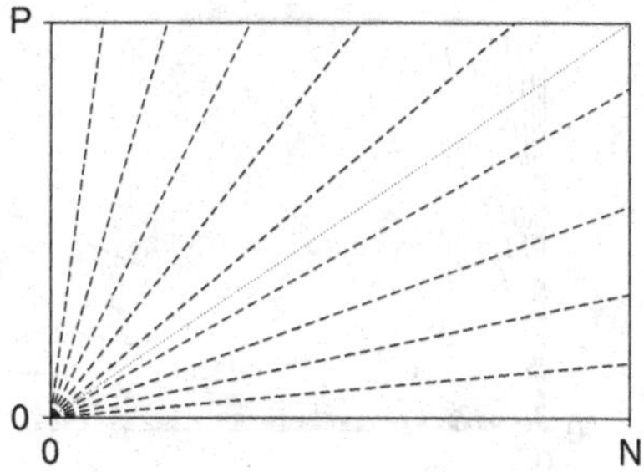

Fig. 7.3 Isometrics for precision

(the P-axis), and like *UncovNeg*, it considers all rules that only cover negative examples as equally bad (the N-axis). All other isometrics are obtained by rotation around the origin $(0, 0)$, for which the value of *Precision* is undefined. Thus, at least near the axes, *Precision* has similar disadvantages as *CovPos* and *UncovNeg*. In particular, it is unable to discriminate between good rules with high and low coverage, as we already discussed in Sect. 2.7.1.

Despite this weakness, precision was directly used in the GREEDY3 (Pagallo & Haussler, 1990) and SWAP-1 (Weiss & Indurkhya, 1991) rule learning algorithms. Several other, seemingly more complex heuristics can be shown to be equivalent to precision. For example, the heuristic that is used for evaluating rules in the pruning phase of RIPPER (Cohen, 1995):

Theorem 7.3.3. RIPPER*'s pruning heuristic* $\frac{\hat{P}-\hat{N}}{\hat{P}+\hat{N}}$ *is equivalent to Precision.*

Proof. $\frac{\hat{P}-\hat{N}}{\hat{P}+\hat{N}} = \frac{\hat{P}}{\hat{P}+\hat{N}} - \frac{\hat{N}}{\hat{P}+\hat{N}} = \frac{\hat{P}}{\hat{P}+\hat{N}} - (1 - \frac{\hat{P}}{\hat{P}+\hat{N}}) = 2 \cdot Precision - 1$ □

Also, the covering ratio of covered positive over covered negative examples is equivalent to precision.

Theorem 7.3.4. *The covering ratio* $\hat{P}/\hat{N}$ *is equivalent to Precision.*

Proof. We show equivalence via compatibility.

$$\frac{\hat{P}_1}{\hat{N}_1} > \frac{\hat{P}_2}{\hat{N}_2} \Leftrightarrow \hat{P}_1 \cdot \hat{N}_2 > \hat{P}_2 \cdot \hat{N}_1$$

$$\Leftrightarrow \hat{P}_1 \cdot \hat{P}_2 + \hat{P}_1 \cdot \hat{N}_2 > \hat{P}_2 \cdot \hat{P}_1 + \hat{P}_2 \cdot \hat{N}_1$$

$$\Leftrightarrow \frac{\hat{P}_1}{\hat{P}_1 + \hat{N}_1} > \frac{\hat{P}_2}{\hat{P}_2 + \hat{N}_2}$$

□

In the next section, we will see that more-complex heuristics, such as entropy and the Gini index, are also essentially equivalent to precision. On the other hand, we will later see that seemingly minor modifications like the Laplace or m-estimates are not.

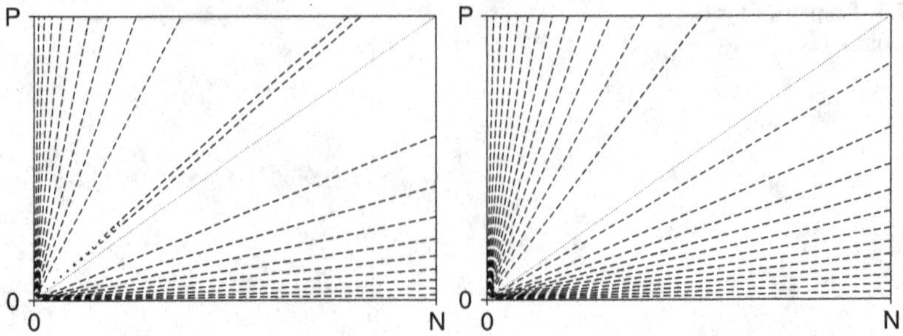

Fig. 7.4 Isometrics for entropy (*left*) and the Gini index (*right*)

7.3.4 Information-Theoretic Measures and Gini Index

Some algorithms (e.g., PRISM; Cendrowska, 1987) use the logarithm of the rule's precision, which measures the *information content* in the classification of the covered examples.

$$\boxed{Info(\mathbf{r}) = -\log_2 \frac{\hat{P}}{\hat{P}+\hat{N}}}$$

Obviously, information content and precision are antagonistic, i.e., a set of rules ordered by ascending information content will exhibit the same order as when ordered by descending precision. Thus the disadvantages of *Precision* apply here as well. The main advantage of using a logarithmic scale is that it tends to assign higher penalties to less-frequent events, which may be particularly important if information content is used as part of more complex functions. For example, *entropy* may be viewed as the weighted average of the information content of the positive and negative class.

$$\boxed{Entropy(\mathbf{r}) = -\left(\frac{\hat{P}}{\hat{P}+\hat{N}} \cdot \log_2 \frac{\hat{P}}{\hat{P}+\hat{N}} + \frac{\hat{N}}{\hat{P}+\hat{N}} \cdot \log_2 \frac{\hat{N}}{\hat{P}+\hat{N}}\right)}$$

Originating from the ID3 decision tree learning system (Quinlan, 1983, 1986), this measure has been used in early versions of the CN2 learning algorithm (Clark & Niblett, 1989). However, it suffers from similar deficiencies as precision and information content and was later replaced by the Laplace estimate (Clark & Boswell, 1991), which is discussed below in Sect. 7.3.6.

Entropy is not equivalent to information content and precision, even though it seems to have the same isometrics as these heuristics (see Fig. 7.4). First, entropy needs to be minimized instead of maximized (negative entropy can be maximized). More importantly, however, the difference is that the isometrics of entropy go through the undefined point $(0, 0)$ and continue on the other side of the 45° diagonal.

The motivation for this is that the original version of CN2 did not assume a positive class, but labeled its rules with the majority class among the examples covered. Thus rules **r** with coverage $(\hat{N}, \hat{P})$ and $\bar{\mathbf{r}}$ with coverage $(\hat{P}, \hat{N})$ can be considered to be of equal quality because one of them will be used for predicting the positive class and the other for predicting the negative class.

Based on this we can, however, prove the following theorem, which essentially shows that negative entropy is equivalent to precision for rules that cover more positive than negative examples.

Theorem 7.3.5. *Entropy and Precision are antagonistic for* $\hat{P} \geq \hat{N}$ *and compatible for* $\hat{P} \leq \hat{N}$.

Proof. $Entropy = -x \cdot \log_2(x) - (1-x) \cdot \log_2(1-x)$ with $x = Precision \in [0, 1]$. This function has its maximum at $Precision = 1/2 \Leftrightarrow \hat{P} = \hat{N}$. From the fact that it is strictly monotonically increasing for $\hat{P} \leq \hat{N}$, it follows that $Precision(\mathbf{r}_1) < Precision(\mathbf{r}_2) \Rightarrow Entropy(\mathbf{r}_1) < Entropy(\mathbf{r}_2)$ in this region. Analogously, $Precision(\mathbf{r}_1) < Precision(\mathbf{r}_2) \Rightarrow Entropy(\mathbf{r}_1) > Entropy(\mathbf{r}_2)$ for $\hat{P} \geq \hat{N}$, where *Entropy* is monotonically decreasing in *Precision*. □

Notice that the isometrics of entropy are symmetrical around the $\hat{P} = \hat{N}$ line. If we want to have the diagonal of the coverage space as the symmetry line, making the measure independent of class distributions, we have to use the *Kullback–Leibler divergence*, or KL-divergence for short.

$$KL(\mathbf{r}) = \frac{\hat{P}}{\hat{P}+\hat{N}} \log_2 \frac{\frac{\hat{P}}{\hat{P}+\hat{N}}}{\frac{P}{P+N}} + \frac{\hat{N}}{\hat{P}+\hat{N}} \log_2 \frac{\frac{\hat{N}}{\hat{P}+\hat{N}}}{\frac{N}{P+N}}$$

The KL-divergence (Kullback & Leibler, 1951) is an information-theoretic measure for the distance between two probability distributions, in this case the a priori distribution of examples and the a posteriori distribution of the examples that are covered by a rule. It is always > 0 except for the diagonal of the coverage space, where it is exactly 0. Thus, it behaves similar to entropy, but has the diagonal of the coverage space as the symmetry. Therefore:

Theorem 7.3.6. *KL and Precision are compatible for Precision* $> \frac{P}{P+N}$ *and antagonistic for Precision* $\leq \frac{P}{P+N}$.

Proof. $KL = Precision \cdot \log_2 \frac{Precision}{P/(P+N)} + (1 - Precision) \cdot \log_2 \frac{(1-Precision)}{1-P/(P+N)}$ with $Precision \in [0, 1]$. This function has its minimum at $Precision = P/(P+N)$. The rest of the proof is analogous to the proof of Theorem 7.3.5. □

The *Gini index* is a very popular heuristic in decision-tree learning (Breiman, Friedman, Olshen, & Stone, 1984). To our knowledge, it has not been used in rule learning, but we list it for completeness:

$$Gini(\mathbf{r}) = -\left(\frac{\hat{P}}{\hat{P}+\hat{N}}\right)^2 - \left(\frac{\hat{N}}{\hat{P}+\hat{N}}\right)^2 \sim \frac{\hat{P} \cdot \hat{N}}{(\hat{P}+\hat{N})^2}$$

It can be seen from Fig. 7.4 that the Gini index has the same isometric landscape as entropy, it only differs in the distribution of the values (hence the lines of the contour plot are a little denser near the axes and less dense near the diagonal). This, however, does not change the ordering of the rules.

Theorem 7.3.7. *Gini and Entropy are equivalent.*

Proof. Like entropy, the Gini index can be formulated in terms of *Precision* ($Gini \sim Precision \cdot (1 - Precision)$). This has the same shape as entropy, i.e., it also grows or falls with *Precision*. □

7.3.5 F- and G-Measures

The F-measure (van Rijsbergen, 1979) has its origin in information retrieval, where it is commonly used to trade off the performance measures recall and precision via their harmonic mean.

$$\boxed{FMeasure(\mathbf{r}) = \frac{(\beta^2+1)\cdot Precision\cdot Recall}{\beta^2\cdot Precision+Recall}}$$

The parameter β performs the trade-off between precision and recall with $\beta = 0$ resulting in *Precision* and $beta \to \infty$ going towards *Recall*. Consequently, the isometrics for $\beta = 0$ must be rotating around the origin. With increasing values of β, the isometrics must approach the parallel horizontal lines of *Recall*. This happens because increasing values of β implicitly move the rotational point of precision more and more towards infinity.

Its isometrics are illustrated in Fig. 7.5. Basically, the isometrics are identical to those of precision, with the exception that the rotational point does not originate in $(0, 0)$ but in a point $(-g, 0)$, where g depends on the choice of β. For $g \to \infty$, the slopes of the isometrics become flatter and flatter and converge towards the axis-parallel lines of *Recall*. On the other hand, for $g = 0$, the measure is obviously equivalent to *Precision*.

The parameter g can also be directly used, resulting in the *generalization quotient* (Gamberger & Lavrač, 2002).

$$\boxed{GQuotient(\mathbf{r}) = \frac{\hat{P}}{\hat{N}+g}}$$

It can be shown with an argument similar to the one used in the proof of Theorem 7.3.4 that *GQuotient* is, in effect, equivalent to $\hat{P}/(\hat{P} + \hat{N} + g)$ (Fürnkranz & Flach, 2005). The special case $g = P$ results in the *G-measure* $\hat{P}/(\hat{N} + P)$, which has been proposed by Flach (2003) as a simplicification of the F-measure with $\beta = 1$, the case where equal weight is given to recall and precision.

Gamberger and Lavrač (2002) also discuss why the generalization quotient is better suited for supervised descriptive rule induction than the linear cost measure.

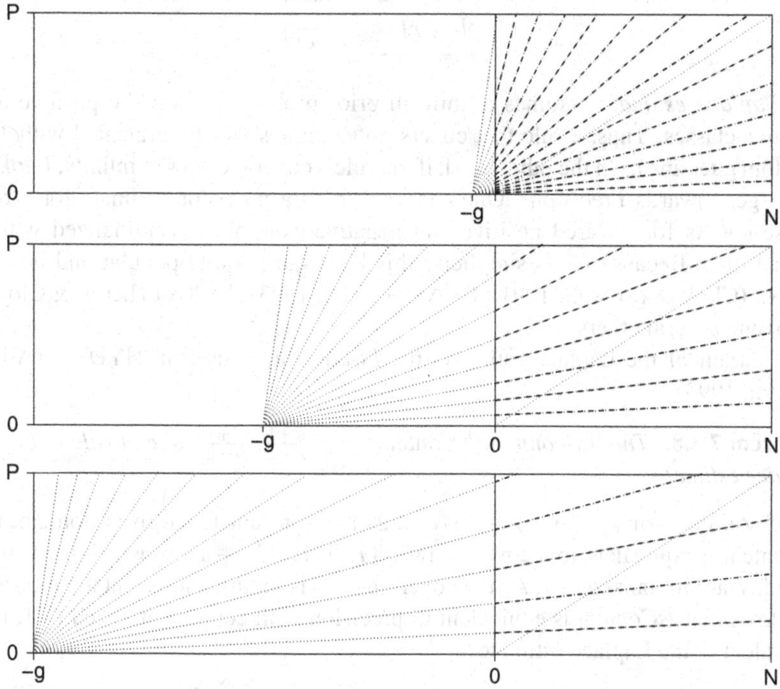

Fig. 7.5 Isometrics for the F-measure and the generalization quotient for increasing values of g

Both heuristics are similar in that they both trade off true positives and false positives, but *GQuotient* does so with a weighted quotient whereas *LinCost* uses a weighted sum. Gamberger and Lavrač argue that the rotational behavior of the generalization quotient is better suited for a top-down beam search because it is flatter in the upper right region of the coverage space and steeper in the lower left region, thereby implementing a gradual shift of focus from generic to specific features as the top-down search progresses.

7.3.6 Laplace and m-Estimate

The Laplace and m-estimates (Cestnik, 1990; Niblett & Bratko, 1987) are very common modifications of *Precision*. Just like *Precision*, both are unbiased estimates of the probability $\Pr(\oplus \mid B)$ of the positive class given the rule body. However, both apply a correction towards the prior probability $\Pr(\oplus)$ of the positive class. This correction will be stronger for rules with low coverage, i.e., those will be closer to $\Pr(\oplus)$, whereas rules with high coverage will be closer to the precision-based estimate of $\Pr(\oplus \mid B)$. In the following, we will discuss both as separate rule evaluation measures, but we note that both can also be used as probability estimates in complex measures (e.g., for replacing *Precision* in *Entropy*).

$$\boxed{Laplace(\mathbf{r}) = \frac{\hat{P}+1}{\hat{P}+\hat{N}+2}}$$

The *Laplace estimate* assumes a uniform prior probability over the positive and negative classes. Thus, a rule that covers no examples will be evaluated with $1/2$ (random guessing). On the other hand, if the rule's coverage goes to infinity, *Laplace* converges towards *Precision*.[3] One can view it as a precision estimate for a rule whose counts for covered positive and negative examples are initialized with 1 instead of 0. Because of its simplicity this heuristic is quite popular and is used in CN2 (Clark & Boswell, 1991), CLASS (Webb, 1993), BEXA (Theron & Cloete, 1996) and several others.

A variant of the Laplace estimate, the ls-content, is used in HYDRA (Ali & Pazzani, 1993).

Theorem 7.3.8. *The ls-content IsContent $= \frac{\hat{P}+1}{P+2} / \frac{\hat{N}+1}{N+2}$ is equivalent to the Laplace estimate.*

Proof. As the terms $(P + 2)$ and $(N + 2)$ are constant for a given domain, the ls-content is equivalent to a covering ratio $(\hat{P} + 1)/(\hat{N} + 1)$, in which the counts $\hat{P}$ and $\hat{N}$ are initialized to 1. As the covering ratio is in turn equivalent to precision, it follows that *IsContent* is equivalent to precision with counts initialized to 1, i.e., equivalent to the Laplace estimate. □

The *m-estimate* generalizes the Laplace so that rules with no coverage will be evaluated with the a priori probability $\pi = P/(P + N)$ of the positive examples in the training set instead of the uniform prior probability $1/2$. It has, e.g., been used in mFOIL (Džeroski & Bratko, 1992).

$$\boxed{MEstimate(\mathbf{r}) = \frac{\hat{P}+m\cdot\pi}{\hat{P}+\hat{N}+m}}$$

The parameter m is a positive number, which trades off the importance between the prior probability π ($m = \infty$) and the posterior probability $\hat{P}/(\hat{P} + \hat{N})$ ($m = 0$). Higher values of m give more weight to the prior probabilities and less to the examples. Higher values of m are thus appropriate for datasets that contain more noise and are more prone to overfitting. The Laplace estimate can be obtained from the m-estimate for $m = 2$ in problems with an equal number of positive and negative examples.

As already mentioned for the Laplace estimate, an alternative interpretation of these measures is to assume that each rule covers a certain number of examples a priori. Both the Laplace and the m-estimate compute a precision estimate, but start to count covered positive or negative examples at a number > 0. With the Laplace estimate, both the positive and negative coverage of a rule are initialized with 1

[3] In its general form the Laplace estimate has the number of classes C in its denominator, so that it will return $1/C$ for rules with no coverage.

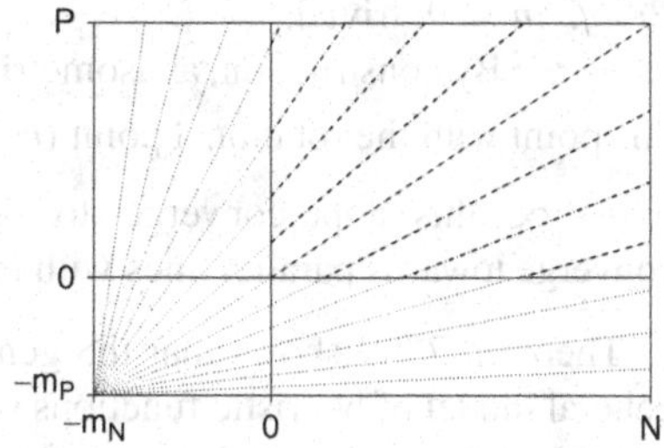

Fig. 7.6 Isometrics for the *m*-estimate

(thus assuming an equal prior distribution), while the *m*-estimate assumes a prior total coverage of *m* examples which are distributed according to the distribution of positive and negative examples in the training set.

In the coverage graphs, this modification results in a shift of the origin of the precision isometrics to the point $(-m_N, -m_P)$, where $m_N = m_P = 1$ in the case of the Laplace heuristic, and $m_P = m \cdot \pi$ and $m_N = m - m_P$ for the *m*-estimate (see Fig. 7.6). The resulting isometric landscape is symmetric around the line that goes through $(-m_N, -m_P)$ and $(0, 0)$. Thus, the Laplace estimate is symmetric around the 45° line, while the *m*-estimate is symmetric around the diagonal of the coverage graph.

Another noticeable effect of the transformation is that the farther the origin $(-m_N, -m_P)$ moves away from $(0, 0)$, the more the isometrics in the relevant window $(0, 0) - (N, P)$ approach parallel lines. For example, the isometrics of the *m*-estimate converge towards the isometrics of weighted relative accuracy for $m \to \infty$ (see Theorem 7.3.9 below).

7.3.7 The Generalized m-Estimate

Based on the above discussion, Fürnkranz and Flach (2005) have proposed the following *generalized m-estimate*, which takes the rotational point of the precision isometrics as a parameter:

$$\text{GenM}(\mathbf{r}) = \frac{\hat{P} + m \cdot \pi}{\hat{P} + \hat{N} + m} = \frac{\hat{P} + m_P}{(\hat{P} + m_P) + (\hat{N} + m_N)}$$

The second version of the heuristic allows us to directly specify the coordinates $(-m_N, -m_P)$ of the rotational point. The first measure specifies the rotational point via its slope $(\frac{c}{1-c})$ and its Manhattan distance (m) from the origin $(0, 0)$. Obviously, both versions of *GenM* can be transformed into each other by choosing $m = m_P + m_N$ and $c = m_P/m$, or $m_P = m \cdot c$ and $m_N = m \cdot (1 - c)$.

Theorem 7.3.9. *For $m = 0$, the generalized m-estimate is equivalent to precision, while for $m \to \infty$, its isometrics converge to the linear cost metric.*

Proof. $m = 0$: trivial.
$m \to \infty$: By construction, an isometric of *GenM* through the point $(\hat{N}, \hat{P})$ connects this point with the rotational point $(c-1)\cdot m, -c \cdot m)$ and has slope $\frac{\hat{P}+c\cdot m}{\hat{N}+(1-c)\cdot m}$. For $m \to \infty$, this slope converges to $\frac{c}{1-c}$ for all points $(\hat{N}, \hat{P})$. Thus all isometrics converge towards parallel lines with the slope $\frac{c}{1-c}$. □

Theorem 7.3.9 shows that the generalized m-estimate may be considered as a general model of heuristic functions with linear isometrics that has two parameters: $c \in [0, 1]$ for trading off the misclassification costs between the two classes, and $m \in [0, \infty]$ for trading off between precision *Precision* and the linear cost metric *LinCost*.[4] Therefore, all heuristics with linear isometrics discussed in this section may be viewed as equivalent to some instantiation of this general model.

7.4 Heuristics with Nonlinear Isometrics

All heuristics considered so far have linear isometrics. This is a reasonable assumption as concepts are typically evaluated with linear cost metrics (e.g., accuracy or cost matrices). However, these evaluation metrics are concerned with evaluating complete rules. As the value of an incomplete rule lies not in its ability to discriminate between positive and negative examples, but in its potential of being *refinable* into a high-quality rule, it might well be the case that different types of heuristics are useful for evaluating incomplete candidate rules. One could, for example, argue that for rules with high coverage on the positive examples, it is less important to exclude negatives than for rules with low coverage, because high-coverage candidates may still be refined accordingly. Similarly, overfitting could be countered by penalizing regions with low coverage. Possibly, these and similar problems could be better addressed with a nonlinear isometric landscape, even though the learned rules will eventually be used under linear cost models.

In this section, we will briefly look at two nonlinear information metrics, the J-measure and the correlation heuristic, which implement quite different strategies. However, note that many of the heuristics in subsequent sections, such as information gain or the Klösgen measures (cf. Sect. 7.5) or the likelihood ratio (cf. Sect. 9.2.4) also have nonlinear isometrics.

[4]The reader may have noted that for $m \to \infty$, *GenM* $\to c$ for all $\hat{P}$ and $\hat{N}$. Thus, for $m = \infty$, the function does not have isometrics because it converges towards a constant function. However, we are not concerned with the isometrics of the function *GenM* at the point $m = \infty$, but with the convergence of the isometrics of *GenM* for $m \to \infty$. In other words, the isometrics of *LinCost* are not equivalent to the isometrics of *GenM* for $m = \infty$, but they are equivalent to the limits to which the isometrics of *GenM* converge if $m \to \infty$.

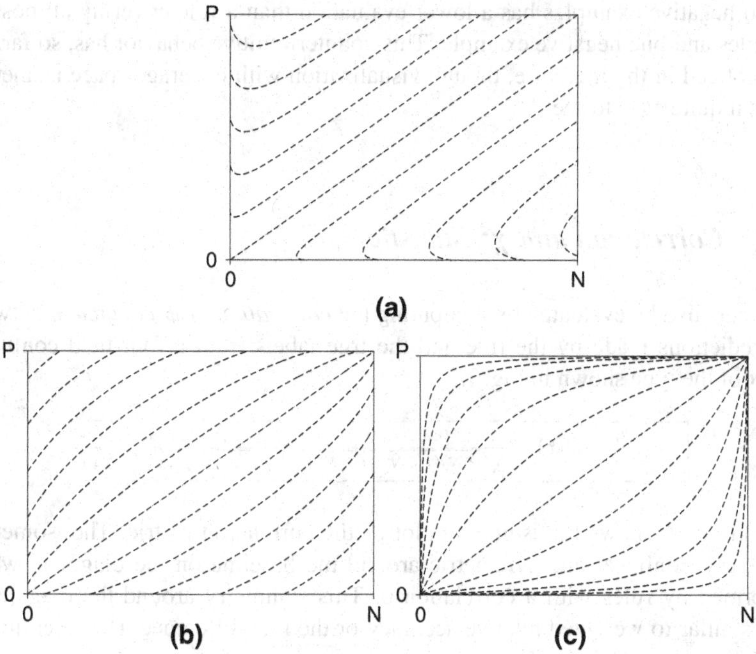

Fig. 7.7 Three evaluation measures with nonlinear isometrics. (**a**) J-measure. (**b**) Correlation. (**c**) Odds ratio

7.4.1 J-Measure

Smyth and Goodman (1991) have proposed to weigh the KL-divergence with the coverage of the rule, i.e., with the percentage of covered examples, resulting in the *J-measure*:

$$JMeasure(\mathbf{r}) = Coverage(\mathbf{r}) \cdot KL(\mathbf{r}) = \frac{1}{P+N}\left(\hat{P} \cdot \log_2 \frac{\frac{\hat{P}}{\hat{P}+\hat{N}}}{\frac{P}{P+N}} + \hat{N} \cdot \log_2 \frac{\frac{\hat{N}}{\hat{P}+\hat{N}}}{\frac{N}{P+N}}\right)$$

Figure 7.7a shows the isometrics of this measure. Interestingly, it turns out that this complex measure is in large regions of the coverage space equivalent to the simpler rate difference or weighted relative accuracy. The only difference is near the P- and N-axes, where the isometrics bend away from the origin. This means that rules that cover no or very few negative examples are actually penalized in comparison to rules that cover the same number of positives but somewhat more negative examples. Consider a point on the P-axis. Moving a little bit to the right now yields a rule that has a higher evaluation although it covers the same number of positive examples but more negative examples. For example, in all problems except those with very few positive examples, a rule covering all positive examples and no negative examples has a lower evaluation than a rule covering all positive examples and one negative example. This counterintuitive behavior has, so far, not been noticed in the literature, though visualization with coverage space isometrics makes it quite easy to see.

7.4.2 Correlation and χ^2 Statistic

Rules can also be evaluated by computing the *correlation* or *ϕ-coefficient* between the predictions made by the rule and the true labels from a four-field confusion matrix of the type shown in Fig. 2.3.

$$\text{Correlation}(\mathbf{r}) = \frac{\hat{P}\cdot\bar{N} - \bar{P}\cdot\hat{N}}{\sqrt{P\cdot N\cdot(\hat{P}+\hat{N})\cdot(\bar{P}+\bar{N})}} = \frac{\hat{P}\cdot N - P\cdot\hat{N}}{\sqrt{P\cdot N\cdot(\hat{P}+\hat{N})\cdot(\bar{P}+\bar{N})}}$$

Figure 7.7b shows the isometric plot of the correlation metric. The isometrics are clearly nonlinear and symmetric around the baseline on the diagonal, which are formed by rules with a correlation 0. This symmetry around the diagonal is again similar to weighted relative accuracy or the rate difference. However, unlike the J-measure, correlation bends the isometrics *towards* the axes. Thus, it has a bias that prefers purer rules (covering fewer negative examples) and more complete rules (covering more positive examples).

Moreover, the linear isometrics that result from connecting the points where the isometrics intersect with the P-axis on the left and the $\hat{P} = P$ line on the top, are *not* parallel to the diagonal (as they are for *WRA*). In fact, it can be shown that the slope of these lines in the area above the diagonal is $\frac{P+c^2\cdot N}{N+c^2\cdot P}$, where $c \in [0, 1]$ is the value of the correlation coefficient. For $c = 0$, the slope is P/N, but for $c = 1$, the slope is 1. Thus, the isometrics of these connecting lines change their slope with growing c, starting with those of *WRA* and slowly approaching those of *Accuracy*.

The correlation heuristic has been suggested for predictive rule induction by Fürnkranz (1994a) and for descriptive rule induction by Piatetsky-Shapiro (1991). Correlation rule discovery has become an important task in descriptive data mining (Brin, Motwani, & Silverstein, 1997). For example, recently Xiong, Shekhar, Tan, and Kumar (2004) proposed a monotonic upper bound for this measure that allows an efficient computation of a large number of item pairs. Similarly, Zhou and Xiong (2011) derive upper bounds that can be used for tracking the correlation in dynamically changing environments.

Note that the four-field correlation coefficient is basically a normalized version of the χ^2 *statistic* over the four events in the confusion matrix.

Theorem 7.4.1. *The χ^2 statistic is equivalent to correlation.*

Proof. One can show that $\chi^2 = (P+N)\cdot \textit{Correlation}^2$ (cf., e.g., Fürnkranz & Flach, 2005). □

7.4.3 *Odds Ratio*

The idea of the odds ratio is to compute the ratio of the odds of a positive example being covered ($\hat{P}$) vs. not being covered ($\bar{P} = P - \hat{P}$) and the odds of a negative example being covered ($\hat{N}$) vs. not being covered ($\bar{N} = N - \hat{N}$).

$$\mathit{OddsRatio}(\mathbf{r}) = \frac{\hat{P}/\bar{P}}{\hat{N}/\bar{N}} = \frac{\hat{P}\cdot\bar{N}}{\hat{N}\cdot\bar{P}}$$

Thus, in a four-field confusion matrix of the type shown in Fig. 2.3, the diagonal elements (the correct predictions) are multiplied and then divided by the product of the off-diagonal elements (the incorrect predictions). The odds ratio has, e.g., been used in the ELEM2 rule learning algorithm (An & Cercone, 1998).

Figure 7.7c shows the isometrics of the odds ratio. Similar to the correlation metric, they bend towards the diagonal, but in a much steeper way. In particular, all isometrics approach the $(0, 0)$-point, so that around this point, the odds ratio behaves quite similar to precision. Unlike precision, however, this rotating behavior is mirrored around the (N, P)-point. As a result of both phenomena, odds ratio has a strong bias towards both rules that do not cover any negative examples (resulting in *overfitting*) and rules that do not exclude any positive examples (resulting in *overgeneralization*) because both types of rules are evaluated with ∞. This undesirable behavior may explain the bad results with this heuristic that have been reported by Janssen and Fürnkranz (2009).

7.5 Gain Heuristics

Gain heuristics compute the difference in the heuristic estimates between the candidate rule $\mathbf{r}$ and typically its immediate predecessor $\mathbf{r}^p$. In other words, they try to estimate how much can be gained by refining $\mathbf{r}^p$ into $\mathbf{r}$. As $H(\mathbf{r}^p)$ will often be constant for all candidate rules (e.g., when using hill-climbing), optimizing $H(\mathbf{r}) - H(\mathbf{r}^p)$ would produce the same behavior as directly optimizing $H(\mathbf{r})$. Thus, the difference is often multiplied with a weighting function, so that the general form of gain-based heuristics is the following:

$$\mathit{Gain}(\mathbf{r}) = W(\mathbf{r}) \cdot (H(\mathbf{r}) - H(\mathbf{r}^p))$$

As mentioned above, the current rule is typically compared to its immediate predecessor. This strategy is particularly well-suited for a greedy top-down search. However, many heuristics also compare the current rule to the universal rule $\mathbf{r}^\top$ predicting the same class as $\mathbf{r}$. This estimates how much information about the class label can be gained by knowing the conditions of $\mathbf{r}$. In fact, we have already seen a few examples of such heuristics. Most notably, weighted relative accuracy computes the weighted difference of the precision of a rule and the precision of the universal rule. We will discuss this in more detail further below. The KL-divergence may also be interpreted as a gain-based heuristic. It is based on the ratio of the rule's precision and the prior probability. Taking the logarithm of these values again results in a difference between a statistic for **r** and a statistic for the universal rule $\mathbf{r}^\top$.

In this section, we discuss two well-known gain-based heuristics in detail, the Klösgen measures (Sect. 7.5.2), which are quite popular in descriptive rule induction, and information gain (Sect. 7.5.3), which is frequently used in predictive rule induction. Before that, we start with the simple precision gain.

7.5.1 Precision Gain, Lift, and Leverage

A simple example for a gain heuristic is *precision gain*, which is defined as the difference between the proportion of positive examples covered by rule **r** and the overall proportion of positive examples.

$$\boxed{\mathit{PrecGain}(\mathbf{r}) = \mathit{Precision}(\mathbf{r}) - \mathit{Precision}(\mathbf{r}^\top) = \frac{\hat{P}}{\hat{P}+\hat{N}} - \frac{P}{P+N}}$$

Precision gain measures the increase in the precision of rule **r** relative to the universal rule $\mathbf{r}^\top$. A rule is only interesting if it improves upon this default precision, measured by the relative frequency of positive examples in the training set.

The same property can also be expressed as a ratio of both probabilities, in which case it is known as *lift*.

$$\boxed{\mathit{Lift}(\mathbf{r}) = \frac{\frac{\hat{P}}{\hat{P}+\hat{N}}}{\frac{P}{P+N}}}$$

Both precision gain and lift are essentially equivalent to precision, because the default precision (the precision of the default rule) is constant for all rules of the same learning problem. Thus, neglecting it does not change the order in which the rules are ranked. However, precision gain is frequently used as a component in other learning heuristics, such as weighted relative accuracy and the Klösgen measures. Lift has been proposed as a pruning criterion for rule sets (Boström, 2004).

A well-known variant of precision gain and lift is the so-called *leverage* of a rule (Piatetsky-Shapiro, 1991), which measures the difference between the actual and the expected probability of the co-occurrence of rule body and rule head, whereas lift measures their ratio.

$$\boxed{\mathit{Leverage}(\mathbf{r}) = \frac{\hat{P}}{P+N} - \frac{P}{P+N} \cdot \frac{\hat{P}+\hat{N}}{P+N}}$$

It is straightforward to see that this measure is equivalent to weighted relative accuracy. In this formulation, we can also see that support forms an upper bound for leverage.

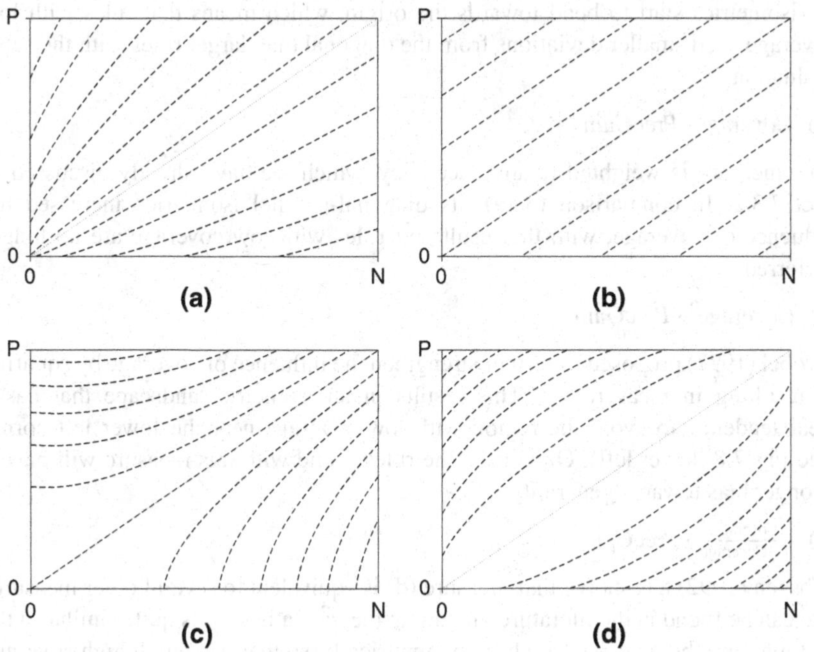

Fig. 7.8 Different ways of trading off coverage and precision gain. (**a**) $\sqrt{Coverage} \cdot PrecGain$. (**b**) $Coverage \cdot PrecGain$. (**c**) $Coverage^2 \cdot PrecGain$. (**d**) $\frac{Coverage}{1-Coverage} \cdot PrecGain$

7.5.2 Klösgen Measures

In the following, we concentrate on a family of well-known evaluation metrics for subgroup discovery (Klösgen, 1992). They have in common that they trade off coverage and precision gain. Klösgen (1996) identified three different variations for combining these two measures, which satisfy a set of four basic axioms proposed by Piatetsky-Shapiro (1991) and Major and Mangano (1995). He further showed that several other measures are equivalent to these. Wrobel (1997) added a fourth measure. All four measures only differ in the way in which they trade off coverage versus precision gain. The isometrics of these measures are shown in Fig. 7.8.

(a) $\sqrt{Coverage} \cdot PrecGain$

This measure was proposed by Klösgen (1992). Its idea is to perform a statistical test on the distribution of precision gain, under the assumption that, if the true precision of the rule were the same as the overall precision in the example set, the observed value for precision gain should follow a binomial distribution around 0. The variance of this distribution brings in the factor $\sqrt{Coverage}$. The isometrics show that the measure has a slight tendency to prefer rules that are near the origin. In that region, the isometrics start to bend towards the origin, which means that rules with low coverage need smaller deviations from the diagonal than larger rules with the same evaluation.

(b) *Coverage · PrecGain*

This measure is weighted relative accuracy, which we have already discussed in Sect. 7.3.2. In comparison to (a), its diagonal-parallel isometrics increased the influence of coverage, with the result that rules with low coverage are no longer preferred.

(c) *Coverage*^{2} · *PrecGain*

Wrobel (1997) proposed to further strengthen the influence of coverage by squaring it, resulting in measure (c). This results in an isometric landscape that has a clear tendency to avoid the region with low coverage near the lower left corner (see Fig. 7.8, lower left). Obviously, the rules found with this measure will have a stronger bias towards generality.

(d) $\frac{\mathit{Coverage}}{1-\mathit{Coverage}} \cdot \mathit{PrecGain}$

Klösgen (1992) has shown that measure (d) is equivalent to several other measures that can be found in the literature, including the χ^2-statistic. It is quite similar to the first measure, but its edges are bent symmetrically, so that rules with high coverage are penalized in the same way as rules with a comparably low coverage.

It is quite interesting to see that in regions with higher coverage, the isometrics of all measures except (d) approach parallel lines, i.e., with increasing rule coverage, they converge towards some measure that is equivalent to weighted relative accuracy. However, measures (a)–(c) differ in their behavior near the low-coverage region of the rule space. Measure (a) makes it easier for rules in the low-coverage region near the origin, (b) is neutral, whereas (c) penalizes this region.

It seems to be the case that two different forces are at work here. On the one hand, rules with low coverage may be more surprising than strong regularities, and it might be reasonable to encourage their discovery, as measure (a) does. On the other hand, low-coverage rules tend to be less reliable because their estimated performance parameters (such as their precision) are associated with a larger variance and larger uncertainty. A simple solution for this problem might be to try to avoid these regions if possible, as measure (c) does. Weighted relative accuracy (b) tries to compromise between these two approaches.

Note that more measures are conceivable in this framework. Janssen and Fürnkranz (2010) investigated all heuristics of the form

$$\boxed{\mathit{Kloesgen}(\mathbf{r}) = \mathit{Coverage}(\mathbf{r})^{\omega} \cdot \mathit{PrecGain}(\mathbf{r})}$$

where ω is a parameter that specifies the trade-off between coverage and precision. For example, setting $\omega = 0$ reduces the measure to precision gain, which in turn is equivalent to precision. On the other hand, large values of ω give more and more weight to coverage, and the isometrics make a sharper and sharper bend away from the origin, so that they eventually approach the parallel lines of the coverage estimate. Empirically, the best-performing parameter setting for predictive rule induction (evaluated by accuracy) was found to be between 0.4 and 0.5, i.e., quite similar to variant (a) above. Thus, avoiding the low-coverage region seems to yield worse performance. This is related to the *small disjuncts problem*: rules with high coverage are responsible for a large part of the overall error of a rule set. Nevertheless, the experiments in (Holte, Acker, & Porter, 1989) suggest that avoiding them entirely is not a good strategy, which is confirmed by the above-mentioned finding.

7.5.3 FOIL's Information Gain

The classical example for a gain heuristic is the *weighted information gain* heuristic used in FOIL (Quinlan, 1990). Here the basic heuristic is information content and the difference is weighted with the number of covered positive examples.[5]

$$\boxed{\mathit{FoilGain}(\mathbf{r}) = \mathit{PosCov}(\mathbf{r}) \cdot (\mathit{Info}(\mathbf{r}) - \mathit{Info}(\mathbf{r}^p)) = \hat{P} \cdot \left(\log_2 \frac{\hat{P}}{\hat{P}+\hat{N}} - \log_2 c\right)}$$

where c is the precision of the predecessor rule $\mathbf{r}^p$. In the following, it will turn out to be convenient to interpret c as a cost parameter taking values in the interval $[0, 1]$.

Figure 7.9 shows the isometrics of *FoilGain* for four different settings of c. Although the isometrics are nonlinear, they appear to be linear in the region above the isometric that goes through $(0, 0)$. Note that this isometric, which we will call the *baseline*, has a slope of $\frac{c}{1-c}$: In graph (a) it is the diagonal, in (b) it has a 45° slope, and in (c) and (d) it coincides with the vertical and horizontal axes, respectively.

The baseline represents the classification performance of the parent rule. In the area below the baseline are the cases where the precision of the rule is smaller than the precision of its predecessor. Such a refinement of a rule is usually not considered to be relevant. In fact, this is also the region where the information gain is negative, an information loss. The points on the baseline have information gain 0, and all points above it have a positive gain.

For this reason, we will focus on the region above the baseline, which is the area where the refined rule $\mathbf{r}$ improves upon its predecessor rule $\mathbf{r}^p$. In this region, the isometrics appear to be linear and parallel to the baseline, just like the isometrics of the linear cost metric *LinCost*. However, closer investigation reveals that this is

[5]This formulation assumes that we are learning in a propositional setting. In a relational setting, FOIL in effect counts the number of proofs (extended tuples) for rule $\mathbf{r}$ when computing *Info*($\mathbf{r}$), but counts the number of unextended tuples for computing *PosCov*($\mathbf{r}$) (Quinlan, 1990).

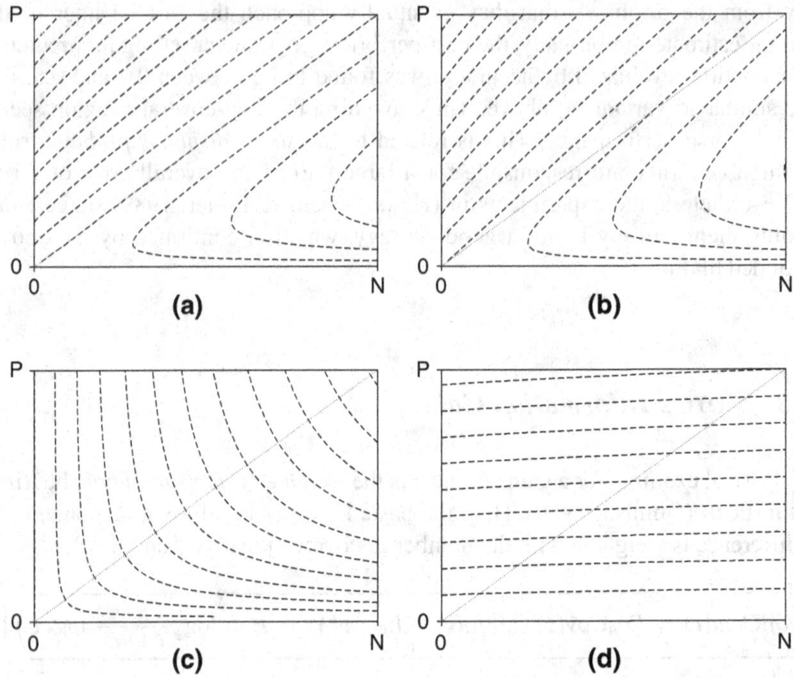

Fig. 7.9 Isometrics for FOIL's information gain for different values of c for the precision of the predecessor rule. **(a)** $c = \frac{P}{P+N}$. **(b)** $c = \frac{1}{2}$. **(c)** $c = 1$. **(d)** $c = 10^{-6}$

not the case. The farther the isometrics move away from the baseline, the steeper they become. Also, they are not exactly linear but are slightly bent towards the P-axis, i.e., purer rules are preferred. Overall, the behavior of FOIL's information gain may be interpreted as an interesting combination of precision and the linear cost metric, where FOIL adapts the cost to the class distribution of the predecessor rule.

This process is illustrated in Fig. 7.10. In graph (a) we see the initial situation: the starting rule covers all examples, i.e., it lies on the upper right corner of the coverage space. Whenever the learner adds a condition to a rule, the resulting rule becomes the new starting point for the next refinement step, and subsequent rules should improve upon this line. The area below this baseline is shown in gray.

After adding a new condition to the rule, the baseline is rotated until it goes through the point corresponding to the new rule. The isometrics of the heuristic rotate with this line and become steeper and steeper the closer the rule moves towards the upper left corner of the coverage space. This rotation of the baseline around the origin has an interesting parallel to the use of precision as a search heuristic. The main difference is that, due to the (almost) parallel lines above the baseline, information gain has a tendency to prefer more general refinements, thus

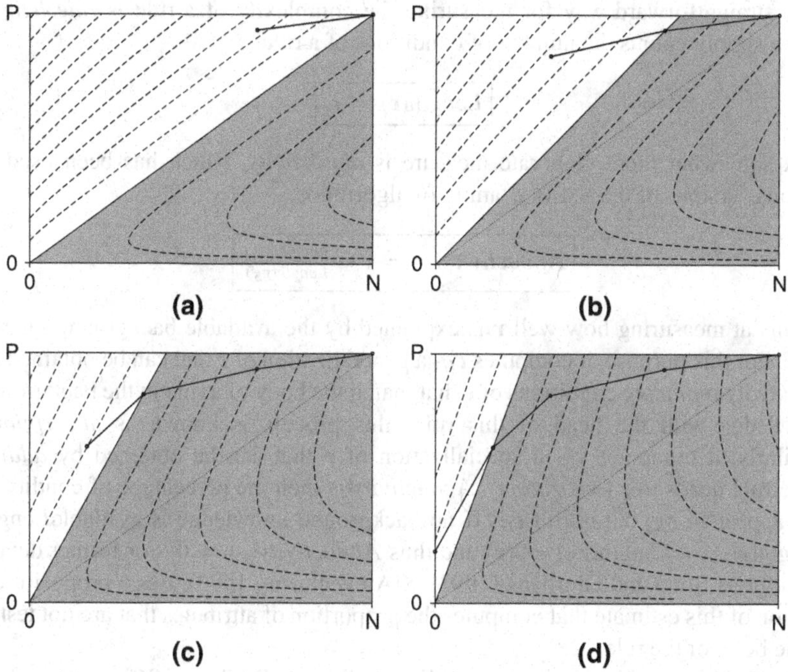

Fig. 7.10 A typical path through the coverage space when FOIL learns a rule. (**a**) Selecting the first condition. (**b**) Selecting the second condition. (**c**) Selecting the third condition. (**d**) Selecting the last condition

trying to stay in the upper regions of the coverage space. This may also be interpreted as a method for implementing 'patient' rule induction, as advocated in Friedman and Fisher (1999).

7.6 Complexity-Based Heuristics

Typically, it is a good idea to bias a learning algorithm towards more general hypotheses, i.e., rules that cover a larger number of examples. The main reason for this is to avoid overfitting. Many algorithms deal with this issue with separate so-called pruning heuristics, which we will discuss in Chap. 9.

However, some algorithms also address the problem by trying to measure the generality of a rule, and incorporate this information into the heuristic. Typically, there is a direct relationship between the complexity of a rule and its generality. Complex rules tend to be very specific and cover only a few examples, whereas short rules tend to be general and cover more examples. In addition to heuristics which try to directly measure the generality (such as support or coverage), there are a variety of heuristics for capturing the complexity of candidate rules.

A straightforward way for measuring the complexity of a rule is *rule length*, which simply counts the number of conditions of a rule.

$$\boxed{\textit{Length}(\mathbf{r}) = L}$$

A somewhat more elaborate measure is *abductivity*, which has been used in various versions of the SMART family of algorithms.

$$\boxed{\textit{Abductivity}(\mathbf{r}) = 1 - \frac{\textit{Length}(\mathbf{r}_G)}{\textit{Length}(\mathbf{r}_S)}}$$

It aims at measuring how well $\mathbf{r}$ is explained by the available background knowledge. For this purpose, it computes $\mathbf{r}_G$, a generalization of $\mathbf{r}$ that can be obtained by repeatedly *replacing* conditions of $\mathbf{r}$ that match the body of a rule in the background knowledge with the head of this rule (this process is known as *absorption*). Similarly, it computes $\mathbf{r}_S$, a specialization of $\mathbf{r}$ that can be obtained by *adding* these rule heads to $\mathbf{r}$ (*saturation*). *Abductivity* is then the percentage of conditions that appear in $\mathbf{r}_S$, but not in $\mathbf{r}_G$. If no background knowledge is available*Length* $\textit{Length}(\mathbf{r}_G) = \textit{Length}(\mathbf{r}_S) = (\mathbf{r})$ and thus $\textit{Abductivity}(\mathbf{r}) = 0$. For further details we refer to Botta and Giordana (1993). SIA (Venturini, 1993) uses a propositional version of this estimate that computes the proportion of attributes that are not tested in the body of the rule.

The *minimum message length (MML)* (Wallace & Boulton, 1968) and its close counterpart the *minimum description length (MDL)* (Rissanen, 1978) principles aim at finding a trade-off between the complexity and the accuracy of a hypothesis (Georgeff & Wallace, 1984). The key idea is to measure the information contained in a set of examples as the amount of information I needed to transmit a rule $\mathbf{r}$ and the amount of information needed to transmit the set of examples $\mathcal{E}$ with the help of this hypothesis. Essentially, this amounts to transmitting the examples that are misclassified by this rule. The goal is to minimize the sum of both, i.e., to minimize the total amount of information that needs to be transmitted.

$$\boxed{\textit{MDL}(\mathbf{r}) = I(\mathbf{r}) + I(\mathcal{E}|\mathbf{r})}$$

Unfortunately, both of the above terms are not computable and have to be approximated. For various computable approximations for rule learning we refer to Kovačič (1994a, 1994b) and Pfahringer (1995a, 1995b). *MDL* can be straightforwardly extended to rule sets, and is often used as a pruning criterion (Cohen, 1995; Quinlan, 1994, cf. also Sect. 9.3.4).

With the exception of various MDL measures, which already incorporate classification accuracy, complexity heuristics are rarely used on their own, but usually form components of more complex heuristics that combine various measures into one evaluation function so that different aspects of a candidate rule can be taken into account. The most popular methods for combining heuristics are described in the following sections.

7.7 Composite Heuristics

Many heuristics are complex combinations of simpler heuristics. For example, several approaches combine discriminative heuristics with heuristics that try to minimize the complexity of a rule. In this section, we will briefly review two ways for combining different heuristics, namely weighted sums and lexicographic evaluation functionals.

7.7.1 Weighted Sum

Many heuristics use weighting functions for combining several basic heuristics H_i. The general form of weighted heuristics is the following:

$$H(\mathbf{r}) = \sum_{i=1}^{n} w_i \cdot H_i(\mathbf{r})$$

One example for a weighted sum of different component heuristics can be found in the SMART family of algorithms (Botta and Giordana, 1993; Botta, Giordana, & Saitta, 1992; Giordana and Sale, 1992).

$$\mathit{Smart}(\mathbf{r}) = a \cdot \mathit{FoilGain}(\mathbf{r}) + b \cdot \mathit{Recall}(\mathbf{r})\mathit{Precision}(\mathbf{r}) + c \cdot \mathit{Abductivity}(\mathbf{r})$$

It computes the weighted average of three terms, the weighted information gain, the product of recall and precision, and the abductivity measure. The weights a, b, and c can be used to trade off the relative importance of the three factors of the evaluation function. SIA (Venturini, 1993) uses a similar measure with $a = 0$.

7.7.2 Lexicographic Evaluation Functionals

Sometimes it is more desirable that one heuristic dominates the others in the sense that the latter are only taken into consideration if the former is not able to discriminate between two rules. Lexicographic evaluation functionals (*lef*s) (Michalski, 1983) are a general mechanism for using a hierarchy of evaluation functions. A *lef* is an ordered set of pairs $(H_i(\mathbf{r}), t_i)$, where the H_i are heuristic functions and the $t_i \in [0, 1]$ are tolerance thresholds. The heuristics are evaluated sequentially. All candidate rules that have an optimal evaluation ($H_i(\mathbf{r}) = H_i(\mathbf{r}_{opt})$) or are within the tolerance threshold ($H_i(\mathbf{r}) \geq t_i \cdot H_i(\mathbf{r}_{opt})$) are evaluated by the next pair $(H_{i+1}(\mathbf{r}), t_{i+1})$. This is continued until a unique best candidate is determined.

In its general form, *lef*s are primarily used in the AQ-family of algorithms. Michalski (1983) suggests the use of not-covered negative examples and covered positive examples as the basic heuristics for a *lef*. The special case where $t_i = 1$

is often used for tie-breaking when two rules are initially judged to be of the same quality. PRISM (Cendrowska, 1987), for example, evaluates rules with a variant of the information content heuristic and breaks ties using positive coverage.

7.8 Practical Considerations and Recommendations

In this chapter, we have provided an elaborate survey of commonly used heuristics for inductive rule learning. While our main analytic tool, visualization in coverage space, gave us a good understanding about the similarities and differences between different heuristics, and also helped to identify some measures which seem to have a counterintuitive behavior (e.g., the J-measure), it has not led to practical recommendations which heuristics should be used in an inductive rule learner. Some intuitions and recommendations are given in the rest of this section.

7.8.1 Predictive Rule Induction

For predictive induction, there are, somewhat surprisingly, relatively few works that empirically compare different rule learning heuristics. For example, Lavrač, Cestnik, and Džeroski (1992a, 1992b) compare several heuristics for inductive logic programming. Most works only perform a fairly limited comparison, which typically introduces a new heuristic and compares it to the heuristic used in an existing system. A typical example for work in this area is Todorovski et al. (2000), where the performance of weighted relative accuracy was compared to the performance of CN2's Laplace-heuristic.

The most exhaustive empirical work in this respect can be found in Janssen & Fürnkranz (2010). There, ten different heuristics were implemented in a simple CN2-like learner that solely relied on its rule selection criterion (no additional pruning algorithms were used). The heuristics included five parameter-free heuristics (*Accuracy*, *WRA*, *Precision*, *Laplace*, *Correlation*) and five parametrized heuristics (*LinCost*, *RelLinCost*, *MEstimate*, *FMeasure*, *Kloesgen*). The experimental procedure used 57 datasets, divided into two groups. The first group was only used for tuning the parameter, and the second group of 30 datasets was used for evaluation.

Figure 7.11 shows a summary of the results in the form of the average rank of the heuristics obtained in the experiments on the 30 held-out datasets. For comparison, the results of two versions of JRIP, a reimplementation of RIPPER in WEKA, are also included (one with and one without post-pruning). The most important conclusions are the following:

- The performance of precision, accuracy, and weighted relative accuracy is the worst. *Precision* overfits, while *Accuracy* and, in particular, *WRA*, overgeneralize.

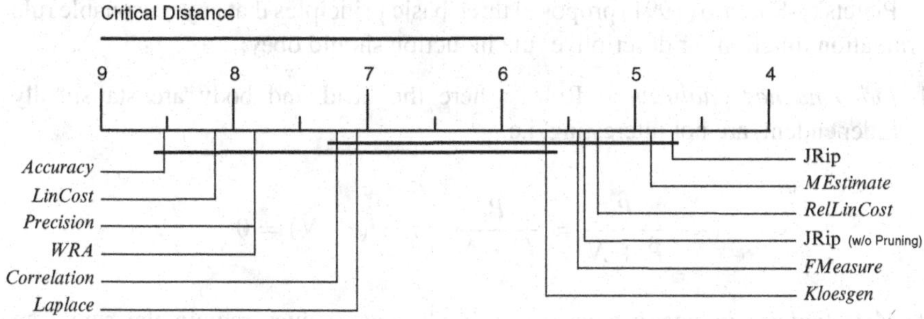

Fig. 7.11 Results of an experimental comparison of ten heuristics and RIPPER with pruning (JRIP) and without pruning (JRIP-P) as a benchmark (Janssen & Fürnkranz, 2010). For each heuristic, the graph shows its average rank in 30 experiments. Groups of heuristics that are not significantly different with a Nemenyi test (at $p = 0.05$) are connected (cf. the footnote on p. 90)

- The four parametrized heuristics *Kloesgen*, *FMeasure*, *LinCost*, and *RelLinCost* outperform the other heuristics by a wide margin. Their performance was almost as good as the performance of JRIP, although the heuristic versions solely relied on a rule selection criterion, while the latter uses extensive pre- and post-pruning algorithms.
- The linear cost heuristic *LinCost* is among the worst heuristics. It differs from the others, in particular from *RelLinCost*, primarily in the fact that it does not normalize its values to the class distribution. This seems to be a very important criterion.

In addition, Janssen & Fürnkranz (2010) also provide optimal parameter settings for the five parametrized heuristics. These were $m = 22.5$ for the m-estimate, $\omega = 0.432$ for the Klösgen measures, $\beta = 0.5$ for the F-measure, $c = 0.437$ for the linear cost measure, and $c_r = 0.342$ for the relative linear cost measure. However, these values have to be interpreted with caution, as additional experiments in different rule learners have shown that optimal values are sensitive to the learning algorithm.

7.8.2 Descriptive Rule Induction

Interestingness measures in descriptive rule induction are often formulated in a more general context, where the head of the rule is not a special target variable, but an arbitrary conjunction of features. Thus, the references that we will discuss below typically do not refer to covered positive or negative examples, but to the examples covered by the head and examples covered by the body of the rule. However, this formulation can be directly translated into the one used throughout this chapter, by realizing that P corresponds to the number of examples covered by the head of the rule, and $\hat{E} = \hat{P} + \hat{N}$ to the number of examples covered by the body of the rule. Thus, for maintaining consistency, we stick to our notation.

Piatetsky-Shapiro (1991) proposed three basic principles that any reasonable rule evaluation function for descriptive rule induction should obey:

1. *Independence calibration:* Rules where the head and body are statistically independent, are not interesting, i.e.,

$$\frac{\hat{P}}{\hat{P}+\hat{N}} = \frac{P}{P+N} \Rightarrow H(\hat{P},\hat{N}) = 0$$

2. *Monotonicity in positive coverage*: If all other values remain the same, an increase in the number of covered positive examples should lead to increase in the value of the rule evaluation function, i.e.,

$$\hat{P}_1 > \hat{P}_2 \Rightarrow H(\hat{P}_1,\hat{N}) > H(\hat{P}_2,\hat{N})$$

3. *Anti-monotonocity in negative coverage*: If all other values remain the same, an increase in the number of covered negative examples should lead to decrease in the value of the rule evaluation function, i.e.,[6]

$$\hat{N}_1 > \hat{N}_2 \Rightarrow H(\hat{P},\hat{N}_1) < H(\hat{P},\hat{N}_2)$$

Major and Mangano (1995) later suggested a fourth principle:

4. *Monotonicity in coverage along precision:* If the percentage of covered positive examples remains constant but higher than the overall percentage of positive examples, an increase in the total number of covered examples should lead to an increase in the rule evaluation, i.e.,

$$\left(\frac{\hat{P}_1}{\hat{P}_1+\hat{N}_1} = \frac{\hat{P}_2}{\hat{P}_2+\hat{N}_2} > 0\right) \wedge \left(\hat{P}_1+\hat{N}_1 > \hat{P}_2+\hat{N}_2\right)$$
$$\Rightarrow H(\hat{P}_1,\hat{N}_1) > H(\hat{P}_2,\hat{N}_2)$$

These principles can be easily checked in coverage space:

1. The diagonal of the coverage space should have a constant value of 0.
2. The measure should increase along the P-axis and its parallels.
3. The measure should decrease along the N-axis and its parallels.
4. The measure should increase along any straight line going out from the origin into the area above the diagonal.

---

[6]Piatetsky-Shapiro (1991) used P or $\hat{E}$ instead of $\hat{N}$ as the basis for the anti-monotonicity. All three versions are equivalent.

It is also easy to check which of the measures discussed in this chapter conform to these principles and which do not. For example, precision violates 1 and 4; precision gain fixes 1 but still violates 4; accuracy satisfies 2, 3, and 4; whereas weighted relative accuracy normalizes accuracy towards the diagonal so that it also satisfies 1. In fact, Lavrač et al. (2004) argue that *WRA* is particularly well-suited for subgroup discovery because a rule's evaluation is proportional to its distance from the diagonal in coverage space. *WRA* is, in fact, the simplest function that satisfies all principles (Piatetsky-Shapiro, 1991). Similar arguments could be brought forward for all variants of the Klösgen measure. In fact, Klösgen (1996) proposed three of the variants in order to satisfy the four axioms above. Less obvious is, e.g., to see that correlation satisfies all criteria, but, despite its bended isometrics (Fig. 7.7b), this is also the case because no line from the origin intersects an isometric twice.

Several studies have tried to identify commonalities and differences between different rule evaluation metrics. Tan et al. (2002) surveyed 21 rule learning heuristics and compared them according to a set of desirable properties. In general, they conclude that the choice of the right interestingness measure is application-dependent, but they also identify situations in which many measures are highly correlated with each other. Bayardo Jr. and Agrawal (1999) analyze several heuristics in support and confidence space, and show that the optimal rules according to many criteria lie on the so-called support/confidence border, i.e., the set of rules that have maximum or minimum confidence for a given support level. Recently, Wu, Chen, and Han (2007) showed that a group of so-called null-invariant measures (measures that are not influenced by the number of records that do not match the pattern) can be generalized into a single parametrized heuristic. Finally, a recent theoretical study by Kralj Novak, Lavrač, and Webb (2009) shows that the heuristics used in various supervised descriptive rule learning settings are compatible and basically optimize a similar trade-off between coverage and precision as *WRA*. This will be discussed in more detail in Sect. 11.5.3 of this book.

7.8.3 Discussion

Naturally, there are some similarities between heuristics used for descriptive and for predictive tasks. For example, Lavrač et al. (1999) derived weighted relative accuracy with the aim of unifying these two realms. Fürnkranz and Flach (2004) analyzed filtering and stopping heuristics and showed that FOIL's information gain search and MDL-based pruning have a quite similar effect as support and confidence thresholds that are commonly used in association rule discovery. Nevertheless, it is important to note that good heuristics for descriptive rule induction are not necessarily well suited for predictive rule induction (weighted relative accuracy is a good example). The key difference is that in the latter case one typically needs to learn an entire rule set, where lack of coverage in individual rules can be corrected by the entire ensemble of rules. Inconsistencies, on the other hand, cannot be corrected by the induction of additional rules (at least not in the case of concept learning). In fact, a good classification rule must optimize or trade off various criteria simultaneously. Among them are:

Consistency: *How many negative examples are covered?*
In concept learning or decision list learning, if a rule covers negative examples, these misclassifications cannot be corrected with subsequent rules. When sets of unordered rules are learned for multiclass problems, such corrections are possible, but obviously one should nevertheless try to find pure rules.

Completeness: *How many positive examples are covered?*
Even though subsequent rules may cover examples that the current rule leaves uncovered, rules with higher coverage are typically preferred because they bring the learner closer to the goal of covering all positive examples.

Gain: *How good is the rule in comparison to other rules (e.g., default rule, predecessor rules)?*
A rule with high consistency may be bad if its predecessor had a higher consistency and vice versa. Thus, many heuristics, such as weighted relative accuracy or information gain, relate the quality of a rule to other rules.

Utility: *How useful will the rule be in the context of the other rules in the theory?*
A rule with high consistency and completeness may nevertheless be a bad addition to the current theory if it does not explain any new examples.

Bias: *How will the quality estimate change on new examples?*
It is well-known that estimates obtained on the training data will be optimistically biased. A good heuristic has to address this problem so that the algorithm will not overfit the training examples.

Potential: *How useful is the rule for obtaining good future refinements?*
An incomplete rule, i.e., a rule that is encountered during the search for a good rule, should not be evaluated by its ability to discriminate between positive and negative examples, but by its potential to be refined into such a rule.

Simplicity: *How complex or understandable is the rule?*
In addition to its predictive quality, rules are often also assessed by their comprehensibility, because this is one of the key factors for preferring rule learning algorithms over competing inductive classification algorithms. As comprehensibility is difficult to measure, it is often equated with simplicity or rule length.

The measures discussed in this chapter address many of these issues directly and simultaneously (consistency, completeness, gain, bias). Some of these criteria are also addressed, at least to some extent, algorithmically. Utility, for example, is addressed in part by the covering loop which removes examples that have been covered by previous rules. Obviously, it is harder to take the context of rules that will be subsequently learned into account. This is particularly the case because most rule learning heuristics only focus on the examples covered by the rule, and not on the examples that are not covered by a rule. This is contrary to decision tree learning heuristics, which consider all possible outcomes of a condition simultaneously.[7] The PART (Frank & Witten, 1998) algorithm may be viewed as an attempt to use the best of both worlds. Similarly, RIPPER's global optimization phase, where rules in a final theory are tentatively relearned in the context of all previous and all subsequent rules, may be viewed as an attempt to address this issue. These techniques will be discussed in more detail in the next chapters.

It is also important to realize that these criteria are not independent. For example, comprehensibility and simplicity are correlated with completeness: simple rules tend to be more general and to cover more examples. Thus, a bias for completeness will automatically correlate with a bias for shorter rules. Similarly, the idea of the Laplace-correction as introduced in CN2 (Clark & Boswell, 1991) was to correct a too-strong bias for consistency over completeness (by, e.g., penalizing pure rules that cover only single examples), and also to try to provide more accurate probability estimates.

Nevertheless, it is not clear whether all these points can or should be addressed simultaneously with a single rule learning heuristic. However, it is the case that common rule learning algorithms essentially assume that all these objectives can be captured into a single heuristic function (only overfitting is frequently addressed by using a separate criterion), so a thorough understanding of existing heuristics is helpful in understanding the behavior of current rule learning algorithms.

7.9 Conclusion

In this chapter, we reviewed and analyzed the most common evaluation metrics for supervised rule learning. We identified coverage spaces as a framework for visualizing the main objective of rule learning heuristics, namely to trade off the two basic quality criteria completeness and consistency. Within this framework, we showed that there is a surprising number of equivalences and similarities among commonly-used evaluation metrics. In fact, we identified two basic prototype metrics, precision and a linear cost metric, and showed that many other heuristics can be reduced to a combination of these two. Other heuristics exhibit a nonlinear isometric structure. In particular, gain heuristics, which do not evaluate the rule by itself but in comparison to a reference rule such as its predecessor or the default rule, often fall into that category. Finally, we discussed the performance of rule learning heuristics in practice, both in descriptive and predictive rule induction, and provided some recommendations about which heuristics to use. We will briefly return to that subject in Chap. 11, where a unifying framework for descriptive rule learning algorithms is developed.

---

[7] Consider, e.g., the difference between the information gain heuristic used in ID3 (Quinlan, 1983) and the information gain heuristic used in FOIL (Quinlan, 1990).

Chapter 8
Learning Rule Sets

The main shortcoming of algorithms for learning a single rule is the lack of expressiveness of a single conjunctive rule. In this chapter we present approaches that construct sets of rules. These sets can be constructed by iterative usage of the algorithms for constructing single rules presented in Chap. 6. We start with a brief discussion of the expressiveness of conjunctive and disjunctive rules, and show that through the use of rule sets we can effectively formulate rules in disjunctive normal form (Sect. 8.1). We then discuss the fundamental covering algorithm for learning rule sets (Sect. 8.2) and its generalization to weighted covering, which is suitable for learning redundant rule sets (Sect. 8.3). In the final two sections of this chapter we discuss approaches that directly learn entire rule sets (Sect. 8.4) and, conversely, algorithms that combine learned rules into global models other than rule sets (Sect. 8.5). In this chapter, we will primarily remain in a concept-learning framework, even though many of the discussed algorithms have been presented in a more general context (later, in Chap. 10, we will demonstrate how a wide variety of learning tasks can be essentially reduced to concept learning tasks).

8.1 Disjunctive Rules and Rule Sets

We will first discuss the necessity for rule sets by showing that individual conjunctive rules are not expressive enough, even for concept learning problems. Thereafter, we will see how rule sets may be interpreted as enriching conjunctive rules with disjunctions, and discuss the duality of representations in conjunctive and disjunctive normal form.

Assume a domain with A mutually independent attributes, each having V possible values. Then we can represent V^A different examples in this domain. Each of these examples can be either positive or negative. Different concepts are defined depending on class assignments for these examples. For V^A examples there are 2^{V^A} different ways in which the subset of positive examples can be selected. This is the

J. Fürnkranz et al., *Foundations of Rule Learning*, Cognitive Technologies,
DOI 10.1007/978-3-540-75197-7_8, © Springer-Verlag Berlin Heidelberg 2012

number of concepts that can be defined over these examples. At the same time, the number of different features that can be constructed for each attribute is proportional to V, the number of different attribute values. This means that the total number of features for all attributes is proportional to $V \times A$, and the total number of possible conjunctive rules is proportional to $2^{V \cdot A}$ (each feature may be part of the rule body or not). Thus, it is clear that the number of concepts that can be represented with a single conjunctive rule is only a small fraction of all possible Boolean concepts.

Expressiveness can be increased if we allow *disjunctions*. In principle, the conditional part of a rule could be any logical combination of one or more features. This means that in some cases the rule body may be a single feature, but that in other cases it may be a complex disjunctive and conjunctive mixture of several features. However, it is generally accepted practice that the rule body is restricted to the conjunctive combination of features only. Disjunctions can only be encoded with multiple rules for the same class. For example, if we have two rules

$$\oplus \leftarrow \mathbf{f}_1 \wedge \mathbf{f}_2$$

$$\oplus \leftarrow \mathbf{f}_3 \wedge \mathbf{f}_4$$

this essentially means that an example is positive if it is covered by features $\mathbf{f}_1$ and $\mathbf{f}_2$ (first rule) or that it is positive if it is covered by features $\mathbf{f}_3$ and $\mathbf{f}_4$ (second rule). This could also be written as a single disjunctive rule

$$\oplus \leftarrow (\mathbf{f}_1 \wedge \mathbf{f}_2) \vee (\mathbf{f}_3 \wedge \mathbf{f}_4)$$

Note that this limited use of disjunctions is, in principle, not a restriction of expressiveness. It is easy to see, that all Boolean concepts over a finite domain can be represented with disjunctions, because one could specify a concept using a disjunction of all examples that belong to the concept. In fact, any logical formula can be transformed into a disjunction of conjunctions, its so-called *disjunctive normal form (DNF)*.

In the context of concept learning, this implies that any possible concept over a given feature set can be represented as a set of conjunctive rules, and that, due to the fact that the conjunctive operator is commutative, the problem of constructing a rule body reduces to the problem of selecting the most relevant subset of features. This syntactic restriction of the hypothesis space has the advantage that it significantly reduces the complexity of the algorithms that have to search through the hypotheses space looking for appropriate combinations of features (cf. Sect. 2.5.1). Also, a decomposition into a set of conjunctions is much more comprehensible for a human than a Boolean formula with an arbitrarily complex structure.

On the other hand, the disjunctive normal form of a concept may be considerably more complex than its most compact representation, so that the rule learner has to learn many complex rules. Imagine a domain in which the following rule describes the positive examples:

$$\oplus \leftarrow \mathsf{Color} = \mathsf{orange} \vee \mathsf{Color} = \mathsf{red} \vee \mathsf{Color} = \mathsf{pink} \vee \mathsf{Color} = \mathsf{violet}$$

Theory for class +	Theory for class -

`IF Color = orange THEN Class = +`
`IF Color = red    THEN Class = +`
`IF Color = pink   THEN Class = +`
`IF Color = violet THEN Class = +`
	`IF      Color not orange`
	`    AND Color not red`
	`    AND Color not pink`
	`    AND Color not violet`
	`THEN Class = -`

Fig. 8.1 Two rule sets describing the same concept in four rules for the positive class (*left*) and one rule for the negative class (*right*)

As shown in Fig. 8.1, a complete theory for the positive class needs four rules, but the negative class can be described with a single conjunctive rule.

It is important to recall at this point that any example that is not covered by a set of rules is classified as not belonging to the concept that is represented by these rules. In this case, any example that is not covered by one of the four rules for the positive class is classified as negative, and, conversely, any example that is not covered by the single rule for the negative class is classified as positive. Note that this completeness assumption effectively turns the implication of the DNF-formula into an equivalence:

$$\oplus \leftrightarrow \textsf{Color} = \textsf{orange} \vee \textsf{Color} = \textsf{red} \vee \textsf{Color} = \textsf{pink} \vee \textsf{Color} = \textsf{violet}$$

This says that the specified conditions are indicative of the positive class ($\leftarrow$) and that the entire positive class is described by these conditions ($\rightarrow$). Logically this means that if we negate the head of the rule (replacing the positive class with the negative class) we also have to negate the body of the rule.

According to De Morgan's laws, any negated complex logical function expression can be transformed into the complementary non-negated logical function with negated arguments. For example, the negation of the function $\neg(a \wedge (b \vee c))$ is equivalent to $\neg a \vee (\neg b \wedge \neg c)$. This also means that any rule or rule set for the positive class can be transformed into an equivalent rule for the negative class by negating its DNF expression and transforming the result into DNF again. In the above example, negating the DNF for the positive class results in

$$\ominus \leftrightarrow \neg(\textsf{Color} = \textsf{orange} \vee \textsf{Color} = \textsf{red} \vee \textsf{Color} = \textsf{pink} \vee \textsf{Color} = \textsf{violet})$$

$$\leftrightarrow \neg(\textsf{Color} = \textsf{orange}) \wedge \neg(\textsf{Color} = \textsf{red}) \wedge \neg(\textsf{Color} = \textsf{pink}) \wedge \neg(\textsf{Color} = \textsf{violet})$$

$$\leftrightarrow \textsf{Color} \neq \textsf{orange} \wedge \textsf{Color} \neq \textsf{red} \wedge \textsf{Color} \neq \textsf{pink} \wedge \textsf{Color} \neq \textsf{violet}$$

which proves the equivalence of the two theories in Fig. 8.1.

Note that a negated DNF expression corresponds to an expression in *conjunctive normal form (CNF)*, in which each term in the DNF and each Boolean operator is negated. Efficient algorithms for transforming DNF to CNF and vice versa have been developed for the design of logic circuitry (Friedman, 1986). There are some algorithms that learn theories in CNF instead of the more common DNF (Mooney, 1995), most notably the CLAUDIEN algorithm for clausal discovery (De Raedt & Dehaspe, 1997). Some systems, such as ICL (De Raedt & Van Laer, 1995), are also able to switch dynamically between learning CNF and learning DNF, because, as we have seen in Fig. 8.1, some concepts are easier to represent in CNF, while others can be better learned in DNF. For a deeper discussion of this and related issues we refer to (De Raedt, 1996; Flach, 1997).

The main point to take from the above discussion is that, due to the fact that we have restricted the rule body to a conjunctive form, we can decompose the learning process into parts, one for each rule, without losing expressiveness. The separate-and-conquer or covering algorithm is the best known approach for that. We will discuss it in detail in the next section.

8.2 The Covering Strategy

In Chap. 2, we defined that the goal of a rule learning algorithm is to find a set of rules that cover all positive examples of a class, and none of the negative examples. In the light of the discussion in the previous section, this means that we aim at finding a DNF expression that separates the positive from the negative examples. An additional constraint is that the expression should be as small as possible, in order to ensure good generalization and to avoid trivial solutions like a disjunction of all positive examples.

8.2.1 The Algorithm

A simple strategy to ensure this is to have a learning algorithm that finds a single conjunctive rule that covers some of the training examples (like the FINDBESTRULE algorithm of Chap. 6) and use this algorithm repeatedly to find a set of rules that collectively cover all examples. The single rule learning algorithm should ensure that no rule covers negative examples (or not many of them). This ensures consistency with the training examples. As it can, in general, not be expected that a single rule covers all examples, we thus have to learn a set of rules in order to ensure completeness.

The main problem that needs to be solved for achieving completeness is to ensure that the learned rules are diverse enough so that they collectively produce a complete theory. Ideally, every training example should be covered by a rule. A simple strategy to ensure this is to simply ignore examples that are already covered by rules, so that the learner can focus on those examples that are not yet handled by the current rule set. The covering or separate-and-conquer algorithm realizes this idea by an iterative algorithm that removes all examples that are covered by a learned rule before it continues to learn subsequent rules.

```
function COVERING(ℰ)

  Input:
    ℰ = 𝒫 ∪ 𝒩: a set of positive and negative examples for a class c,
               represented with a set of features ℱ

  Algorithm:
    // initialize an empty rule set
    ℛ := ∅
    ℰ^cur := ℰ

    //loop until all positive examples are covered
    while 𝒫^cur ≠ ∅ do
      //find the best rule for the current examples
      r := FINDBESTRULE(ℰ^cur)

      // check if we need more rules
      if THEORYSTOPPINGCRITERION(ℛ, ℰ^cur)
      then  break while

      // add rule to rule set and adjust example set
      ℛ := ℛ ∪ r
      ℰ^cur := ADJUSTEXAMPLES(ℛ, ℰ^cur)

    endwhile

    // post-process the rule set
    ℛ = POSTPROCESSRULES(ℛ, ℰ)

  Output:
    ℛ the learned rule set
```

Fig. 8.2 The generic COVERING algorithm for incrementally learning a complete set of rules

Figure 8.2 shows a generic separate-and-conquer or covering rule learning algorithm that calls various subroutines which can be used to instantiate the generic algorithm into specific algorithms known from the literature. COVERING starts with an empty theory. If there are any positive examples in the training set it calls the subroutine FINDBESTRULE (Fig. 6.1) for learning a rule that will cover a subset of the positive examples. ADJUSTEXAMPLES will then adapt the training set to the new situation, typically by removing all covered examples or only the covered positive examples. Thereafter, the learned rule is added to the theory, and another rule is learned from the remaining examples. Rules are learned in this way until no positive examples are left or until the THEORYSTOPPINGCRITERION fires. Often the resulting theory undergoes some post-processing (POSTPROCESSRULES). This can, e.g., be a separate optimization phase as in RIPPER (cf. Sect. 9.5).

All separate-and-conquer algorithms share the basic structure of this simple algorithm. Their variants mostly differ in choices for the FINDBESTRULE procedure. We have already discussed these in detail in Chap. 6. Other common modifications try to avoid the induction of complete and consistent theories, which can lead to overfitting. These are realized via the THEORYSTOPPINGCRITERION or POSTPROCESSRULES routines. These are discussed in more detail in Chapter. 9.

A minor issue is the realization of the ADJUSTEXAMPLES procedure. In principle, we have two choices here:

1. Remove all covered examples
2. Remove only the positive covered examples

There is no commonly accepted recommendation for this procedure. Option 1 mirrors the situation for learning a decision list: once a rule is added to a theory, it will classify all examples that it covers as positive. Thus, one can argue that negative examples that are covered by this rule should not influence the learning of subsequent rules. On the other hand, one can argue that negative examples that are erroneously covered by one rule should still be considered by subsequent rules in order to avoid adding more evidence for the wrong class. This may be particularly important if unordered rule sets for multiple classes are learned, where the final prediction may depend crucially on the total voting mass that the rules give to each class. A third choice might be to not remove any examples at all, but to adjust their weight. This will be discussed further below, in Sect. 8.3. First, however, we will take a look at the behavior of the covering algorithm in coverage space.

8.2.2 *The Covering Strategy in Coverage Space*

In Chap. 7, we demonstrated how to visualize the behavior of heuristics for evaluating single rules via their isometrics in coverage space. In this section, we take a look at the evaluation of sets of rules.

Recall the discussion of Sect. 3.4 where we have seen that the successive addition of rules to an initially empty theory results in a so-called coverage path through coverage space. Of particular interest for the covering approach is the property that coverage graphs reflect a change in the total number or proportion of positive (P) and negative (N) training examples via a corresponding change in the relative sizes of the P- and N-axes. As a consequence, the coverage space PN_i for a subset $\mathcal{E}_i$ can be drawn directly into the coverage graph of the entire training set $\mathcal{E}$. In particular, the sequence of training sets that are produced by the recursive calls of the covering strategy—after each new rule all covered examples are removed and the learner calls itself on the remaining examples—can be visualized by a nested sequence of coverage graphs, as shown in Fig. 3.3 (left).

Note that we assume here that ADJUSTEXAMPLES removes all the examples and not only the positives (cf. the discussion in the previous section). The latter case corresponds to reducing only the height of the coverage space, but not its width. If we evaluate a rule in this reduced coverage space, the result cannot be immediately transferred back into the original coverage space because computing the rule's position in the original space requires us to subtract the number of previously and newly covered negative examples from the total count of covered negative examples in the height-reduced space, so that these examples are not counted twice. This is

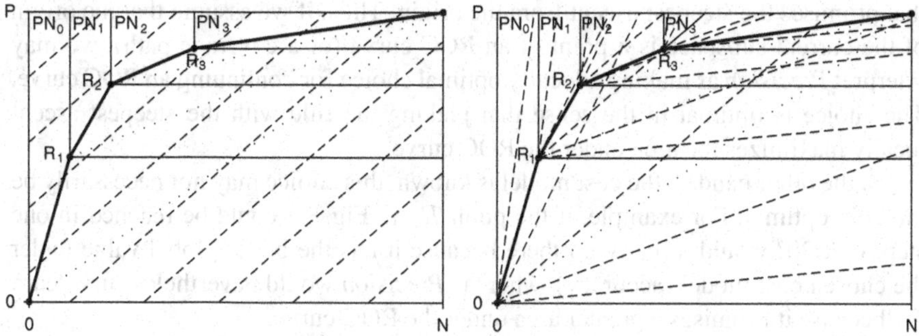

Fig. 8.3 Isometrics for accuracy and precision in nested coverage spaces

not necessary if negative examples are removed because then all previously covered negative examples are taken care of by the reduction of the coverage space. Thus, only newly covered examples are counted in the reduced coverage space.

8.2.3 Global and Local Optima

The nesting property implies that evaluating a single rule in the reduced dataset $(N - \hat{N}_i, P - \hat{P}_i)$ amounts to evaluating the point in the subspace PN_i that has the point $(\hat{N}_i, \hat{P}_i)$ as its origin. Moreover, the rule that maximizes the chosen evaluation function in PN_i is also a maximum point in the full coverage space if the the evaluation metric does not depend on the size or the shape of the coverage graph.

The linear cost metric *LinCost* is such a heuristic: a local optimum in the subspace PN_i is also optimal in the global coverage space because all isometrics are parallel lines with the same angle, and nested coverage spaces (unlike nested ROC spaces) leave angles invariant. *Precision*, on the other hand, cannot be nested in this way. The evaluation of a given rule depends on its location relative to the origin of the current subspace PN_i.

This is illustrated in Fig. 8.3. The subspaces PN_i correspond to the situation after removing all examples covered by the rule set $\mathcal{R}_i$. The left graph of Fig. 8.3 shows the case for accuracy: the accuracy isometrics are all parallel lines with a 45° slope. The right graph shows the situation for precision: each subspace PN_i evaluates the rules relative to its origin, i.e., *Precision* always rotates around $(\hat{N}_i, \hat{P}_i)$ in the full coverage space, which is $(0,0)$ in the local space.

However, local optimization is not necessarily bad. At each point $(\hat{N}, \hat{P})$, the slope of the line connecting $(0,0)$ with $(\hat{N}, \hat{P})$ is $\hat{P}/\hat{N}$.[1] *Precision* picks the rule

---

[1]One may say that *Precision* assumes a different cost model for each point in the space, depending on the relative frequencies of the covered positive and negative examples. Such local changes of cost models are investigated in more detail by Flach (2003).

that promises the steepest ascent from the origin. Thus, if we assume that the origin of the current subspace is a point of an ROC curve (or a coverage path), we may interpret *Precision* as making a locally optimal choice for continuing an ROC curve. The choice is optimal in the sense that picking the rule with the steepest ascent locally maximizes the area under the ROC curve.

On the other hand, if the cost model is known, this choice may not necessarily be globally optimal. For example, if the point $\mathcal{R}_2$ in Fig. 8.3 could be reached in one step, *CovDiff* would directly go there because it has the better global value under the chosen cost model (accuracy), whereas *Precision* would nevertheless first learn $\mathcal{R}_1$ because it promises a greater area under the ROC curve.

In brief, we may say that *Precision* aims at optimizing under unknown costs by (locally) maximizing the area under the ROC curve, whereas *LinCost* tries to directly find a (global) optimum under known (or assumed) costs.

8.3 Weighted Covering

A key problem for constructing a theory out of a set of rules is that the rules are learned in isolation, whereas they will be used in the context of other rules. The covering strategy partially addresses this problem by learning rules in order: all examples that are covered by previous rules are removed from the training set before a new rule is learned. This guarantees that the new rule will focus on new, unexplored territory. However, it also ignores the evidence contained in the removed examples, and the successive removal of training examples eventually leads to training sets with a very skewed class distribution, and possibly isolated, scattered examples.

As a remedy for the above-mentioned problem, several authors (Cohen & Singer, 1999; Gamberger & Lavrač, 2000; Weiss & Indurkhya, 2000) have independently proposed the use of *weighted covering*. The weighted covering algorithm modifies the classical covering algorithm in such a way that covered examples are not deleted from the current training set. Instead, in each run of the covering loop, the algorithm stores with each example a count indicating how often (with how many rules) the example has been covered so far. These weights are used in the computation of the rule learning heuristics, and therefore influence the selection of the next condition.

Figure 8.4 shows a high-level depiction of the algorithm. The key difference to conventional covering is the use of the example weights which need to be specified by appropriate choices of the functions INITIALIZEEXAMPLEWEIGHT and ADJUSTEXAMPLEWEIGHT. Different algorithms use different weight adaptation formulas, ranging from simple 'lightweight' weight updates (Gamberger & Lavrač, 2000; Weiss & Indurkhya, 2000) to complex error-based procedures motivated by boosting (Cohen & Singer, 1999; Dembczyński et al., 2010).

We will briefly describe the technique proposed by Lavrač et al. (2004), which has a particularly clear interpretation. Initially, all positive examples have a weight

```
function WEIGHTEDCOVERING(ℰ)
  Input:
    ℰ = 𝒫 ∪ 𝒩: a set of positive and negative examples for a class c,
               represented with a set of features ℱ

  Algorithm:
    // initialize rule set and example weights
    ℛ ← ∅
    for e ∈ ℰ do
        w(e) := INITIALIZEEXAMPLEWEIGHT(ℰ)
    endfor

    // loop for a fixed number of iterations I
    for i = 1...I do
        // find the best rule for the current example weights
        r := FINDBESTRULE(ℰ)
        // add the rule to the classifier
        ℛ := ℛ ∪ r

        // reduce the weight of covered examples
        for e ∈ ℰ do
            w(e) := ADJUSTEXAMPLEWEIGHT(ℛ, ℰ)
        endfor
    endfor
  Output:
    ℛ the learned rule set
```

Fig. 8.4 The weighted covering algorithm

of $w(e) = 1.0$, which denotes that the example has not been covered by any rule, meaning *'please cover this example, since it has not been covered before'*. The weights of examples that are covered by a rule will not be set to 0.0 (which is the equivalent to removing them from the training set), but instead their weight will only be reduced to a value between 0 and 1, meaning *'do not try too hard on this example'*. Consequently, the examples already covered by one or more constructed rules decrease their weights, whereas the weights of uncovered target class examples remain unchanged, which means that they have a greater chance to be covered in the following iterations of the algorithm. This ensures that their influence on the evaluation of subsequent rules is reduced, but not entirely eliminated.

For the adaptation of weights, Lavrač et al. (2004) propose two variants, a multiplicative and an additive weighting scheme. In the *multiplicative weighting scheme*, weights decrease multiplicatively. For a given parameter $0 < \gamma < 1$, the weight of a rule that is covered by C rules is determined by

$$w(e) = \gamma^C$$

Note that the weighted covering algorithm with $\gamma = 1$ would result in finding the same rule over and over again, whereas with $\gamma = 0$ the algorithm would perform the same as the standard covering algorithm. In the *additive weighting scheme*, the weight of an example that is covered by C rules is determined by

$$w(e) = \frac{1}{C + 1}$$

In the first iteration all target class examples contribute the same weight $w(e) = 1$, while in the following iterations the contributions of examples are inversely proportional to their coverage by previously induced rules.

Note that example weights also need to be appropriately incorporated into the heuristics used for rule quality evaluation inside the FINDBESTRULE algorithm. In principle, we can reuse all heuristics discussed in Chap. 7, with the main difference that all counts need to be replaced with the accumulated weights. For example, the total number of examples $E = P + N$ needs to be replaced with the total sum of all weights

$$E \leftarrow \sum_{e \in \mathcal{E}} w(e)$$

or the number of covered positive examples $\hat{P}$ needs to be replaced with the sum of weights of all covered positive examples

$$\hat{P} \leftarrow \sum_{e \in \hat{\mathcal{P}}} w(e)$$

Most weighted covering algorithms also adopt a very simple stopping criterion, namely they simply learn a fixed number of rules. Diversity of the rules is encouraged by the reweighting of the examples, but it is no longer strictly enforced that each example is covered by a rule. Also, the number of learned rules is typically higher, which has the effect that most examples will be covered by more than one rule. Thus, weighted covering algorithms have two complementary advantages: on the one hand they may learn better rules because the influence of previously covered examples is reduced but they are not entirely ignored; on the other hand they will produce a better classifier by combining the evidence of more rules, thus exploiting the redundancy contained in an ensemble of diverse rules (Dietterich, 2000).

Weighted covering is also an alternative for ensuring the diversity of discovered rules in subgroup discovery (Gamberger & Lavrač, 2002; Lavrač et al., 2004). The main reason is that removing examples during training, as done by the covering algorithm, distorts the training set statistics and introduces order dependencies between rules. For instance, the last rule learned is heavily dependent on the previous rules and the positives they cover, and it may not be meaningful (or statistically significant) when interpreted individually. Weighted covering, on the other hand, ensures that subsequently induced rules (i.e., rules induced in the later stages) also represent interesting and sufficiently large subgroups of the population.

8.4 Nonreductive Rule Learning Algorithms

All the algorithms that we covered so far learn rule sets by reducing the problem to learning individual rules, essentially via calls to the FINDBESTRULE algorithm. An alternative approach is to try to directly learn a complete rule set. Among the approaches found in the literature, we will discuss here algorithms that combine rule learning and instance-based learning (Sect. 8.4.1), tree-based approaches (Section 8.4.2), and algorithms that perform an exhaustive or stochastic search for complete theories (Sect. 8.4.3).

8.4.1 Generalization of Nearest Neighbor Algorithms

There are several algorithms that use a bottom-up learning strategy for generalizing the given examples into a set of rules. They typically start with a set of rules, each representing one example, and successively generalize the entire rule set. Thus, they may be viewed as generalizations of instance-based or nearest-neighbor learning algorithms.

A prototypical example is the incremental learning algorithm NEAR (Salzberg, 1991). It learns so-called *hyper-rectangles*, which are essentially rules in a numerical feature space. The key idea is to classify each new example with the nearest rule (measured to the border of the closest border of the hyper-rectangle). If this classification is correct, this rule is minimally generalized to include the new example, otherwise the example is converted into a new rule.

Probably the best-known variant of such algorithms is RISE (Domingos, 1996b), a slightly generalized variant of which is shown in Fig. 8.5. The algorithm starts with a set of rules, where each rule is the bottom rule $\mathbf{r}_e$ for this example e, i.e., the most specific rule that covers this example. These rules are then successively generalized by computing the lgg of the rule with its nearest (positive) example that is not yet covered by this rule. More precisely, the lgg is computed between the rule and the bottom rule of the selected examples, but the pseudocode omits this operation for simplicity of presentation. This rule then replaces the original rule in a new rule set $\mathcal{R}'$, and both rule sets, the original $\mathcal{R}$ and the new $\mathcal{R}'$, are evaluated on the training data. If $\mathcal{R}'$ is no less accurate than $\mathcal{R}$, $\mathcal{R}'$ replaces $\mathcal{R}$. The classification accuracy of the rule set $\mathcal{R}$ is estimated in a leave-one-out fashion where for each example e its own bottom rule $\mathbf{r}_e$ is ignored for classification should it still be contained in the rule set. Examples that are not covered by any rule are classified by the closest rule (cf. Sect. 10.3.2), where ties are broken by using the rule with the highest Laplace-corrected precision (cf. Sect. 10.3.1). Rule generalization is repeated until one iteration through the rules no longer yields any improvement. Domingos (1996b) has named his method 'conquering without separating'.

```
function RISE(ℰ)
  Input:
    ℰ = 𝒫 ∪ 𝒩: a set of positive and negative examples for a class c,
                represented with a set of features ℱ

  Algorithm:
    // initialize rule set and examples
    ℛ ← BottomRules(ℰ)

    repeat
       // loop through all rules
       for r ∈ ℛ do
          // choose closest positive example not yet covered by r
          e ← SelectClosestExample(𝒫, r)

          // generalize r so that it covers e
          r′ ← lgg(r, e)

          //keep the new rule if it does not decrease accuracy
          ℛ′ ← ℛ \ r ∪ r′
          if (Accuracy(ℛ′, ℰ) ≥ Accuracy(ℛ, ℰ))
          then
                  ℛ = ℛ′
          endif
       endfor
    until ℛ has not changed

  Output:
    ℛ the learned rule set
```

Fig. 8.5 The RISE Algorithm (Domingos, 1996b)

Many close variants of this algorithm can be found in the literature, in particular in inductive logic programming. For example, the algorithm CHILLIN (Zelle et al., 1994) generalizes rules by forming their *rlgg* and, if necessary, successively specializes them using top-down hill-climbing as in FOIL. CWS (Domingos, 1996a) interleaves the induction of different rules by starting to induce the next rule in the same cycle as the second condition of the current rule is learned. The PLCG algorithm uses similar ideas to combine and generalize multiple rules via clustering (Widmer, 2003).

8.4.2 Tree-based Approaches

In Sect. 1.5, we have already briefly discussed decision-tree learning and noted that a decision tree may be considered as a set of nonoverlapping rules. An explicit conversion of a tree into a rule set is straightforward, but more elaborate algorithms exist for this purpose. Most notably, the C4.5 decision tree induction algorithm has an option, C4.5RULES, that allows us to generate compact decision lists from decision trees by turning a decision tree into a set of nonoverlapping rules, pruning these rules, and ordering the resulting overlapping rules into a decision list. The key steps of the C4.5RULES algorithm are (Quinlan, 1987a, 1993):

1. Convert the tree to rules
2. Form blocks of rules, one for each class label
3. Generalize the rules via post-pruning (Sect. 9.3)
4. Identify a subset of rules for each block that minimizes a MDL-based criterion (Sect. 9.3.4)
5. Rank the rule blocks so that they minimize the number of false positives

An attempt to directly combine the advantages of decision-tree and rule-based learning is the PART algorithm (Frank & Witten, 1998), which does not learn the next rule in isolation but (conceptually) learns a global model in the form of a decision tree. From this tree, a single path is selected as the next local pattern that can be added to the theory. The implementation of this phase can be optimized so that the selected branch can be grown directly, without the need of growing an entire tree first. These partial trees are repeatedly learned in a covering loop, i.e., after the selection of a rule, the covered examples are removed and a new tree is learned.

8.4.3 *Direct Search for Theories*

Just as we have defined the learning of single rules as a search in the space of all possible rules (Chap. 6), we can also define the learning of a rule set as search in the space of all possible rule sets. Recently, Rijnbeek and Kors (2010) proposed EXPLORE, a general framework for the exhaustive search of DNF formulas in propositional logic. Earlier, the ITRULE algorithm (Smyth & Goodman, 1991) implemented an efficient exhaustive search for a fixed number of rules with a fixed maximum length. In inductive logic programming, there has been some work on *theory revision*, i.e., on refinement operators for entire theories instead of individual rules (Badea, 2001; Esposito, Semeraro, Fanizzi, & Ferilli, 2000; Wrobel, 1996). A particularly interesting proposal for refining entire theories is the HYPER system (Bratko, 1999), which assumes that the number of rules is fixed.

The main problem with these approaches is that the hypothesis of rule sets is potentially infinite so that most algorithms pose a fixed limit on the size of the concepts. Even then, exhaustive search is infeasible for larger problems and is frequently replaced with a stochastic search. For example, Rückert and Kramer (2003) have proposed to learn rule sets using stochastic local search (Hoos & Stützle, 2004), a particularly efficient variant of stochastic hill-climbing with multiple restarts, which allows to learn an entire theory in parallel. Their algorithm aims at finding a rule set of a user-specified maximum size, which is consistent with the training examples. In a follow-up work, Rückert and De Raedt (2008) showed that a variant of this algorithm is able to learn simpler theories than conventional covering algorithms, while maintainin a similar level of predictive accuracy.

Pfahringer, Holmes, & Wang (2005) take the stochastic approach to the extreme and propose to classify examples with a large set of randomly generated rules. This sacrifices the comprehensibility of the resulting rule set, but gains in accuracy due to the boosting effect of ensemble techniques.

8.5 Combining Rules to Arbitrary Global Models

All of the above-mentioned algorithms combine individual rules into a rule set. Rules, however, are quite flexible and can be used as input to arbitrary learning algorithms, including statistical and probabilistic algorithms. The key idea is to treat the body of a discovered rule as a *local pattern* and use these patterns as binary features. The input data can then be recoded in terms of its local patterns, and an arbitrary learning algorithm can be used for learning a *global model* from these data. Essentially, this is the key idea of the LEGO framework for inducing global models from local patterns (Fürnkranz and Knobbe, 2010; Knobbe, Crémilleux, Fürnkranz, & Scholz, 2008).

In this context, a *local pattern* refers to a regularity that holds for a particular part of the data (Morik, Boulicaut, & Siebes, 2005). The term local refers to the fact that it captures some aspect of the data, without providing a complete picture of the database. Local patterns do not necessarily represent exceptions in the data (Hand, 2002), but rather fragmented and incomplete knowledge, which may be fairly general. Although this is not necessarily the case, we will think of a local pattern as a single rule.

Figure 8.6 shows the key steps of the LEGO framework:

Local Pattern Discovery: This phase is responsible for producing a set of candidate patterns by means of an exploratory analysis of a search-space of patterns, defined by a set of inductive constraints provided by the user. As such, this phase can be seen as an automated instance of the feature construction phase in the KDD process. Patterns are typically judged on qualities such as their frequency or predictive power with respect to some target concept.

Pattern Set Selection: This phase considers the potentially large collection of patterns produced in the preceding phase, and selects from those a compact set of informative and relevant patterns that shows little redundancy. This phase is the counterpart of the feature selection phase in the KDD process.

Global Modeling: This phase is responsible for turning the condensed set of relevant patterns into a well-balanced global model. It either treats each local pattern as a constructed feature, and applies an existing inductive method, or applies some pattern combination strategy that is specific to the class of patterns discovered.

A prototypical instantiation of this framework is *associative classification*, as exemplified by the CBA rule learning algorithm (Liu, Hsu, & Ma, 1998; Liu, Ma, & Wong, 2000). This type of algorithm typically uses a conventional association rule

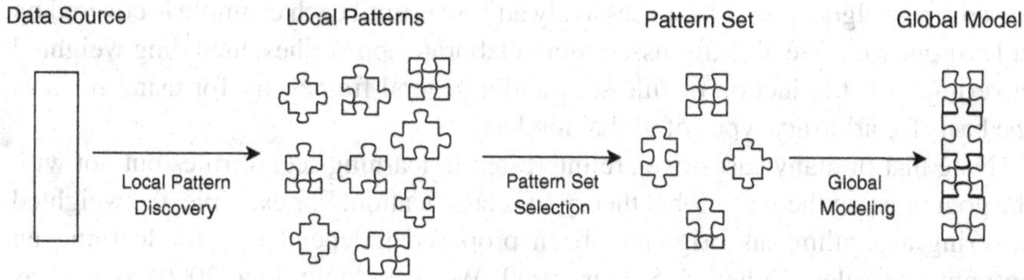

Fig. 8.6 The LEGO framework (Knobbe et al., 2008)

discovery algorithm, such as APRIORI, to discover a large number of patterns. From these, all patterns that have the target class in the head are selected, and only those are subsequently used for inducing a global theory. The global theory is typically a disjunction of patterns, found by a variant of the COVERING algorithm: patterns are sorted according to some heuristic function and the best one is repeatedly added to the disjunction. Variations in the global model may use decision lists or redundant rule sets. A variety of successor systems have been proposed that follow the same principal architecture (e.g., Bayardo Jr., 1997; Jovanoski & Lavrač, 2001; Li, Han, & Pei, 2001; Mutter, Hall, and Frank, 2004; Yin & Han, 2003). Sulzmann & Fürnkranz (2008) compare various approaches for combining association rules into a rule-based theory. Azevedo & Jorge (2010) propose to generate an ensemble of rule sets instead of a single rule set. A good survey of associative classification and related algorithms can be found in (Bringmann et al., 2009).

Note that the separation of the phases does not have to be as clear as it is in these algorithms. It is also useful to view the covering and weighted covering algorithms within this framework (Fürnkranz, 2005). In these algorithms, the local pattern discovery phase focuses on finding a single best global pattern, which is subsequently added to the growing rule set. Thus, the phases of local pattern discovery, pattern set discovery, and global modeling are tightly interleaved. Dembczyński et al. (2010) describe a unifying approach of weighted covering algorithms in the context of the LEGO framework, which is particularly interesting because it demonstrates how different loss functions for the performance of the global model can be decomposed into impurity measures that can be optimized during the construction of local patterns in the form of individual classification or regression rules.

8.6 Conclusion

In this chapter, we saw how single rules, which only apply locally to parts of the instance space, can be combined into complete global models, which can be used to obtain a prediction for any point in the instance space. The simplest approach, the covering algorithm, is to successively add rules until each example is covered by at least one rule. We also discussed more elaborate approaches, including weighted covering, direct induction of rule sets, and a general framework for using rules as the basis for arbitrary types of global models.

Note that in many cases, we are interested in learning sets of rules but not with the goal of using them as global theory for classification. For example, the weighted covering algorithm has originally been proposed independently for learning an ensemble of rules (Cohen & Singer, 1999; Weiss & Indurkhya, 2000) as well as for discovering a set of interesting subgroups (Gamberger & Lavrač, 2000). The problem of discovering the set of the best subgroups is also known as *top-k rule discovery* (cf., e.g., Scheffer & Wrobel, 2002). We will return to subgroup discovery in Chap. 11.

Chapter 9
Pruning of Rules and Rule Sets[†]

Overfitting is a key problem for all learning algorithms. It describes the phenomenon that a very good fit of the learned rule set to the training data may result in inaccurate predictions on unseen data. Typically, this problem occurs when the hypothesis space is too large. Prepruning and post-pruning are two standard techniques for avoiding overfitting. Prepruning deals with it during learning, while post-pruning addresses this problem after an overfitting rule set has been learned.

In this chapter, we will discuss a variety of pruning techniques for rule learning algorithms. We will start with a brief discussion of the problem of overfitting (Sect. 9.1), and then focus on the two principal approaches, prepruning (Sect. 9.2) and post-pruning (Sect. 9.3). State-of-the-art approaches typically integrate post-pruning into the rule learning phase, which is discussed in Sect. 9.4. Finally, we will also briefly summarize alternative methods for optimizing a learned rule set (Sect. 9.5).

9.1 Overfitting and Pruning

The central problem of any machine learning process, including rule learning, is that learning is performed on a restricted set of training examples and that the learned hypothesis is expected to present relevant dependencies in the whole domain, including unseen instances that may be encountered in the future. Learning is based on the assumption that patterns that hold on the training set are also valid in the entire domain. However, this assumption does not necessarily hold. There are many possible patterns and theories that can be constructed for the training set, and there is always a possibility that the constructed hypothesis will be of excellent quality on the training data and very poor quality on the complete domain. This effect is called *overfitting*.

[†]Parts of this chapter are based on Fürnkranz and Flach (2005) and Fürnkranz (1997).

J. Fürnkranz et al., *Foundations of Rule Learning*, Cognitive Technologies,
DOI 10.1007/978-3-540-75197-7_9, © Springer-Verlag Berlin Heidelberg 2012

Overfitting is a common problem for all learning algorithms. The reasons for overfitting are manifold, ranging from example representations that do not capture important domain characteristics to inadequate hypothesis spaces. One particular problem is that real-world databases very often contain *noise*, i.e., erroneous or incomplete instances. There may be many reasons for noise in databases: manual data collection and data entry are error-prone, measuring devices may have a natural variance in their precision or may fail occasionally, some attribute values may not always be available, judgments obtained from human experts are often inconsistent, and many more.

Overfitting typically has the effect that a too-complex theory will be learned, and that the size of this theory will increase with the increasing size of the training set. Consider a propositional dataset consisting of E examples, each described with F features. The examples are evenly distributed, so that each feature covers about half of the examples. Further assume that a constant fraction of the training examples are incompressible noise; these examples cannot be explained by a nontrivial pattern or rule in the data. In this idealized noisy dataset, it is easy to see the following:

Theorem 9.1.1 (Complexity of an Overfitting Rule Set). *The size of a rule set that explains all examples in a noisy dataset is* $\Theta(E \cdot \log E)$.

Proof. A constant fraction of the examples in a noisy dataset cannot be explained by a pattern. Thus, $\Theta(E)$ examples can only be encoded with $\Theta(E)$ rules. Furthermore, each of these rules requires $\Theta(\log E)$ tests for discriminating the noisy example from other examples, because each test for a feature will exclude about half of the instances. Thus, the size of the overfitting theory is $\Theta(E \cdot \log E)$ conditions. □

Note that the factor of $\log E$ goes away with increasing E, because at some point $F < \log E$, and F is the maximum length of a rule. Nevertheless, we see that the size of an overfitting theory grows at least linearly with E. This is typical for learning from noisy datasets. On the other hand, in a noise-free dataset, once the underlying theory has been found, additional training examples should not lead to further growth in theory size.

Overfitting is typically avoided by biasing the learner towards simpler concept descriptions to explicitly counter this growth and force the theory to stabilize. Figure 9.1 shows typical curves of the accuracy of a rule set and its complexity (number of features in the learned rule set) over a parameter that controls the complexity of a rule set, in this case the cutoff parameter of the FOSSIL rule learning algorithm, which will be described in more detail in Sect. 9.2.5.[1] The most accurate

[1] The learning task was to discriminate between legal and illegal chess positions in a king–rook–king chess endgame. The training set consisted of 500 training positions, 10 % of which were corrupted with noise.

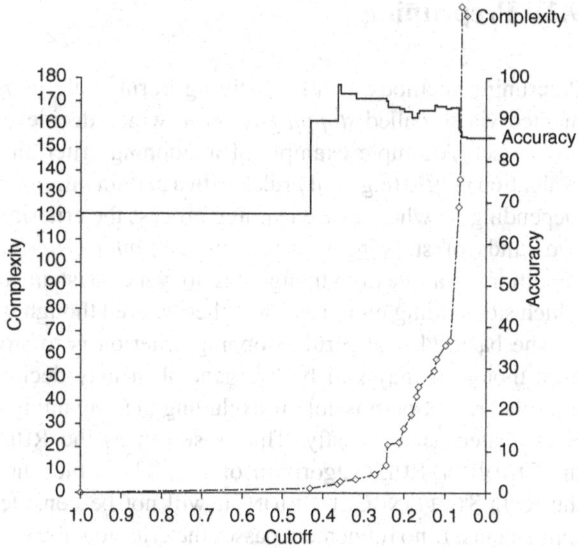

Fig. 9.1 Accuracy and complexity *vs.* pruning level in a noisy domain

theories are found for cutoff values between approximately 0.25 and 0.35. Higher cutoff values result in too simple theories (*underfitting*), while lower settings of the cutoff obviously result in overfitting of the data.

In Chap. 7, we have already seen that many commonly used rule learning heuristics prefer rules that cover a few exceptions over rules that do not make any mistakes on the training data. However, many rule learning algorithms try to separate this problem to some extent from the selection heuristic and use a separate pruning heuristic for controlling the complexity of the resulting target theory. There are two fundamental approaches for achieving this goal: *prepruning* in the form of stopping and filtering criteria that decide when a rule should no longer be specialized, and *post-pruning*, which simplifies overfitting theories by generalizing them as long as their performance on an independent part of the data increases. These two approaches, which we will discuss in detail in Sects. 9.2 and 9.3, have complementary advantages: prepruning approaches are fast but turn out to be inferior in practice, whereas post-pruning algorithms are flexible and accurate, but are very inefficient. Another successful strategy integrates the best of both approaches: it uses post-pruning at the rule level, which effectively results in prepruning at the level of entire rule sets. This algorithm, known as *incremental reduced error pruning*, is described in more detail in Sect. 9.4. It is used in RIPPER (Cohen, 1995), still one of the most powerful rule learning algorithms available today.

9.2 Prepruning

Prepruning methods avoid overfitting during rule construction. They are implemented via so-called *stopping criteria*, which decide when to stop refining a rule or rule set. A simple example of a stopping criterion is a threshold upon a rule evaluation metric (e.g., only rules with a certain minimum precision are acceptable). Depending on where in the learning process the criterion is applied, we can discern two kinds of stopping criteria: *rule stopping criteria*, which stop adding more conditions to a rule even though it is not yet consistent, and *theory stopping criteria*, which stop adding more rules to a theory, even though it may not yet be complete.

The basic idea of a rule stopping criterion is to stop the refinement of a rule even though it may still be overgeneral. It may decide that some refinements of the rule are not permissible if excluding the remaining covered negative examples is estimated as too costly. This is shown as the RULESTOPPINGCRITERION in the FINDBESTRULE algorithm of Fig. 6.1. If the current refinement $\mathbf{r}'$ satisfies the RULESTOPPINGCRITERION, it will not be considered as one of the possible refinements. If no refinement passes the criterion, the refinement process is stopped.

Conversely, the decision whether to further refine a rule set by adding more rules is made by a so-called *theory stopping criterion*. This is shown by the call to THEORYSTOPPINGCRITERION in the COVERING algorithm of Fig. 8.2. If the rule found by FINDBESTRULE satisfies this criterion, it is assumed that no further rule can be found that covers the remaining positive examples and satisfies the theory stopping criterion. Consequently, the incomplete theory without the rule that triggered the criterion is returned as the final theory. The remaining positive examples are thus considered to be noisy and will be classified as negative by the returned theory.

Note that in the following we do not discriminate between rule and theory stopping criteria because most criteria can be used in both roles. For example, a simple threshold criterion, which only accepts a rule if it has a certain minimum heuristic value, can be used as a rule stopping criterion filtering out candidate refinements that do not satisfy this constraint, or as a theory stopping criterion stopping the learning process as soon as the best found rule does not meet this quality threshold.

In the following, we illustrate prominent filtering and stopping criteria for greedy specialization: minimum coverage constraints, support and confidence, significance tests, encoding length restrictions, and correlation cutoff. As in Chap. 7, we use coverage space to analyze stopping criteria, by visualizing regions of acceptable hypotheses.

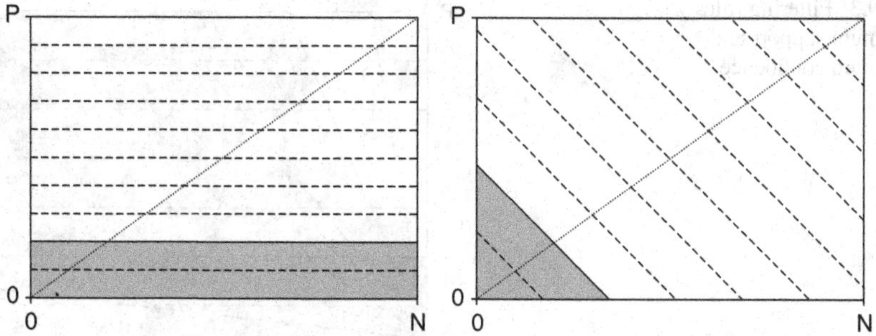

Fig. 9.2 Thresholds on minimum coverage of positive examples (*left*) and total number of examples (*right*)

9.2.1 *Minimum Coverage and Minimum Precision Constraints*

The simplest form of overfitting avoidance is to disregard rules with low coverage. For example, one could require that a rule covers a certain minimum number of examples or a minimum number of positive examples. These two cases are illustrated in Fig. 9.2. The graph on the left shows the requirement that a minimum fraction (here 20 %) of the positive examples in the training set are covered by the rule. All rules in the gray area are thus excluded from consideration. The right graph illustrates the case where a minimum fraction (here 20 %) of examples needs to be covered by the rule, regardless of whether they are positive or negative. Changing the size of the fraction will cut out different slices of the coverage space, each delimited with a coverage isometric (−45° lines). Clearly, in both cases, the goal is to fight overfitting by filtering out rules whose quality cannot be reliably estimated because of the small number of training examples they cover. As with the linear cost evaluation metric (Sect. 7.3.2), different misclassification costs can be modeled in this framework by changing the slope of the coverage isometrics.

Similar to minimum coverage constraints, some rule learning systems also use a minimum precision or minimum rule accuracy constraint. This simple criterion requires that a certain percentage of the examples covered by the learned rules is positive. It is, for example, used in the SFOIL algorithm (Pompe, Kovačič, & Kononenko, 1993) as a termination criterion for the stochastic search. In FOIL (Quinlan, 1990) this criterion is used as a theory stopping criterion: when the best rule is below a certain purity threshold (usually 80 %), it is rejected and the learned theory is considered to be complete.

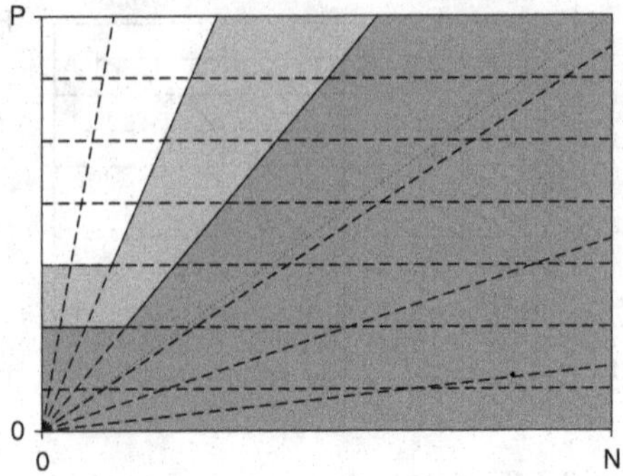

Fig. 9.3 Filtering rules with minimum support and minimum confidence

9.2.2 *Multiple Threshold Constraints*

There is no reason why only a single measure should be used for filtering out unpromising rules. The most prominent example for combining multiple estimates are the thresholds on support and confidence that are used mostly in association rule mining algorithms, but also in classification algorithms that obtain the candidate rules for the covering loop in an association rule learning framework (Jovanoski & Lavrač, 2001; Liu, Hsu, & Ma, 1998; 2000).

Figure 9.3 illustrates the effect of thresholds on support and confidence in coverage space. Together, both constraints specify an area for valid rules around the $(0, P)$-point. Rules in the gray areas will be filtered out. The dark gray region shows a less restrictive combination of the two thresholds, and the light gray region a more restrictive setting. In effect, confidence constrains the quality of the rules, whereas support aims at ensuring a minimum reliability by filtering out rules whose confidence estimate originates from too few covered positive examples.

9.2.3 *Encoding Length Restriction*

To decide when to stop refining the current rule, Quinlan (1990) suggested a criterion based on the minimum description length principle (Rissanen, 1978; Wallace & Boulton, 1968), similar to the *MDL* heuristic (Sect. 7.6), The criterion tries to avoid learning complicated rules that cover only a few examples by making sure that the number of bits that are needed to encode a rule is less than the number of bits needed to encode the instances covered by the rule.

More precisely, for transmitting the information which of the $P + N$ training examples are covered by a rule, one needs:

$$\boxed{ECL(\mathbf{r}) = \log_2(P+N) + \log_2 \binom{P+N}{\hat{P}}}$$

bits. The idea behind this estimate is that, in order to keep the transmission time from the sender to the receiver of the information short, first the sender has to transmit how many of the $P + N$ examples are covered by the rule ($\log_2(P + N)$ bits). The receiver can then construct all possible subsets of the transmitted size $\hat{P}$ in an order that has been agreed upon by the sender and the receiver. The sender then only has to transmit which one of these subsets is the correct one. This is the second term, the $\log_2$ of the number of possible subsets of size $\hat{P}$ in a set of $P + N$ elements. Note that this derivation is based on the assumption that the rule does not cover any negatives (i.e., $\hat{P} + \hat{N} = \hat{P}$).

The resulting number is then compared to the number of bits that are needed to encode the rule itself. The idea is that if the receiver can also use the rule to determine which examples are positive. A rule is encoded by encoding each feature that appears in the rule body. Quinlan (1990) uses relational features and suggests to encode them by specifying the relation, a suitable variabilization for this relation, and whether the relation is negated or not. In our feature-based framework, this simply amounts to specifying which of the F features we use. Thus, the encoding length for a rule of length $L_{\mathbf{r}}$ is Pfahringer (1995a):

$$ERL(\mathbf{r}) = \log_2 \binom{F}{L_{\mathbf{r}}}$$

If transmitting the rule is cheaper than transmitting the information of which examples are covered by the rule, the rule is acceptable; otherwise the rule will be rejected. When no feature can be added without exceeding the limit imposed by *MDL*, the incomplete rule is no longer refined, and it is added to the theory if a certain percentage (usually 80 %) of the examples it covers are positive.

For the purposes of our analysis, we interpret *MDL* as a heuristic that is compared to a variable threshold, whose size depends on the length $L_{\mathbf{r}}$ of the current rule $\mathbf{r}$. If $ECL(\mathbf{r}) < ERL(\mathbf{r})$, i.e., if the encoding of the rule is longer than the encoding of the examples themselves, the rule is rejected. As $ERL(\mathbf{r})$ depends solely on $L_{\mathbf{r}}$, and $ECL(\mathbf{r})$ depends on the number of covered positive examples $\hat{P}$, FOIL's stopping criterion depends on the size of the training set: the same rule that is too long for a smaller training set might be good enough for a larger training set, in which it covers more examples.

Figure 9.4 shows the behavior of *ECL* in coverage space. The isometric landscape is equivalent to the minimum support criterion, namely parallel lines to the N-axis. This is not surprising, considering that *ECL* is independent of $\hat{N}$, the number of covered negative examples.

For $P < N$, *ECL* is monotonically increasing with $\hat{P}$. If we see this in connection with FOIL's search heuristic *FoilGain*, we note that while the rule refinement process rotates the 0-gain line towards the P-axis (Fig. 7.10), the *ECL* metric steadily increases a minimum support constraint. Thus, the combined effect is the same as the effect of support and confidence constraints (Fig. 9.3) with the difference that FOIL chooses the thresholds dynamically based on the quality and length of the rules.

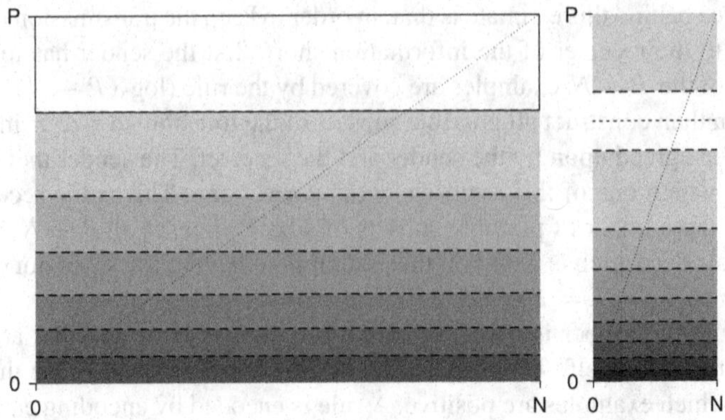

Fig. 9.4 Illustration of FOIL's encoding length restriction for domains with $P < N$ (*left*) and $P > N$ (*right*). Lighter shades of *gray* correspond to larger encoding lengths for the rule

Particularly interesting is the case $P > N$ (right graph of Fig. 9.4): the isometric landscape is still the same, but the corresponding values are no longer monotonically increasing. In fact, *ECL* has a maximum at the point $\hat{P} = (P + N)/2$. Below this line (shown dashed in Fig. 9.4), the function is monotonically increasing (as above), but above this line it starts to decrease again. Thus, one can formulate the following theorem (Fürnkranz & Flach, 2005):

Theorem 9.2.1. *ECL is compatible with CovPos iff* $\hat{P} \leq (P+N)/2$, *and it is antagonistic to CovPos for* $\hat{P} > (P + N)/2$.

Proof. (a) $\hat{P} \leq \frac{P+N}{2}$:

$$ECL = \log_2(P+N) + \log_2\binom{P+N}{\hat{P}}$$

$$\sim \log_2 \frac{\prod_{i=1}^{\hat{P}}(P+N+1-i)}{\prod_{i=1}^{\hat{P}} i}$$

$$= \sum_{i=1}^{\hat{P}} \log_2(P+N+1-i) - \sum_{i=1}^{\hat{P}} \log_2 i$$

$$= \sum_{i=1}^{\hat{P}} (\log_2(P+N+1-i) - \log_2 i)$$

The terms inside the sum are all independent of the number of covered examples. Thus, for two rules $\mathbf{r}_1$ and $\mathbf{r}_2$ with $0 \leq CovPos(\mathbf{r}_1) < CovPos(\mathbf{r}_2) \leq \frac{P+N}{2}$, the corresponding sums only differ in the number of terms. As all terms are > 0, $CovPos(\mathbf{r}_1) < CovPos(\mathbf{r}_2) \Leftrightarrow (\mathbf{r}_1) < ECL(\mathbf{r}_2).ECL$

(b) $\hat{P} > \frac{P+N}{2}$:

$$ECL \sim \log_2 \binom{P+N}{\hat{P}} = \log_2 \binom{P+N}{P+N-\hat{P}}$$

Therefore, each rule $\mathbf{r}$ with coverage $\hat{P}$ has the same evaluation as some rule $\mathbf{r}'$ with coverage $P+N-\hat{P} < (P+N)/2$. As the transformation $\hat{P} \to P+N-\hat{P}$ is monotonically decreasing:

$$\begin{aligned} ECL(\mathbf{r}_1) > ECL(\mathbf{r}_2) &\Leftrightarrow ECL(\mathbf{r}_1') > ECL(\mathbf{r}_2') \Leftrightarrow \\ &\Leftrightarrow CovPos(\mathbf{r}_1') > CovPos(\mathbf{r}_2') \\ &\Leftrightarrow CovPos(\mathbf{r}_1) < CovPos(\mathbf{r}_2) \end{aligned}$$

□

Corollary 9.2.1. *ECL is equivalent to CovPos iff* $P \le N$.

Proof. ECL reaches its maximum at $\hat{P} = \frac{P+N}{2}$, which is inside coverage space only if $P > N$. □

Thus, for skewed class distributions with many positive examples, there might be cases where a rule $\mathbf{r}_2$ is acceptable, while a rule $\mathbf{r}_1$ that has the same encoding length ($ERL(\mathbf{r}_1) = ERL(\mathbf{r}_2)$) covers the same number or fewer negative examples ($\hat{N}(\mathbf{r}_1) \le \hat{N}(\mathbf{r}_2)$), but more positive examples ($\hat{P}(\mathbf{r}_1) > \hat{P}(\mathbf{r}_2)$) is *not* acceptable. For example, in the example shown on the right of Fig. 9.4, we assumed $P = 48$ and $N = 20$. A rule $\mathbf{r}_1 = (0, 48)$ that covers all positive examples and no negative examples would be allowed approximately 62.27 bits, whereas a rule $\mathbf{r}_2 = (20, 34)$ that covers only 34 positive examples but all 20 negative examples would be allowed approximately 70.72 bits. The reason is that ECL is independent of the number of covered negative examples and that $\hat{P} = 34 = (P+N)/2$ maximizes the second term in ECL. Thus, if we assume that both rules have the same encoding length (e.g., $ERL(\mathbf{r}_1) = ERL(\mathbf{r}_2) = 65$ bits), $\mathbf{r}_2$ would still be acceptable, whereas the perfect rule $\mathbf{r}_1$ would not. This is very counterintuitive and sheds some doubts upon the suitability of FOIL's encoding length restriction for such domains. We refer to Sect. 9.3.4 for an alternative formulation of an MDL-based heuristic that has primarily been used for pruning.

9.2.4 Likelihood Ratio Significance Test

CN2 filters rules for which there is no statistically significant difference between the distribution of the covered examples and the distribution of examples in the full data set. To this end, it computes the *likelihood ratio statistic*, which we already briefly mentioned in Sect. 2.7.2:

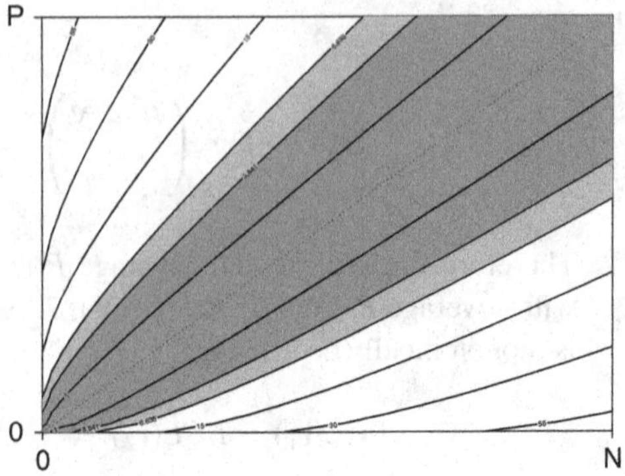

Fig. 9.5 Illustration of CN2's significance test. Shown are the regions that would not pass a 95 % (*dark gray*) and a 99 % (*light gray*) significant test

$$LRS(\mathbf{r}) = 2 \cdot \left(\hat{P} \cdot \log \frac{\hat{P}}{\mathbb{E}\hat{P}} + \hat{N} \cdot \log \frac{\hat{N}}{\mathbb{E}\hat{N}}\right) = 2 \cdot \left(\hat{P} \cdot \log \frac{\frac{\hat{P}}{\hat{P}+\hat{N}}}{\frac{P}{P+N}} + \hat{N} \cdot \log \frac{\frac{\hat{N}}{\hat{P}+\hat{N}}}{\frac{N}{P+N}}\right)$$

Here, $\mathbb{E}\hat{P} = (\hat{P} + \hat{N})\frac{P}{P+N}$ and $\mathbb{E}\hat{N} = (\hat{P} + \hat{N})\frac{N}{P+N} = (\hat{P} + \hat{N}) - \mathbb{E}\hat{P}$ are the number of positive and negative examples, respectively, which one could expect the rule to cover if the $\hat{P} + \hat{N}$ examples covered by the rule were distributed in the same way as the $P + N$ examples in the full data set.

The likelihood ratio statistic *LRS* is distributed as χ^2 with one degree of freedom. Only rules that pass the test at a predefined significance level are considered to be valid. Typical choices for the significance level are 95 %, which corresponds to a value $LRS(\mathbf{r}) = 3.841$, or 99 %, which corresponds to $LRS(\mathbf{r}) = 6.635$.

This form of significance testing for prepruning was first used in the propositional CN2 induction algorithm (Clark & Niblett, 1989) and later in the relational learner *m*FOIL (Džeroski & Bratko, 1992). In BEXA (Theron & Cloete, 1996) this test is also used for comparing the distribution of instances covered by a rule to that of its direct predecessor. If the difference is insignificant, the rule is discarded.

Figure 9.5 illustrates CN2's filtering criterion in coverage space. The dark gray area shows the location of the rules that will be filtered out because it cannot be established with 95 % confidence that the distribution of their covered examples is different from the distribution in the full dataset. The light gray area shows the set of rules that will be filtered out if 99 % confidence in the difference is required. The area is symmetric around the diagonal, which represents all rules that have the same distribution as the full example set.

Note that the shape of the likelihood ratio isometrics does not depend on the size of the training set. Any graph corresponding to a bigger training set with the same class distribution will contain this graph in its lower left corner. Therefore, a rule that is not significant for a small dataset may become significant if the dataset grows. The reason is that if the size of the training set grows, the ratios $\hat{P}/P$, $\hat{N}/N$, and also $\hat{P}/(\hat{P} + \hat{N})$ and $P/(P + N)$ should remain constant, but the absolute numbers of covered examples $\hat{P}$ and $\hat{N}$ grow. Thus, whether a rule is significant according to *LRS* depends not only on the class distribution of the examples it covers, but also on their absolute numbers.

The isometric structure of the likelihood ratio statistic is quite similar to those of precision or the m-estimate, with the difference that the isometrics are not linear but are bent towards the origin. Thus, the significance test has a tendency to prefer purer rules. As small pure rules are often the result of overfitting, it is questionable whether this strategy is reasonable if the primary goal of the significance test is to counter overfitting. The main purpose is to filter out uninteresting rules, i.e., rules for which the class distribution of the covered examples does not differ significantly from the a priori distribution.

This similarity between search and stopping heuristics also poses the question why different heuristics are used. It seems to be unlikely that CN2's behavior would be much different if the likelihood statistic is directly used as a search heuristic. Conversely, thresholds upon the m-estimate would have a similar effect as CN2's significance test (although the semantics of the associated significance levels would be lost).

9.2.5 Correlation Cutoff

Fürnkranz (1994a) proposes a threshold upon the absolute value of the correlation heuristic $|Correlation|$ (cf. Sect. 7.4.2). Only rules that evaluate above this threshold (also known as the *cutoff*) are admitted. The use of the absolute value allows us to evaluate both, a feature and its negation, in a single computation: a high negative correlation value for a candidate feature implies a high positive correlation for the negation of the feature. Thus, only the strength of the correlation is of interest; its sign is only used for selecting the sign of the feature. For $cutoff = 0$, all rules will be admitted (no prepruning), whereas $cutoff = 1$ means that FOSSIL will learn an empty theory (maximum prepruning). Values between 0 and 1 trade off the two extremes. For the correlation heuristic, a value of 0.3 has been shown to yield good results at different training set sizes and at differing levels of noise (Fürnkranz) as well as across a variety of test domains (Fürnkranz, 1997).

Figure 9.6 shows the isometric landscape of the correlation heuristic and the effect of a cutoff of 0.3. FOSSIL does not return the rule with the highest evaluation, but it continues to add conditions until the stopping criterion is satisfied (cf. Sect. 9.2.6). Thus, the cutoff line—the line separating the white from the gray region in Fig. 9.6—may be viewed as a minimum quality line: learning stops as soon as the path of the learner crosses this line (from the acceptable region to the inacceptable region).

It can be clearly seen that, like CN2, FOSSIL filters out uninteresting rules, i.e., rules whose example distribution does not deviate much from the example distribution in the full training set. Similar to CN2's likelihood ratio statistic, rules

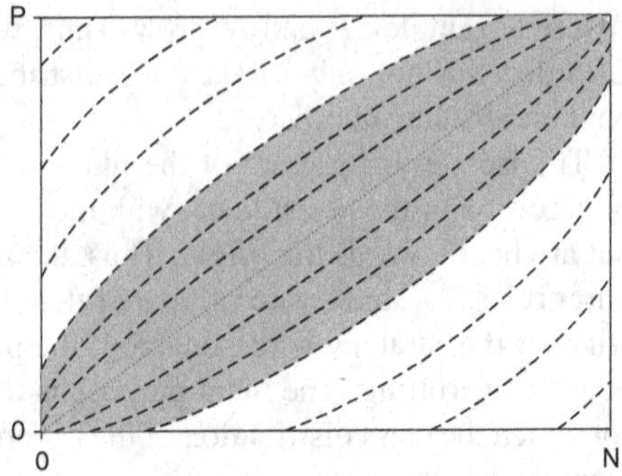

Fig. 9.6 Illustration of the correlation cutoff criterion

that cover few negative examples are preferred by the bent shape of the isometric lines of *Correlation*. In fact, the correlation cutoff criterion can also be interpreted as a statistical χ^2-test, quite similarly to CN2's significance testing. In the case of a two-by-two contingency table, both $(P + N) \cdot \mathit{Correlation}^2$ (cf. Theorem 7.4.1) and *LRS* are distributed as χ^2 with one degree of freedom. Note, however, this does not imply that they are equivalent as heuristics. In fact, there are some interesting differences between the isometric landscapes in Figs. 9.5 and 9.6.

As we have seen in the previous section, whether a rule is significant or not according to *LRS* depends not only on the class distribution of the covered examples, but also on their absolute numbers. We call this *absolute significance*. In contrast, *Correlation* always fits the same isometric landscape into coverage space. As a result, the evaluation of a point $(\hat{N}, \hat{P})$ only depends on its relative location $(\frac{\hat{N}}{N}, \frac{\hat{P}}{P})$. In this case, the same rule will always be evaluated in the same way irrespective of the size of the training set, as long as it covers the same fraction of the training data. We call this *relative significance*. It is still an open question whether absolute or relative significance is preferable in practice, but there is considerable evidence that FOIL's and CN2's prepruning heuristics are inefficient in preventing overfitting in the presence of noisy data (Fürnkranz, 1994a; Todorovski, Flach, & Lavrač, 2000).

9.2.6 *Stopping Versus Filtering*

Strictly speaking, we can discriminate between two slightly different points of view on the above-mentioned criteria: *stopping criteria* determine when the refinement process should stop and the current best candidate should be returned, whereas *filtering criteria* determine regions of acceptable performance. As an illustration, let us compare the strategies used by FOIL and CN2.

FOIL, which forms the basis of many rule learning algorithms, most notably RIPPER (Cohen, 1995), does not evaluate rules on an absolute scale but instead relative to their respective predecessors. Hence, the evaluation of two rules with different predecessors are not directly comparable. For this reason, FOIL-like algorithms always return the last rule searched (cf. Fig. 7.10). As FOIL will (almost) always find an improvement over its predecessor—because the point (0, 1), covering no negative examples and a single positive example, is an improvement over all predecessor rules that cover at least one negative example—FOIL's refinement process will usually continue until a pure rule is found. As this typically leads to overfitting, it is crucial to have a good *stopping* criterion, which determines when the refinement process should terminate.

On the other hand, CN2 evaluates rules using either entropy or the Laplace-estimate. These measures evaluate rules on an absolute scale, i.e., the evaluations of all searched rules are comparable. As a consequence, CN2 continues to search until no further refinements are possible, and the *best* rule (the one with the highest evaluation) encountered during this search is returned. In this case, the role of a stopping criterion is replaced with a *filtering* criterion that filters out unpromising candidates, but does not directly influence the choice of the best rule (Clark & Boswell, 1991).

In any case, filtering and stopping criteria are closely related. In particular, filtering criteria can also be used as stopping criteria: if no further rule can be found within the acceptable region of a filtering criterion, the learned theory is considered to be complete. Similarly, if no further refinement for a single rule can be found within the acceptable region, the rule is considered to be complete, and the specialization process stops. For this reason, we decided to use the term stopping criterion instead of filtering criterion because this terminology is more established.

9.3 Post-pruning

While prepruning approaches try to avoid overfitting during rule generation, *post-pruning* approaches at first ignore the problem of overfitting and learn a complete and consistent theory, which is then simplified in a second pass. A frequently used approach is to use the second pass to maximize the predictive accuracy on a separate set of data that has not been available to the learner during theory construction.

Figure 9.7 shows such an algorithm, which was first suggested in (Pagallo & Haussler, 1990) based on similar algorithms for pruning decision trees (Quinlan, 1987b). Before learning a complete and consistent concept description, the training set is split into two subsets: a *growing set* $\mathcal{G}$ (usually two third) and a *pruning set* $\mathcal{H}$ (one third). LEARNRULEBASE is then used for learning a complete and consistent concept description $\mathcal{R}$ from the growing set, which is subsequently pruned by PRUNERULEBASE in order to optimize its performance on the pruning set.

Most algorithms use this general strategy in one way or another. Examples that follow the above framework include *error complexity pruning* (Breiman, Friedman, Olshen, & Stone, 1984), *iterative pruning* (Gelfand, Ravishankar, & Delp, 1991) and

---

function PostPruning($\mathcal{E}, s$)

Input:
$\mathcal{E}$: a set of training examples
s: a ratio for splitting the training set

Algorithm:
// split examples into a growing set $\mathcal{G}$ and a pruning set $\mathcal{H}$
$(\mathcal{G}, \mathcal{H})$:= SplitExamples($\mathcal{E}, s$)

// learn a theory on $\mathcal{G}$
$\mathcal{R}$:= LearnRuleBase($\mathcal{G}$)

// prune the theory to optimize performance on $\mathcal{H}$
$\mathcal{R}$:= PruneRuleBase($\mathcal{R}, \mathcal{H}$)

Output:
$\mathcal{R}$ the learned rule set

Fig. 9.7 Post-pruning

critical value pruning (Mingers, 1989a). Other algorithms, such as *minimal error pruning* (Niblett & Bratko, 1987) or *pessimistic error pruning* (Quinlan, 1987b), analyze the performance of induced hypothesis on the same data set from which they have been learned.

Post-pruning approaches have been commonly used in decision tree learning algorithms (Breiman et al., 1984; Niblett & Bratko, 1987; Quinlan, 1987b). Mingers (1989a) and Esposito, Malerba, and Semeraro (1993) present an overview and comparison of various approaches, including those mentioned above. However, post-pruning was also used early on in rule learning algorithms. The key idea of removing unnecessary rules and conditions in a post-processing phase has been used in various versions of the AQ algorithm (Michalski, Mozetič, Hong, & Lavrač, 1986). The basic idea is to test whether the removal of a single condition or even of an entire rule would lead to a decrease in the quality of the concept description, usually measured in terms of classification accuracy on the training set. If this is not the case, the condition or rule will be removed.

This framework was generalized in the POSEIDON system (Bergadano, Matwin, Michalski, & Zhang, 1992). POSEIDON can simplify a complete and consistent concept description, which has been induced by AQ15 (Michalski et al., 1986), by removing conditions and rules and by contracting and extending intervals and internal disjunctions. POSEIDON successively applies the operator that results in the highest coverage gain as long as the resulting theory improves according to some quality criterion. One of the best-known pruning techniques is reduced error pruning, which we will discuss in the next section.

```
function REDUCEDERRORPRUNING(R, H)
  Input:
    R: a set of previously learned rules
    H: a set of examples on which to optimize the performance
  Algorithm:
    R̂ := R
    v̂ := EVALUATE(R̂, H)
    // loop until best pruned theory decreases quality on pruning set
    repeat
        // remember best rule set from previous iteration
        R := R̂
        v := v̂
        // compute all possible applications of a pruning operator
        T := APPLYPRUNINGOPERATORS(R)
        // evaluate them and find the best one
        for R' ∈ T do
            v' := EVALUATE(R', H)
            if v' > v̂
            then R̂ := R'
                 v̂ := v'
            endif
        endfor
    until v̂ < v
  Output:
    R the pruned rule set
```

Fig. 9.8 Reduced error pruning

9.3.1 Reduced Error Pruning

Reduced error pruning (REP) follows the basic layout of Fig. 9.7. It first learns a complete, overfitting theory, and then tries to identify and remove unnecessary conditions and rules from the rule set. Figure 9.8 shows the key step, namely the algorithm for pruning $\mathcal{R}$ on the pruning set $\mathcal{H}$. REDUCEDERRORPRUNING is meant to be called in place of PRUNERULEBASE in the POSTPRUNING algorithm of Fig. 9.7.

At the heart of this algorithm is the function APPLYPRUNINGOPERATORS, which applies several pruning operators to $\mathcal{R}$ and returns the resulting simplified theories in the set $\mathcal{T}$. Among them, the algorithm selects the one with the highest accuracy on the pruning set, which is then pruned in the next iteration. This process is repeated until the accuracy of the best pruned theory is below that of its predecessor.

APPLYPRUNINGOPERATORS may implement various pruning operators. The most commonly used operators are:

DELETEANYCONDITION: This operator considers each single condition of $\mathcal{R}$ for deletion. It is very expensive to use, because it needs to evaluate one new theory for each condition in $\mathcal{R}$.

DELETEWORSTCONDITION: This operator considers all conditions of $\mathcal{R}$ for deletion, and selects the condition that results in the best performance on the training data. In many cases, but not always, this will be the last condition added.

DELETELASTCONDITION: This operator only tries to delete the last condition of each rule in $\mathcal{R}$. Brunk and Pazzani (1991) argue that this more efficient operator is suitable for separate-and-conquer rule learning algorithms, because the order in which the conditions of a rule are considered for pruning should be inverse to the order in which they have been learned.

DELETEFINALSEQUENCE: This operator, suggested by Cohen (1993), considers all theories that result from deleting a sequence of conditions of arbitrary length from the end of each rule in $\mathcal{R}$. Each iteration is thus equally expensive as with the DELETEANYCONDITION operator, but the algorithm may converge faster, because this operator can prune several conditions in the same iteration.

DELETERULE: This simple operator examines all theories that result from deleting one rule from $\mathcal{R}$. As it specializes the theory, it should only be used in connection with one of the three other operators that generalize the theory. It is also necessary to remove rules that have become redundant because of the generalization of other rules.

Reduced error pruning in its original formulation for rule learning (Brunk & Pazzani, 1991) employs DELETELASTCONDITION and DELETERULE, so that in each iteration, exactly two new theories per rule have to be evaluated.

Some algorithms also employ operators that modify conditions instead of removing them entirely. Examples are finding the best replacement for a condition (Weiss & Indurkhya, 1991), or extending and contracting internal disjunctions and intervals (Bergadano et al., 1992).

9.3.2 *Problems with Reduced Error Pruning*

Empirically, REP has been shown to perform quite well in terms of predictive accuracy. For example, Brunk and Pazzani (1991) observed that REP learns more accurate theories than FOIL's prepruning approach in an artificial domain for a variety of different noise levels. However, this straightforward adaptation of REP to rule learning is not entirely unproblematic, as we will discuss in the following.

Complexity. As already discussed in the previous section, reduced error pruning is very expensive because it has to evaluate one theory for each literal in the original, overfitting theory $\mathcal{R}$. Moreover, this has to be repeated iteratively, in the worst case until all literals have been deleted. In fact, it has been shown that, in the scenario assumed in Theorem 9.1.1, REP's complexity is at least E^4 for noisy data (Cameron-Jones, 1996; Cohen, 1993).

Theorem 9.3.1 (Complexity of REP). *The worst-case complexity of the pruning phase of reduced error pruning is* $\Omega(E^4)$.

Proof. From Theorem 9.1.1, we know that the overfitting theory $\mathcal{R}$ will have $\Theta(\log E)$ rules with $\Theta(\log E)$ conditions each. Let us assume we use the two cheapest operators DELETELASTCONDITION and DELETERULE, as suggested in Brunk and Pazzani (1991). Thus, in each step of the pruning phase, each of the $\Theta(E)$ rules can be simplified in two ways, resulting in $\Theta(E)$ simplifications. Each simplification has to be tested on the pruning set in order to select the one with the highest accuracy. For the $\Theta(E)$ examples of the pruning set that will be classified as negative by a theory, at least the first conditions of all $\Theta(E)$ rules have to be evaluated. So, the complexity of testing a theory is $\Omega(E^2)$, and the total complexity of testing one iteration is $\Omega(E^3)$.

If we further assume that REP works as intended, it should prune the overfitting theory to the correct theory, whose size should be independent of the size of training set (provided that we have sufficiently many examples for identifying the correct theory). Therefore, REP must at least remove all but a constant number of the $\Theta(E)$ rules of the overfitting theory, i.e., it has to loop $\Omega(E)$ times. When it frequently prunes single conditions there may be considerably more iterations. Thus, we get a total cost of $\Omega(E^4)$.[2] □

With similar arguments, Cohen (1993) has derived a complexity bound of $\Omega(E^2 \log E)$ for the initial growing phase. This lower bound is tight, because each of the $\Theta(E \cdot \log E)$ conditions in the overfitting theory has been tested on at most $O(E)$ examples of the growing set. Thus, Theorem 9.3.1 shows that the costs of pruning will outweigh the costs of generating the initial overfitting set of rules $\mathcal{R}$, which are in turn already an upper bound for the costs of using a prepruning algorithm (because in the worst case, the prepruning algorithm will not make any pruning decision and learn the same theory $\mathcal{R}$).

Bottom-up hill-climbing. REP employs a greedy hill-climbing strategy. Conditions and rules will be deleted from the concept definition so that predictive accuracy on the pruning set is greedily maximized. When each possible operator leads to a decrease in predictive accuracy, the search process stops at this local maximum. This is essentially the greedy bottom-up learning algorithm discussed in Sect. 6.4.2, where we also noted that it has a fundamental problem in some cases, when deleting a condition does not increase its coverage. This problem is not as bad if we restrict ourselves to the DELETELASTCONDITION operator, because then it can be argued that it actually makes sense to delete the final condition of a rule that was only added to exclude some negative examples on the training set, but does not seem to be necessary on the test set.

---

[2]One may argue that the number of simplifications and the testing costs decrease when the theories get simpler, so that the total cost will be lower. However, if we assume that at least half of the rules (or any fixed percentage of the rules) have to be deleted, we still have $\Theta(E)$ simplifications for $\Omega(E)$ iterations.

However, we already noted that in noisy domains the theory $\mathcal{R}$ that has been generated in the growing phase will be very complex (see, e.g., Fig. 9.1). In order to recover the target theory, REP has to prune a large portion of $\mathcal{R}$ and has ample opportunity to err on its way. Therefore, we can also expect REP's complex-to-simple search to be not only slow, but also inaccurate on noisy data.

Rule overlap in decision lists. On the other hand, pruning a condition from a rule means that the rule is generalized, i.e., it will cover more positive and negative instances. In a decision list, the first rule that fires classifies the example: if this rule is generalized, it will classify more examples. Thus, some examples that have been previously classified by subsequent rules are now classified by this generalized rule. Consequently, pruning a decision list may lead to unexpected behavior when using a decision list for classification.

Thus, the same rule **r** that was chosen by the covering strategy on the training data may suddenly look very bad, because the set of examples that remain uncovered by the unpruned rules at the beginning of a theory may yield a different evaluation of candidate conditions for subsequent rules than the set of examples that remain uncovered by the pruned versions of these rules. Thus, if the covering algorithm had known to which extent **r** would eventually be pruned, it might have learned a different set of subsequent rules. However, a wrong choice of a condition during growing cannot be undone by pruning. This is a key motivation for the incremental reduced error pruning algorithm that we will discuss in Sect. 9.4.1, as well as for algorithms for optimizing rule sets, which we will discuss in Sect. 9.5.1.

9.3.3 Reduced Error Regrowth

To solve some of the problems mentioned in the previous section, in particular efficiency, Cohen (1993) has proposed *reduced error regrowth*, a top-down post-pruning algorithm based on a technique used by Pagallo and Haussler (1990). Like REP, the GROW algorithm shown in Fig. 9.9 first learns a rule set $\mathcal{R}$ that overfits the data. But instead of pruning this intermediate theory bottom-up until any further deletion results in a decrease of accuracy, it uses it to *grow* a pruned theory.

For this purpose, GROW first forms the set $\mathcal{R}'$ which contains generalizations of all rules of $\mathcal{R}$. Cohen (1993) experimented with different pruning operators for producing these generalizations. The first one was the DELETEFINALSEQUENCE operator, which deletes each possible final sequence of conditions from a rule. Better performance was achieved via a repeated application of the DELETEWORSTCONDITION operator (cf. Sect. 9.3.1).

After the set $\mathcal{R}'$ has been formed, rules are greedily selected from this expanded set and added to an initially empty theory $\hat{\mathcal{R}}$. This is repeated as long as the error on the *growing* set decreases. When no further rule improves the predictive accuracy of $\hat{\mathcal{R}}$ on the pruning set, GROW stops.

```
function GROW(R, H)
  Input:
    R: a set of previously learned rules
    H: a set of examples on which to optimize the performance

  Algorithm:
    // form a set of all refinements of the rules in R
    R' = R
    for r ∈ R do
        R' := R' ∪ APPLYPRUNINGOPERATORS(r)
    endfor

    R̂ = ∅
    v̂ := EVALUATE(R̂, H)

    // loop until best theory does not improve quality on pruning set
    repeat
        // find the best simplification of the rule set
        for r ∈ R' do
            v' := EVALUATE(R̂ ∪ r, H)
            if v' > v̂
            then R̂ := R̂ ∪ r
                 v̂ := v'
            endif
        endfor
    endloop

  Output:
    R the pruned rule set
```

Fig. 9.9 Reduced error regrowth

Thus, GROW improves upon REP by replacing the bottom-up hill-climbing search of REP with a top-down approach. Instead of removing the most useless rule or condition from the overfitting theory, it adds the most promising generalization of a rule to an initially empty theory. As a result, there is no need to apply the DELETERULE operator, which is necessary in REDUCEDERRORPRUNING.

It has been experimentally confirmed that this results in a significant gain in efficiency, along with a slight gain in accuracy (Cohen, 1993). However, Cameron-Jones (1996) has demonstrated that the asymptotic time complexity of the GROW post-pruning method is still above the complexity of the initial rule growing phase. The explanation for the speed-up that can be gained with the top-down strategy is that it starts from the empty theory, which in many noisy domains is much closer to the final theory than the overfitting theory. Nevertheless, it still suffers from the inefficiency caused by the need of generating an overly specific theory in a first pass.

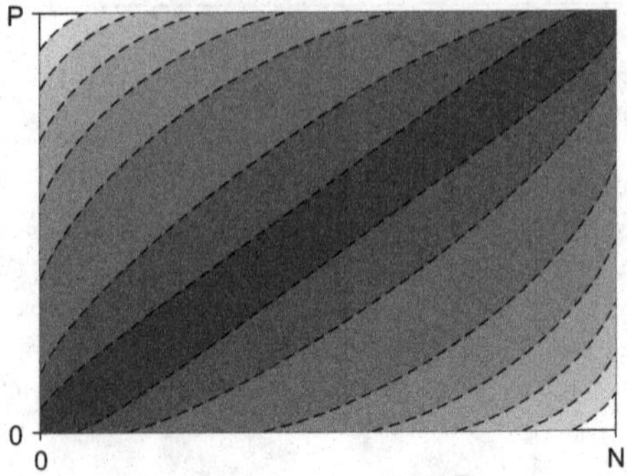

Fig. 9.10 Isometrics of MDL-based pruning using *SimpleMDL* for a constant rule length L_r

9.3.4 MDL-Based Pruning

All of the above approaches have in common that they use a separate pruning set for evaluating the pruned theories in order to get an estimate that is independent of the training phase. This ensures that rules and conditions that only fit the growing set but do not generalize are removed. An alternative approach is to explicitly try to trade off the length of a theory with its accuracy of fit. This is the key idea of all approaches based on the minimum description length principle. We have already discussed FOIL's MDL-based prepruning criterion in Sect. 9.2.3, where the key idea is to compare the description length of a theory to the description length of the examples that are covered by the theory.

An alternative approach is to use a computable approximation of the description length *MDL*, which we already mentioned in Sect. 7.6, to guide the pruning process. A particularly simple approximation of MDL for rule learning is due to Pfahringer (1995a):

$$\boxed{\mathit{SimpleMDL}(\mathbf{r}) = \log_2 \binom{F}{L_r} + \log_2 \binom{\hat{E}}{\hat{N}} + \log_2 \binom{\bar{E}}{\bar{P}}}$$

The first term is the encoding length of the theory $ERL(\mathbf{r})$, which we have already seen in Sect. 9.2.3. The second and third terms calculate the encoding lengths for the false positives $\hat{N}$ and false negatives $\bar{P}$, respectively. The former have to be identified among all covered examples $\hat{E}$; the latter have to be identified among all not-covered examples $\bar{E}$. The idea of encoding the error costs in this way goes back to Quinlan (1993), who later suggested several improvements (Quinlan, 1994, 1995).

Figure 9.10 shows the isometrics of *SimpleMDL* for a constant rule length L_r. It can be seen that their shape is quite similar to those of *Correlation* (Fig. 9.6). However, this figure does not capture the trade-off between the theory costs and the exception costs.

The use of MDL in pruning algorithms like REP or GROW is straightforward. In addition to using a heuristic like *SimpleMDL* in place of the *Evaluate* routines in these algorithms, one only has to modify the general POSTPRUNING algorithm (Fig. 9.7) so that it does not split the examples, but calls the respective pruning method with the learned set of rules $\mathcal{R}$ and the original set of examples $\mathcal{E}$. The key advantage of this approach is, of course, that all examples in the training set can be used for both learning and pruning.

9.4 Integrating Pre- and Post-pruning

Experimental results show complementary advantages of the two basic pruning strategies: While prepruning tends to be fast because it addresses pruning during the learning phase, post-pruning appears to be more accurate, but also too expensive. In particular, the intermediate theory resulting from the initial overfitting phase can be much more complex than the final theory. Post-pruning is very inefficient in this case, because most of the work performed in the learning phase has to be undone in the pruning phase. The algorithms discussed in this section try to bring these complementary advantages together.

9.4.1 Incremental Reduced Error Pruning

Incremental reduced error pruning (I-REP) (Fürnkranz & Widmer, 1994) addresses this problem by pruning each individual rule right after it has been learned. Figure 9.11 shows pseudocode for this algorithm. The algorithm is very similar to REP (Fig. 9.8), but, unlike REP, I-REP will not be called from the generic POSTPRUNING algorithm (Fig. 9.7), but will be called as a replacement for FINDBESTRULE, in the simplest case from the standard COVERING algorithm (Fig. 8.2). Thus, just as in prepruning, noise handling is directly integrated into the rule learning phase.

As in REP, the current set of training examples $\mathcal{E}$ is split into a growing set $\mathcal{G}$ and a pruning set $\mathcal{H}$. However, this does not happen at the outer level, but happens inside the subroutine for finding the best rule. As a consequence, each rule is learned from a different split of the data. A rule **r** is learned by calling FINDBESTRULE on $\mathcal{G}$. Subsequently, conditions are deleted from this rule in a greedy fashion until any further deletion would decrease the evaluation of this rule on the pruning set. Again, we can use all rule pruning operators discussed in Sect. 9.3.1. The original version of I-REP evaluates rules using *Accuracy*, and employs the pruning operator DELETEANYCONDITION, which is too expensive to be used in REP, but can be afforded in I-REP (Fürnkranz & Widmer, 1994).

The best rule found by repeatedly pruning the original rule is returned by I-REP. The COVERING algorithm adds the found rule to the concept description

---

function I-REP($\mathcal{E}, s$)

Input:
$\mathcal{E}$: a set of training examples
s: a ratio for splitting the training set

Algorithm:

```
// split examples into a growing set G and a pruning set H
(G, H) := SPLITEXAMPLES(E, s)

// learn a single rule on G and evaluate it on H
r̂ := FINDBESTRULE(G)
v̂ := EVALUATERULE(r̂, H)

// loop until the best rule decreases quality on pruning set
repeat
    // remember best rule from previous iteration
    r := r̂
    v := v̂

    // compute all possible applications of a pruning operator to the rule r
    T := APPLYPRUNINGOPERATORS(r)

    // evaluate them and find the best one
    for r' ∈ T do
        v' := EVALUATERULE(r', H)
        if v' > v̂
        then r̂ := r'
             v̂ := v'
        endif
    endfor
until v̂ < v
```

Output:
r: the best pruned rule for the given set of examples

Fig. 9.11 Reduced error pruning integrated into FINDBESTRULE

and removes all covered positive and negative examples from the training set $\mathcal{E}$. The next rule is then learned with another call to I-REP, which redistributes the remaining examples into a new growing and a new pruning set. This redistribution of the growing and pruning set after each learned rule ensures that each of the two sets contains the predefined percentage of the remaining examples.

The covering loop can be stopped using any theory stopping criterion. The original version of I-REP suggested to stop when the evaluation of the pruned rule is below the evaluation of the empty rule $\mathbf{r}_\perp$. In this case, the evaluation of the pruned rules on the pruning set also serves as a theory stopping criterion.

Boström (2004) proposed a version of the I-REP algorithm that aims at optimizing rule sets instead of decision lists.

9.4.2 *Complexity of* I-REP

I-REP addresses the problems we discussed in Sect. 9.3.2. First, it can be shown to be considerably more efficient than reduced error pruning or the GROW algorithm.

Theorem 9.4.1 (Complexity of I-REP). *The worst-case complexity of* I-REP *using* DELETEANYCONDITION *as a pruning operator is* $O(E \cdot \log^2 E)$.

Proof. We analyze the same scenario as in Theorems 9.1.1 and 9.3.1. Recall that the cost of growing one rule in REP is $\Theta(E \cdot \log E)$ because $\Theta(\log E)$ conditions are tested against $\Theta(E)$ examples. I-REP considers *every* condition in the rule for pruning. Therefore, each of the $\Theta(\log E)$ conditions has to be evaluated on the $\Theta(E)$ examples in the pruning set. This is repeated until the final rule has been found, which, in the worst case, is after all conditions have been pruned, i.e., at most $O(\log E)$ times. Thus, the cost of pruning one rule is $O(E \cdot \log^2 E)$. Assuming that I-REP stops when the correct theory of constant size has been found, the overall cost is also $O(E \cdot \log^2 E)$. □

This is significantly lower than the cost of growing an overfitting theory, which has been shown to be $\Omega(E^2 \cdot \log E)$ under the same assumptions (Cohen, 1993), and thus also significantly lower than the cost of reduced error pruning and the cost of the GROW algorithm, which are both higher than the initial rule growing phase (Cameron-Jones, 1996). Note that the original version of I-REP used the most expensive pruning operator, the use of DELETELASTCONDITION would further decrease the complexity to $O(E \cdot \log E)$, but may result in a worse predictive performance.

As a result, I-REP has been shown to be considerably more efficient than post-pruning without losing its accuracy. For example, Fig. 9.12 shows the results for the accuracy and the run-time of various algorithms on an artificial task where 10 % of the examples had their signs inverted in order to simulate noise. Shown are the results of the two prepruning algorithms FOIL and FOSSIL, the two post-pruning algorithms REP and GROW, and I-REP. A basic learning algorithm without using any pruning, is also shown for comparison.

It can be seen that I-REP shares its efficiency with the prepruning algorithms, which are all significantly faster (hardly visible near the bottom of the graph) than the post-pruning algorithms. They are even faster (and certainly much more accurate) than no pruning, because they avoid learning the overfitting theory that is the starting point for the two post-pruning algorithms. Conversely, I-REP is significantly more accurate than the prepruning algorithms, and even appears to have slight advantages over the post-pruning algorithms. Thus, it effectively combines the advantages of both basic pruning algorithms.

The key factor responsible for the good performance is that I-REP, like GROW, uses a top-down approach instead of REP's bottom-up search: the final theory is not found by removing unnecessary rules and conditions from an overly complex theory, but by repeatedly adding rules to an initially empty theory. However, while GROW first generates an overfitting theory and thereafter selects the best generalized

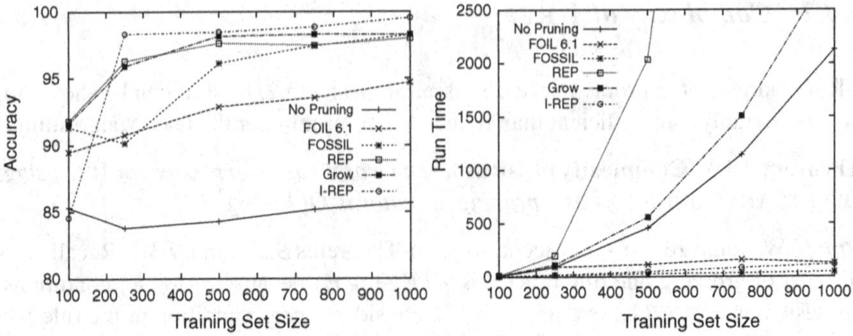

Fig. 9.12 Accuracy and run-time of different pruning algorithms in the KRK domain (10 % noise), for different training set sizes (Fürnkranz, 1997)

rules from this theory, I-REP selects the best generalization of a rule right after the rule has been learned.

Finally, we also note that the problem with overlapping rules (cf. Sect. 9.3.2) is solved by I-REP, because I-REP learns the rules in the order in which they will be used in the decision list. Each rule is completed—learned *and* pruned—before subsequent rules are learned. Therefore, subsequent rules are learned exactly in the same context in which they will be used, and not, as is the case for REP and GROW, in the context of overly specific earlier rules.

9.4.3 Pruning in **RIPPER**

I-REP has been shown to outperform various other pruning algorithms in a variety of noisy domains, in particular in terms of efficiency (Fürnkranz, 1997; Fürnkranz & Widmer, 1994). However, it has several weaknesses, which have been addressed in subsequent work on the RIPPER rule learning algorithm (Cohen, 1995).

First, Cohen (1995) has shown that I-REP's use of the *Accuracy* heuristic for evaluating rules results in a high variance for low-coverage rules. Therefore I-REP is likely to stop prematurely and to overgeneralize in domains that are susceptible to the *small disjuncts problem*, i.e., to the problem that rules with low coverage are responsible for a high percentage of the error of a theory, but that ignoring these rules will also not be helpful (Holte, Acker, & Porter, 1989).

Second, Cohen (1995) has argued that the use of *Accuracy* as pruning heuristic, which is equivalent to *CovDiff* (Theorem 7.3.1), can lead to undesirable choices. For example, *Accuracy* would prefer a rule that covers 2,000 positive and 1,000 negative instances over a rule that covers 1,000 positive and only 1 negative instance. As an alternative, Cohen suggests the ratio between the difference and the sum of covered positive and negative examples, which is essentially equivalent to *Precision* (cf. Theorem 7.3.3). While we have argued that optimizing precision by specializing rules on the training set is bound to lead to overfitting, optimizing precision on an independent validation set by generalizing rules is an entirely different story. Cohen's experimental evidence indicates that this choice leads to significantly better results.

Finally, Cohen (1995) used an alternative rule stopping criterion based on minimal description length that was originally proposed by Quinlan (1995). The key idea is to compute a description length for both the theory and their exceptions (i.e., the misclassified examples), and to greedily search for a rule set that minimizes the sum of the two. It also has a parameter `-S` that allows one to multiply the theory cost with a user-provided factor, thereby increasing or decreasing the estimated theory costs, which leads to heavier or less pruning, respectively (an example can be seen in Fig. 12.1).

Maybe the most significant addition was a novel optimization technique that allows the algorithm to relearn rules in the context of the entire theory. We will discuss this in more detail in Sect. 9.5. A brief summary of RIPPER can be found in Sect. 2.10.5.

9.4.4 Other Combinations of Pre- and Post-pruning

Other pruning algorithms that try to combine the advantages of pre- and post-pruning have been investigated in the literature. For example, a straightforward combination of a prepruning algorithm and a subsequent post-pruning phase has been explored by Cohen (1993), who proposed combining the efficient post-pruning algorithm GROW (see Sect. 9.3.3) with some weak prepruning heuristics that speed up the learning phase. Similar ideas have been tried in the BEXA algorithm (Theron & Cloete, 1996). BEXA uses significance testing as a prepruning criterion and performs an additional post-pruning phase where conditions and rules are pruned in the way described in Quinlan (1987a). The key challenge with these approaches is that the initial prepruning phase should be aggressive enough in order to reduce the amount of overfitting and thus to make the subsequent post-pruning phase more efficient, but also conservative enough to still profit from the flexibility of post-pruning.

Fürnkranz (1994b, 1997) proposed an automated approach to identify the right level of prepruning in such a setting. The *top-down pruning* (TDP) algorithm generates all theories that can be learned with different settings of the cutoff parameter of FOSSIL's cutoff stopping criterion. This series of theories is generated in a top-down fashion. The most complex theory within one standard error of the most accurate theory is selected as a starting point for the post-pruning phase.[3]

---

[3]This method is inspired by the approach taken in the decision tree learner CART (Breiman et al., 1984) where the most general decision tree within this standard error margin is selected as a final theory.

The hope is that this theory will not be an over-simplification (it is more complex than the most accurate theory found so far), but will also be close to the intended theory (its accuracy is still close to the best so far). Thus only a limited amount of post-pruning has to be performed. TDP's implementation made use of several optimizations, so that finding this theory is often cheaper than completely fitting the noise.

Just as I-REP avoids learning an overfitting theory by pruning individual rules instead of entire theories, Fürnkranz (1995) investigated a way of taking this further by pruning directly during the selection of the conditions in order to avoid learning overfitting rules. The resulting algorithm, I^2-REP, splits the training set into two sets of equal size, selects the best condition for each of them, and chooses the one that has the higher accuracy on the entire training set. This procedure is quite similar to a twofold cross-validation, which has been shown to give reliable results for ranking classifiers at low training set sizes (Weiss & Indurkhya, 1994). Conditions are added to the rule until the best choice does not improve the accuracy of the rule, and rules are added as long as this improves the accuracy of the theory. I^2-REP seems to be a little more stable than I-REP with small training sets, but no significant run-time improvements can be observed. It even appears to be a little slower than I-REP, although asymptotically both algorithms exhibit approximately the same subquadratic behavior.

I-REP and TDP have been deliberately designed to closely resemble the basic post-pruning algorithm, REP. In particular, the choice of *Accuracy* as an evaluation metric for pruned rules was mainly motivated by the desire to allow a fair comparison to REP that concentrates on the methodological differences between post-pruning and I-REP's integration of pre- and post-pruning. An important advantage of post-pruning methods is that the way of evaluating theories (or rules in I-REP's case) is entirely independent from the basic learning algorithm. Other pruning and stopping criteria can further improve the performance and eliminate weaknesses. We have already mentioned that RIPPER was able to improve upon the results of I-REP by changing the evaluation heuristic from accuracy to precision. However, an exhaustive experimental study of different evaluation heuristics cannot be found in the literature. For example, in light of our discussion in Chap. 7, it would be interesting to see whether versions of the m-estimate would be able to outperform their basic constituents, accuracy and precision.

9.5 Optimizing Rule Sets

In addition to post-pruning, there are several alternative methods that can be used in place of POSTPROCESSRULES for optimizing a learned set of rules in the COVERING algorithm (Fig. 8.2). Popular methods are to relearn, to reorder, or to reweight (some of) the rules in the rule set.

9.5.1 Relearning Rules in Context

One of the techniques that contributes to RIPPER's excellent performance is a rule optimization phase that aims at improving the learned rule set $\mathcal{R} = \{\mathbf{r}_1 \ldots \mathbf{r}_R\}$ by relearning some of the rules. The key idea behind this approach is that while rules are generally learned in the context of previously learned rules (because the examples that are covered by these rules are removed), they do not see subsequently learned rules. Thus, it might be a good idea to relearn individual rules once a complete theory has been found.

In particular, Cohen (1995) suggests to consider each rule $\mathbf{r}_i$ in turn. For each rule $\mathbf{r}_i$, three different variants are computed:

1. The original rule $\mathbf{r}_i$
2. The replacement rule $\dot{\mathbf{r}}_i$, which is formed by relearning $\mathbf{r}_i$
3. The revised rule $\mathbf{r}'_i$, which is obtained by further specialization and subsequent pruning of $\mathbf{r}_i$

Steps two and three learn the new rule on the set of examples that are uniquely covered by $\mathbf{r}_i$, which is split into a growing and a pruning set. Relearning and specialization are performed on the growing set, and additional pruning on the pruning set.

These three candidate rules are then added into the set $\mathcal{R}$ in place of the original rule $\mathbf{r}_i$. The three resulting theories are then compared with an MDL measure and the best-performing variant is maintained. This is repeated for all rules $\mathbf{r}_i$ in order, and the entire process is iterated a fixed number of times (the default is 2).

9.5.2 Reordering Rule Sets

If we are learning a decision list, the order of the rules within the list is crucial because the first rule that matches a given example will make the prediction. In particular for multiclass problems, the order of the rules may be crucial. We will return to this issue in Chap. 10. In this section, we will focus on work that considers to reorder rules within a single class.

We have seen in Sect. 3.4 that learning a set of rules draws a coverage path through coverage space starting at the point $\mathcal{R}^{\perp} = (0,0)$ (the theory covering no examples) and ending at $\mathcal{R}^{\top} = (N, P)$ (the theory covering all examples). Each intermediate rule set $\mathcal{R}_i$ is a potential classifier, and one of them has to be selected as the final classifier. We also discussed in Sect. 8.2.3 that *Precision* always picks the steepest continuation, thereby implicitly optimizing the area under the ROC curve by making the coverage path as convex as possible.

Nevertheless, the coverage path will frequently not be convex. The left graph of Fig. 9.13 shows a case where the rule set $\mathcal{R}_2$ consisting of rules $\{\mathbf{r}_1, \mathbf{r}_2\}$ is not on the convex hull of the classifiers. Thus, only $\mathcal{R}_1 = \{\mathbf{r}_1\}$ and $\mathcal{R}_3 = \{\mathbf{r}_1, \mathbf{r}_2, \mathbf{r}_3\}$ are potential candidates for the final classifier.

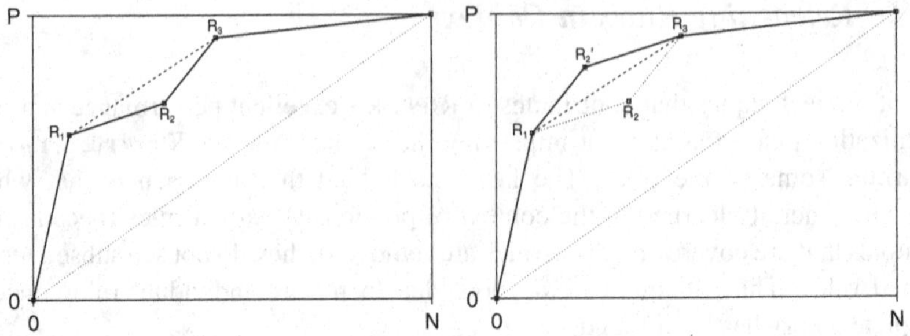

Fig. 9.13 A concavity in the path of a rule learner (*left*), and the result of a successful fix by swapping the order of the second and third rule (*right*)

Interestingly, in some cases it may be quite easy to obtain better classifiers by simply reordering the rules. The right graph of Fig. 9.13 shows the best result that can be obtained by swapping the order of rules $\mathbf{r}_2$ and $\mathbf{r}_3$, resulting in the classifier $\mathcal{R}_2' = \{\mathbf{r}_1, \mathbf{r}_3\}$. Note, however, that this graph is drawn under the assumption that rules $\mathbf{r}_2$ and $\mathbf{r}_3$ cover disjoint sets of examples, in which case the quadrangle $\mathcal{R}_1 - \mathcal{R}_2 - \mathcal{R}_3 - \mathcal{R}_2'$ is a parallelogram. In general, however, $\mathbf{r}_3$ may cover some of the examples that were previously already covered by $\mathbf{r}_2$. This may change the shape of the quadrangle considerably, depending on whether the majority of the examples in the overlap are positive or negative examples. Thus, it is not guaranteed that swapping the rules will indeed increase the area under the ROC curve or make the curve convex.

Flach and Wu (2005) suggested a practical approach in a somewhat more general context, where individual predictions can be reordered. However, in the special case of rule learning, this essentially corresponds to reordering rules. The key idea is to first identify concavities in the ROC curves (such as the one in Fig. 9.13), and then compare a reordering (in our case $\mathcal{R}_2'$) to the original ordering ($\mathcal{R}_2$) on a validation set. Experiments with a decision tree learner and a naïve Bayes classifier confirmed both the utility of the method and the need for a separate validation set in order to make decisions that generalized well.

Related to this approach is also the technique suggested by Ferri, Flach, and Hernández (2002). Their key idea, again adapted to the context of rule learning, is to sort relabel decision trees in the following way. First, all k leaves (or branches) are sorted in decreasing order according to *Precision*. Then, a valid labeling is obtained by splitting the ordering in two, and labeling all leaves before the split point positive and the rest negative. This results in $k + 1$ labelings, and the corresponding ROC curve is necessarily convex. An optimal point can be chosen in the usual way, once the cost model is known. In our context, ordering branches corresponds to ordering rules, and relabeling corresponds to picking the right subset of the learned rules: labeling a rule as positive corresponds to including the rule in the theory, while labeling a rule as negative effectively deletes the rule because it will be subsumed by the final default rule that always predicts the negative class. However, the chosen subset is only guaranteed to be optimal if the rules are mutually exclusive.

9.5.3 Reweighting Rules

Many algorithms learn weights for their rules, which in some way indicate the importance of a rule. Such rule weights can be used for several purposes, such as estimating the reliability of a prediction, or combining the predictions of multiple rules. We will return to this issue in Sect. 10.3. In this section, we briefly discuss different ways for obtaining or correcting rule weights.

A simple approach is to weight the rule with an estimate for the probability Pr(H|B) that the head of the rule is true given that the body of the rule is true, yielding a *probabilistic rule*. A simple estimate for this probability is the relative frequency of the head in all covered examples. For binary data, this would essentially amount to computing the precision heuristic. As this will typically overestimate the true probability because of overfitting, it is more customary to use Laplace-corrected or m-estimates instead.

In machine learning, this problem of correcting overly optimistic probability estimates is also known as *calibrating probabilities* (Niculescu-Mizil & Caruana, 2005b; Rüping, 2006; Zadrozny & Elkan, 2001). The best-known approaches are *isotonic regression* (Fawcett & Niculescu-Mizil, 2007; Zadrozny & Elkan, 2002) and *Platt scaling* (Lin, Lin, & Weng, 2007; Platt, 1999). The problem has, e.g., been studied for decision-tree learning (Hüllermeier & Vanderlooy, 2009; Provost & Domingos, 2003), boosting (Niculescu-Mizil & Caruana, 2005a), and support vector machines (Platt, 1999), but has received considerably less attention in inductive rule learning.

Shrinkage is a general framework for smoothing probabilities, which has been successfully applied in various research areas. It is, e.g., regularly used in statistical language processing (Chen & Goodman, 1998; Manning & Schütze, 1999). Its key idea is to "shrink" probability estimates of rules towards the estimates of some of their generalizations. On the one hand, these estimates suffer less from variance because the generalizations cover more examples, but on the other hand they are, of course, also biased. The Laplace- and m-estimates may be interpreted as shrinking the empirical probability estimate towards the prior probabilities of the data. The main differences are that the general case combines multiple probabilities, and that the weights for the combinations are learned from the data whereas they are fixed for the Laplace correction, and are user-specified for the m-estimate. While this method has worked well for decision trees (Wang & Zhang, 2006), a straightforward adaptation to rule learning was not able to outperform its simpler competitors (Sulzmann & Fürnkranz, 2009).

Another line of research is concerned with learning weights that minimize a given loss function on the data. Prominent examples include boosting-based rule learning approaches such as RULEFIT (Friedman & Popescu, 2008) or REGENDER (Dembczyński, Kotłowski, & Słowiński, 2008). A particularly interesting approach is the RUMBLE algorithm (Rückert & Kramer, 2008), which optimizes the weights for a given set of rules using a margin-based approach.

9.6 Conclusion

In this chapter, we have discussed different pruning techniques for rule learning algorithms. Prepruning heuristics do not seem to be as well explored as, for example, the rule evaluation metrics that we have studied in Chap. 7. There are only a few commonly used criteria, and we have seen that they have some interesting deficiencies. Overall, we believe that we are still far from a systematic understanding of prepruning heuristics. The fact that, unlike decision-tree algorithms, most state-of-the-art rule learning algorithms use variants of post-pruning for noise-handling may not necessarily be a strong indicator for the superiority of this approach, but may also be interpreted as an indicator of the inadequacy of currently used stopping and filtering criteria.

Post-pruning, on the other hand, is fairly well understood. Its simple approach of first ignoring the problem entirely and learning an overfitting theory, which is afterwards refined in a separate pruning phase, is well-tested in decision-tree learning and also seems to be quite accurate in rule learning. However, the approach is quite inefficient, and we also pointed out that post-pruning should not be performed on entire theories, but instead on individual rules.

Incremental reduced error pruning realizes this by integrating pre- and post-pruning into one algorithm. Instead of post-pruning entire theories, each rule is pruned right after it has been learned. Thus, on the level of individual rules, it works like post-pruning, but on the level of theories it may be considered as prepruning. As real-world databases are typically large and noisy, they require learning algorithms that are both efficient and noise-tolerant. I-REP seems to be an appropriate choice for this purpose, and variants of it are commonly used in state-of-the-art rule learning algorithms such as RIPPER.

Finally, we have also discussed approaches for optimizing rules and rule sets in a post-processing phase. Relearning individual rules in the context of the entire theory seems to be an adequate technique for ensuring that rules are optimized to the particular subset of examples that they uniquely cover. Reordering rule sets may be helpful for improving the ranking quality, which in turn can lead to a better selection of a suitable subset of the learned rules. Finally, reweighting rules may lead to improved classification performance. This is particularly important for cases when the predictions of several rules have to be combined, as is the case for some of the techniques discussed in the next chapter.

Chapter 10
Beyond Concept Learning

So far, we have mostly assumed a concept learning framework, where the learner's task is to learn a rule set describing the target concept from a set of positive and negative examples for this concept. In this chapter, we discuss approaches that allow to extend this framework. We start with multiclass problems, which commonly occur in practice, and discuss the most popular methods for handling them: one-against-all classification and pairwise classification. We also discuss error-correcting output codes as a general framework for reducing multiclass problems to binary classification. As many prediction problems have complex, structured output variables, we also present label ranking and show how a generalization of pairwise classification can address this problem and related problems such as multilabel, hierarchical, and ordered classification. General ranking problems, in particular methods for optimizing the area under the ROC curve, are also addressed in this section. Finally, we briefly review rule learning approaches to regression and clustering.

10.1 Introduction

Previous chapters addressed concept learning tasks with known positive and negative classes where the learning goal is to find a high-quality description of the positive class which discriminates it from the examples of the negative class. Although concept learning might also be interpreted as a classification task in a domain with two classes, there are some important differences. The most important differences are that the classification task can have more than two classes, and that the decision-making process may have different overall goals for which the induction of a rule set is only a first step. To stress these differences, we call the set of rules induced as the result of the concept learning task the *rule set*, while the set of rules constructed for classification is called the *rule base*.

---

†Parts of this chapter are based on Fürnkranz (2002b), Park and Fürnkranz (2009) and Fürnkranz and Hüllermeier (2010b).

J. Fürnkranz et al., *Foundations of Rule Learning*, Cognitive Technologies, DOI 10.1007/978-3-540-75197-7_10, © Springer-Verlag Berlin Heidelberg 2012

A rule base may have various forms. It may consist of a simple set of rules (Fig. 2.4a), a single multiclass decision list (Fig. 2.4b), or, as we will see later in this chapter, of multiple rule sets whose predictions have to be aggregated at prediction time. The key difference between a rule set and a rule base is that the latter may be a structured collection of rules, and that it always has a concrete decision procedure associated with the rule base.

In order to clarify the differences between concept learning and classification, we will first discuss binary classification, i.e., classification for two classes (Sect. 10.2). We will there see several problems that may occur when predicting with rule bases, which are then discussed in Sect. 10.3. Thereafter, we will devote our attention to classification problems with more than two classes (Sect. 10.4). One way of addressing such problems is to directly learn a decision list that interleaves rules of different classes. Alternatively, we discuss various ways of addressing multiclass problems by reducing them to learning a collection of concept learning tasks.

We continue the chapter with an overview of recent work on preference learning and ranking via pairwise classification for solving prediction problems with structured output spaces (Sect. 10.5). The *learning by pairwise comparison* paradigm is the natural machine learning counterpart to the relational approach to preference modeling and decision making. From a machine learning point of view, it is especially appealing as it decomposes a possibly complex prediction problem into a number of concept learning problems. This methodology can be used, with minor adaptations, for addressing common machinelearning problems such as multilabel, ordered, and hierarchical classification, as well as multipartite ranking. We also briefly review work on learning rule sets that maximize the area under ROC curve.

Finally, in Sects. 10.6 and 10.7, we briefly discuss work on rule learning approaches to regression and clustering. Although they have not received as much attention in the rule learning literature, these problems appear quite frequently in real-world data, so that rule-based solutions are of great practical interest.

10.2 Binary Classification

The simplest classification tasks are *binary*, i.e., they involve only two classes. In many cases, such problems can be formulated as concept learning tasks. A good example is a medical domain in which one class is patients with some diagnosed disease and the other class is healthy persons. A natural choice is that ill persons represent the target (positive) class. There are two reasons for this choice. The first is that analysis of rules describing ill persons is much easier for medical experts than analysis of rules describing healthy persons in contrast to ill persons with some disease. The second reason is that induced rules for ill persons and expert reasoning based on them can be directly compared and combined with existing expert knowledge and rules describing other diseases.

In other cases, the choice is, in principle, arbitrary, but practical arguments will favor one or the other class. Consider, for instance, a binary classification problem

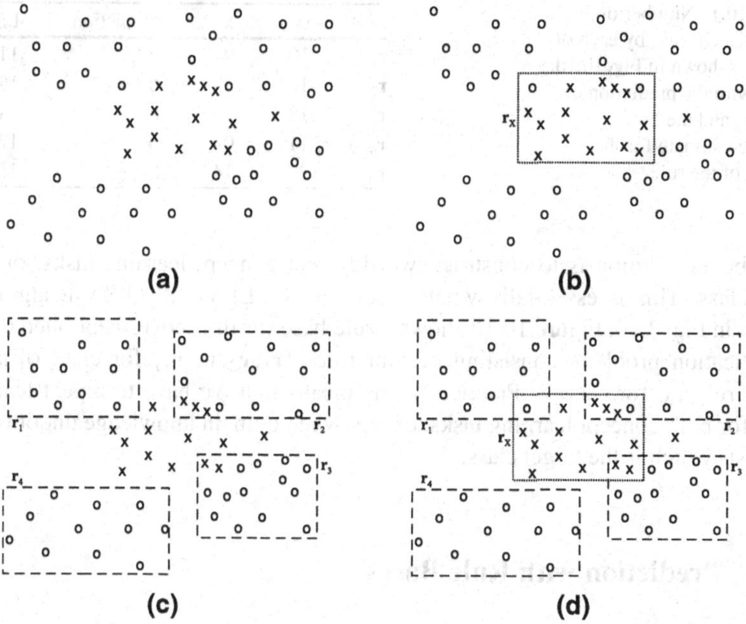

Fig. 10.1 Binary classification problem with several solutions. (**a**) Binary classification problem. (**b**) Learning a rule for class x. (**c**) Learning rules for class. o (**d**) Learning both classes

shown in Fig. 10.1a. If one chooses to learn rules for class x, the resulting rule set may consist of a single rule, as shown in Fig. 10.1b. If, however, one chooses to learn rules for class o first, the solution is not as simple. Figure 10.1c shows a possible solution with four rules. This example demonstrates that for the same domain there can be large differences in the complexity of induced rule sets, depending on the class. Thus, one approach could be to repeat concept learning tasks for differently defined positive classes and to select the result leading to the simpler and more comprehensive rule base, in this case, a single rule for class x.

Such a strategy, however, has the problem that the two classes are treated asymmetrically, because rules are always learned for one of the two classes, and the other class is treated as a *default class*. In many cases, such a choice is not obvious. As an example, consider the *congressional voting dataset* available in the UCI repository of machine learning databases (Frand & Asuncion, 2010). The learning task is to find rules that allow to classify members of the U.S. House of Representatives Congressmen into *democrats* and *republicans* based on their voting records. One could formulate this as a concept learning problem by arbitrarily picking democrats as the concept, learn a set of rules for recognizing democrats, and classify all other representatives as republican. Conversely, one could learn rules for republicans, and classify everybody else as democrats. However, proceeding in either way seems unfair because it treats the two parties in different ways, learning rules for one party and using the other as a default class.

Table 10.1 Number of examples covered by each of the rules shown in Fig. 10.1d, along with the prediction of the rule, and the Laplace-estimate for the quality of the rule

	o	x	Prediction	Laplace
$\mathbf{r}_1$	10	0	o	11/12
$\mathbf{r}_2$	12	3	o	13/17
$\mathbf{r}_3$	13	2	o	14/17
$\mathbf{r}_4$	11	0	o	12/13
$\mathbf{r}_X$	2	14	x	15/18

A better solution is to construct two different concept learning tasks, one for each class. This is essentially what is done in the LEARNRULEBASE algorithm shown in Fig. 2.10. Figure 10.1d shows a rule base for the above-mentioned binary classification problem, consisting of four rules $\{\mathbf{r}_1, \mathbf{r}_2, \mathbf{r}_3, \mathbf{r}_4\}$ for class o, and a single rule $\mathbf{r}_X$ for class x. Practically this means that we have to repeat learning rules for both concept learning tasks, except when domain knowledge undoubtedly suggests which is the target class.

10.3 Prediction with Rule Bases

We already noted in Sect. 2.8 that the learning of rules for multiple classes causes additional problems that we have to deal with during classification, most notably the problems of conflict resolution in the case of multiple, possibly contradicting rules that cover the same example, and the need for classification of uncovered examples. These will be discussed in this section.

10.3.1 Conflict Resolution

Rule bases with rule sets for different classes consist of many rules, and at classification time it often happens that more than one rule fires. A standard procedure for dealing with this problem is to assign a weight to each rule, as discussed in Sect. 9.5.3. Examples can then be classified using the rule with the highest weight. Essentially, this strategy corresponds to converting the rule set into a decision list, in which the rules are sorted by decreasing weight. This method is used in HYDRA (Ali & Pazzani, 1993) where the ls-content of a rule (see Sect. 7.3.6) is used as a weighting heuristic. The RIPPER rule learning algorithm has a mode for learning unordered rule sets, in which it uses the Laplace-value as a weight for making the final prediction.

For example, Table 10.1 shows the number of covered examples of classes o and x, the predicted class, and the Laplace value for each rule of the binary problem shown in Fig. 10.1d. The order of the rules according to the Laplace value is:

$$\mathbf{r}_4 > \mathbf{r}_1 > \mathbf{r}_X > \mathbf{r}_3 > \mathbf{r}_2$$

Thus, an example in the intersection of $\mathbf{r}_1$ and $\mathbf{r}_X$ will be predicted as o, whereas an example in the intersection of $\mathbf{r}_2$ and $\mathbf{r}_X$ will be predicted as x.

Early versions of C4.5RULES (Quinlan, 1987a) sort the rules by their *advantage*, i.e., by the number of examples that would be erroneously classified after deleting the rule. In later work, Quinlan (1993) replaced this strategy by a scheme that groups the rules according to the classes they predict and orders these groups using a heuristic based on the *minimum description length principle* (Rissanen, 1978). This increases the comprehensibility of the learned concepts. Pazzani et al. (1994) present an algorithm that orders a set of learned rules with respect to a minimization of the expected misclassification cost.

Of course, the weights of all rules covering an example can also be combined in more elaborate ways than simply using the rule with the maximum weight. In Sect. 2.8 we already introduced the weighting strategy of CN2. Essentially, it adds up the class distributions of each rule covering an example in order to get an overall class distribution, and then predicts the most frequent class. In this case, the examples in the intersection of $\mathbf{r}_1$ and $\mathbf{r}_X$ will be classified as x because the total coverage of each rule is $10+2 = 12$ examples of class o and $0+14 = 14$ examples of class x.

A more recent version of CN2 can give probabilistic classifications, i.e., it will predict class o with probability $12/26$, and class x with probability $14/26$. While such an approach makes clear that the prediction for x is not very confident, it also clarifies that this approach is not theoretically grounded, because it estimates the probability $\Pr(\mathsf{x} \mid \mathbf{r}_1 \wedge \mathbf{r}_X)$ that the class is x given that the example is covered by both $\mathbf{r}_1$ and $\mathbf{r}_X$ from the number of training examples covered by either of the two rules, where examples in the intersection are counted twice.

A sound approach is to use a Bayesian approach for estimating the probability $\Pr(\mathsf{x} \mid \mathcal{R})$ that the class is x given the rule set $\mathcal{R} = \{\mathbf{r}_1, \mathbf{r}_2, \mathbf{r}_3, \mathbf{r}_4, \mathbf{r}_X\}$.

$$\Pr(\mathsf{x} \mid \mathcal{R}) = \frac{\Pr(\mathcal{R} \mid \mathsf{x}) \cdot \Pr(\mathsf{x})}{\Pr(\mathcal{R})} = c \cdot \Pr(\mathsf{x}) \cdot \Pr(\mathcal{R} \mid \mathsf{x})$$

where $c = 1/\Pr(\mathcal{R})$ is a normalizing constant. By making the naïve Bayes assumption, we can estimate the above term by

$$c \cdot \Pr(\mathsf{x}) \cdot \Pr(\{\mathbf{r}_1, \ldots, \mathbf{r}_X\} \mid \mathsf{x}) = c \cdot \Pr(\mathsf{x}) \cdot \prod_{i \in \{1,\ldots,X\}} \Pr(\mathbf{r}_i \mid \mathsf{x})$$

where $\Pr(\mathbf{r}_i|\mathsf{x})$ is the probability that an example of class x is covered by rule $\mathbf{r}_i$. Consequently, $1 - \Pr(\mathbf{r}_i|\mathsf{x})$ is the probability that an example of class x is not covered by rule $\mathbf{r}_i$. If we estimate all probabilities with the Laplace correction (as in Table 10.1) we obtain $\Pr(\mathbf{r}_1|\mathsf{x}) = \frac{1}{17}$, $\Pr(\mathbf{r}_2|\mathsf{x}) = \frac{4}{17}$, $\Pr(\mathbf{r}_3|\mathsf{x}) = \frac{3}{17}$, $\Pr(\mathbf{r}_4|\mathsf{x}) = \frac{1}{17}$, $\Pr(\mathbf{r}_X|\mathsf{x}) = \frac{15}{17}$, and $\Pr(\mathsf{x}) = \frac{16}{63}$. For an example that is covered by rules $\mathbf{r}_1$ and $\mathbf{r}_X$, we thus obtain an estimate:

$$\Pr(\mathsf{x} \mid \mathcal{R}) = c \cdot \frac{1 \cdot 13 \cdot 14 \cdot 16 \cdot 2}{17^5} \cdot \frac{16}{63} = c \cdot 0.00104173.$$

Analogously we obtain:

$$\Pr(\mathsf{o} \mid \mathcal{R}) = c \cdot \frac{11 \cdot 25 \cdot 24 \cdot 26 \cdot 3}{48^5} \cdot \frac{47}{63} = c \cdot 0.00150727$$

Normalizing with $1/c = 0.00104173 + 0.00150727$ yields $\Pr(\mathsf{x}|\mathcal{R}) = 0.409$ and $\Pr(\mathsf{o}|\mathcal{R}) = 0.591$, i.e., the prediction for an example that is covered by rules $\mathbf{r}_1$ and $\mathbf{r}_X$ is class o.

Making the naïve Bayes assumption as above simplifies the estimation considerably, but the assumption is typically not justified. In the context of conflict resolution, the assumption means that whether an example of a given class is covered by a rule $\mathbf{r}_i$ is independent of it being covered by any of the other rules. This assumption clearly does not hold. For example, if an example is covered by rule $\mathbf{r}_X$ it is less likely that it is also covered by any of the other rules. One can try to solve this problem by using general Bayesian networks. Davis et al. (2004) investigate the use of the tree-augmented naïve Bayes (Friedman, Geiger, & Goldszmidt, 1997), a version that weakens this assumption by admitting one rule as a context. A later system tightly integrates a Bayesian conflict resolution strategy into the learning of the individual rules (Landwehr, Kersting, & De Raedt, 2007).

In principle, any classifier could be used for making a final prediction based on the prediction of individual rules. An interesting variant named *double induction* has been proposed by Lindgren and Boström (2004). They suggest to learn a separate classifier for resolving rule conflicts, which is trained on all training examples that are covered by multiple rules. This classifier is then employed whenever a new instance cannot be uniquely classified by the original rule set.

10.3.2 Classifying Uncovered Examples

As can be seen in Fig. 10.1d, the union of the five rules does not cover the entire example space. Even one of the training examples—the example classified as x between $\mathbf{r}_2$ and $\mathbf{r}_3$—cannot be classified by the learned rules. If such an example has to be classified, we have to come up with additional techniques.

The most commonly used technique is the use of a *default rule*, a rule that is employed whenever none of the regular rules cover the example. Essentially, such a rule has already been used in concept learning, where all examples that are not covered by the rules for the positive class are classified as negative. The key difference is that in binary classification, the default rule predicts a class which has previously also been covered by additional rules. Thus, these previous rules are, in principle, redundant, but are nevertheless learned for reasons sketched above.

Typically, the default rule predicts the largest class in the training examples, as shown in the LEARNRULEBASE algorithm of Fig. 2.10. However, other sensible choices are also possible. In particular, one may choose to predict the majority class among all examples that are not covered by any of the other rules. In the example of

Fig. 10.1, the former approach would result in a default rule predicting the class o, whereas the latter approach would choose the class x.

Of course, more elaborate algorithms for making predictions for uncovered examples have been proposed in the literature. An obvious idea is to use a nearest-neighbor prediction for uncovered examples. This has been proposed by Salzberg (1991), who suggests to use a distance measure between rules and examples in order to find the nearest rule to an example, and use this for classifying the example. Essentially, the distance between a rule and an example is defined as the distance to the closest point that is still covered by the rule. In Fig. 10.1d, the point x that is not covered by $\mathbf{r}_X$ is closest to $\mathbf{r}_2$, and would thus be classified as o.

Along similar lines, Eineborg and Boström (2001) suggest *rule stretching* which minimally generalizes each rule so that it now covers the example, and then classifies the example using the resulting generalized rule set. Generalization could, e.g., be performed using least general generalizations (lggs), and for conflict resolution any of the methods discussed above can be used.

10.3.3 Abstaining Classifiers

Up to now we have assumed a classification setting in which we have to make a decision in every situation. However, there are domains which require very reliable decision making. A first step towards more reliable decision making is to construct rule bases that do not have to give decisive classification in every situation, but instead can *abstain* from making a classification. By allowing that the classification may be also indecisive, it is possible to achieve that all decisive predictions are more reliable (Pietraszek, 2007).

A simple way to achieve abstention is to simply not make a prediction in the case when no example covers a rule. This case has been studied by Blaszczynski, Stefanowski, and Zajac (2009), who showed that ensembles of abstaining rule sets can improve the prediction quality. They also treat conflicting predictions in the same way as predictions for uncovered examples. Hühn and Hüllermeier (2009a) discuss the FR3 algorithm, a version of RIPPER which embeds conflicting predictions (*conflict*) and predictions for uncovered examples (*ignorance*) into a fuzzy preference model (Fodor & Roubens, 1994). In the case of confirmation rule induction (Gamberger, Lavrač, & Krstačić, 2002), predictions are only made if a predefined number of rules cover an example. Obviously, increasing this number leads to more indecisive predictions but the decisive ones get more reliable, which is demonstrated in a real life medical domain.

10.4 Multiclass Classification

Many real-world problems have multiple classes, i.e., the task is to assign one of a set of $C > 2$ classes c_i to any given example. Figure 10.2 shows a sample problem with six classes in a two-dimensional space. The dotted lines show a

Fig. 10.2 A multiclass classification problem

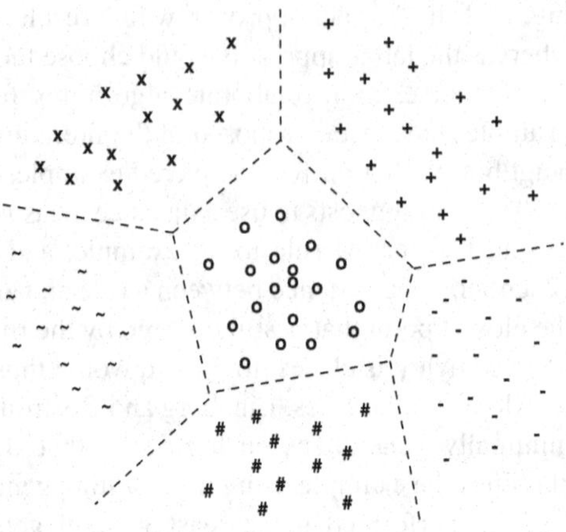

possible decision boundary that could be found by a multiclass learning algorithm. Rule learning algorithms are adapted to this scenario by transforming the original problem into a series of binary classification problems, a process that is also known as *class binarization*.

Definition 10.4.1 (class binarization, decoding). *Class binarization* is a mapping of a multiclass learning problem to several two-class learning problems in a way that allows a sensible *decoding* of the prediction, i.e., it allows the derivation of a prediction for the multiclass problem from the predictions of the set of two-class classifiers.

In inductive rule learning, class binarization can be performed at multiple levels: in the simplest case, class binarization is integrated into the covering loop so that a two-class problem is defined for each individual rule. The resulting theory is a decision list that interleaves rules for different classes. We will discuss this in more detail in Sect. 10.4.1. In the remaining sections, we discuss several approaches that learn several rule sets for binary classification problems, whose predictions are then combined into an overall prediction. In particular, we will describe the one-against-all approach (Sect. 10.4.2), ordered class binarization (Sect. 10.4.3), pairwise classification (Sect. 10.4.4), and, as a general framework, error-correcting output codes (Sect. 10.4.5).

10.4.1 Learning Multiclass Decision Lists

The covering algorithm can naturally be adapted to learning a decision list, as shown in Fig. 10.3. In this case, rules are not learned for one particular class, but all classes are tried successively, and the best resulting rule is chosen. The resulting rule list may interleave rules with different classes in the rule heads.

```
function MULTICLASSCOVERING(𝓔)
Input:
𝓔 set of training examples
Procedure:
𝓡 := ∅
repeat
    r := ∅
    for each class c_i, i = 1 to C do
        𝓟_i := {subset of examples in 𝓔 with class label c_i}
        𝓝_i := {subset of examples in 𝓔 with other class labels}
        ṙ = LEARNONERULE(c_i, 𝓟_i, 𝓝_i)
        if ṙ better than r according to quality criterion
        then r = ṙ
        endif
    endfor
    𝓡 := 𝓡 ∪ r
    𝓔 := 𝓔 \ COVERED(r, 𝓔)
until 𝓡 satisfies a quality threshold or 𝓔 is empty
𝓡 := 𝓡 ∪ default rule (c_max ← true)
    where c_max is the majority class in 𝓔.
Output: 𝓡 the learned decision lists
```

Fig. 10.3 The covering algorithm for decision lists

In effect, the inner loop of this MULTICLASSCOVERING algorithm performs a class binarization inside the covering loop. Note that in some cases, this inner **for**-loop need not be performed explicitly, but can be integrated into the LEARNONERULE function, which will then evaluate all possible refinements for all possible classes in the rule head. Some rule quality measures can be adapted to directly evaluate a multiclass class distribution, so that only one evaluation of a refinement has to be performed. For example, in the decision list version of CN2, the following generalized version of the *Entropy* heuristic is used for evaluating a rule body.[1]

$$\text{Entropy}(\mathbf{r}) = \sum_{i=1}^{C} \hat{\pi}_i \cdot \log_2 \hat{\pi}_i \quad .$$

The entropy is minimal if all covered examples belong to the same class, and maximal if they are distributed evenly among the classes. Once the best rule body has been found, the rule head is formed using the majority class among the examples covered by the body.

[1]Recall that $\hat{\pi}_i = \frac{\hat{P}_i}{\hat{E}}$ is the proportion of covered examples of class c_i.

Note that when learning a decision list one typically deletes all examples (both positive and negative) covered by the rule from the current example set, while in the process of learning unordered set of rules one often deletes only positive examples. The reason is that with an ordered decision list the first rule which covers the example fires, hence there is no need to try to cover false negatives by another rule assigning the correct class. As a consequence, decision lists are often shorter than rule sets because this elimination of both positive and negative examples may lead to simpler rules. On the other hand, the resulting theories are typically more difficult for human understanding, because the order of evaluation must be taken into account.

Classification with a decision list is quite straightforward. When classifying a new instance, the rules are sequentially tried and the first rule that covers the instance is used for classification/prediction. If no induced rule fires, the default rule is invoked.

Like for the case of learning a rule set, a default rule is added at the end which applies if none of the induced rules fire. However, in the ordered case the default class assigned by the default rule is the majority class among all noncovered training examples (and not the majority class in the entire training set, as with the unordered algorithm).

CN2 orders rules in the order they have been learned. This seems to be a natural strategy, because most search heuristics tend to learn more general rules first. However, it has been pointed out by Webb and Brkič (1993) that prepending (adding to the beginning) a new rule to the previously learned rules can produce simpler concepts. The intuition behind this argument is that there are often simple rules that would cover many of the positive examples, but also cover a few negative examples that have to be excluded as exceptions to the rule. Placing this simple general rule near the end of the rule list allows us to handle these exceptions with rules that are placed before the general rule and keep the general rule simple. This hypothesis has been empirically confirmed in Webb (1994) and Mooney and Califf (1995).

The BBG algorithm (Van Horn & Martinez, 1993) tries to be less greedy by inserting a learned rule at appropriate places in the decision list and evaluating the quality of the resulting list on the entire training set. Thus, it always evaluates the entire model instead of individual rules.

10.4.2 One-Against-All Learning

The most popular class binarization technique is the unordered or one-against-all class binarization, where one takes each class in turn and learns binary concepts that discriminate this class from all other classes. Thus, the multiclass learning problem is reduced to a series of concept learning problems, one for each class.

Definition 10.4.2 (one-against-all class binarization). The *one-against-all class binarization* transforms a C-class problem into C two-class problems, which use

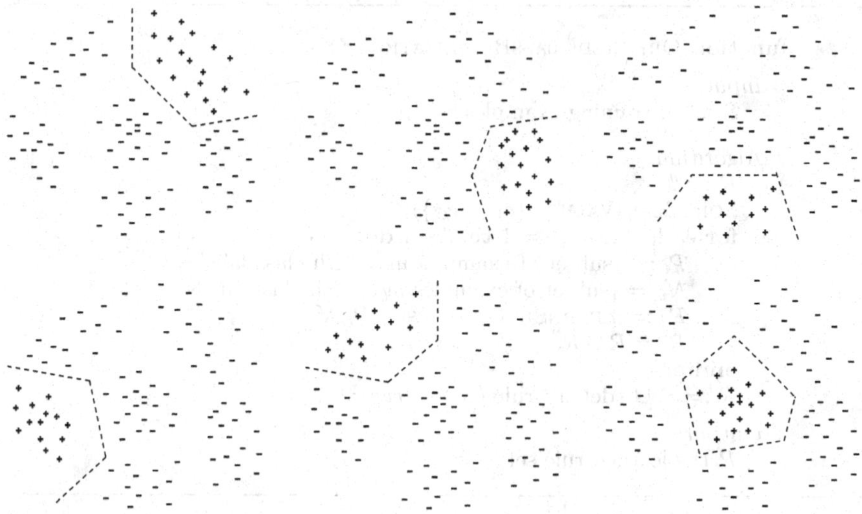

Fig. 10.4 The six binary learning problems that are the result of one-against-all class binarization of the multiclass dataset of Fig. 10.2

the examples of class c_i as the positive examples and the examples of classes c_j $(j = 1 \ldots C,\ j \neq i)$ as the negative examples.

The LEARNRULEBASE algorithm of Fig. 2.10 shows an implementation of this strategy. The basic idea is to learn a separate rule set $\mathcal{R}_i$ for each class c_i, taking the examples of this class as positive and all other examples as negative examples (hence the name *one-against-all* classification). Figure 10.4 shows the six training sets that result from a one-against-all binarization of the multiclass dataset of Fig. 10.2.

For each of these binary problems, a separate rule set $\mathcal{R}_i$ is learned. At classification time, each rule set is queried and predicts its positive class c_i if the example is covered by $\mathcal{R}_i$, or abstains from making a prediction if $\mathcal{R}_i$ does not cover the example. Just as with binary classification (Sect. 10.2), an example may be covered by multiple rules or by no rule at all. All strategies for prediction with rule bases (cf. Sect. 10.3) can be straightforwardly adapted to the multiclass case.

Of course, the ability to use a concept learner for multiclass learning comes at a price: we now have to train on C times as many examples as in the original problem.

Theorem 10.4.1 (complexity of one-against-all binarization). *The total number of training examples generated by one-against-all class binarization is $C \cdot E$.*

Proof. Each of the E examples appears in all of the C training sets. □

10.4.3 Ordered Class Binarization

In rule learning, the one-against-all strategy is also known as *unordered* class binarization because it is typically used for sinducing unordered rule sets

```
function ORDEREDCLASSBINARIZATION(ℰ)
  Input:
    ℰ set of training examples
  Algorithm:
    ℛ := ∅
    SORTCLASSVALUES({c_1 ... c_C})
    for each class c_i, i = 1 to C − 1 do
        𝒫_i := {subset of examples in ℰ with class label c_i}
        𝒩_i := {subset of examples in ℰ with class labels c_j, j > i}
        ℛ_i := LEARNSETOFRULES(c_i, 𝒫_i, 𝒩_i)
        ℛ := ℛ ∪ ℛ_i
    endfor
    ℛ := ℛ ∪ {default rule (c_C ← true)}
  Output:
    ℛ the learned rule set
```

Fig. 10.5 Ordered class binarization for multiclass rule learning

(Clark & Boswell, 1991). This terminology helps to clarify the difference to a variant of the one-against-all strategy, that has its origins in separate-and-conquer rule learning. It goes back to later versions of CN2 (Clark & Boswell) and is, e.g., the default strategy in the RIPPER rule learner. The key idea is that instead of discriminating one class from all other classes, each class is only discriminated from all subsequent classes.

Definition 10.4.3 (ordered class binarization). The *ordered class binarization* transforms a C-class problem into $C - 1$ binary problems. These are constructed by using the examples of class c_i ($i = 1 \dots C - 1$) as the positive examples and the examples of classes $j > i$ as the negative examples.

Figure 10.5 shows pseudocode for this class binarization strategy. The key difference to the one-against-all strategy (Fig. 2.10) is that the classes are first sorted (SORTCLASSVALUES) and then used in order. In each iteration, we only discriminate class c_i from all classes that come after it in the ordering. Thus, the positive examples $\mathcal{P}_i$ for learning the rule set $\mathcal{R}_i$ consist of all examples of class c_i, and the negative examples $\mathcal{N}_i$ of all classes c_j with $j > i$. Consequently, the last rule set that is learned discriminates the largest class c_C from the second-largest class c_{C-1}.

As a result, we obtain one rule set $\mathcal{R}_i$ for each of the classes $c_i, i = 1 \dots C - 1$. The order of the classes has to be adhered to at classification time: the first rule set $\mathcal{R}_1$ discriminating class c_1 from classes $\{c_2 \dots c_C\}$ has to be called first. If $\mathcal{R}_1$ classifies the example as belonging to class c_1, no other classifier is called. If not, we conclude that the example cannot belong to class c_1, and $\mathcal{R}_2$ is called in order to check whether the example belongs to class c_2. If the last classifier $\mathcal{R}_{C-1}$ concludes that the example does not belong to class c_{C-1}, it is automatically assigned to class c_C. As the order of the rules within a block of rules that predict the same class does not matter, we may concatenate the rule sets $\mathcal{R}_i, i = 1 \ldots C-1$ into a single decision list. The final rule in the list is the default rule that predicts the last class c_C.

Ordered class binarization is considerably cheaper than one-against-all class binarization.

Theorem 10.4.2 (complexity of ordered class binarization). *The total number of training examples generated by one-against-all class binarization is* $E \cdot \left(\sum_{i=1}^{C} i \cdot \gamma_i - \gamma_C\right)$ *where* γ_i *is the fraction of examples of class* c_i.

Proof. The examples of the first class c_1 are only contained in the training set for rule set $\mathcal{R}_1$, the examples of the 2nd-largest class are contained only in the first two training sets (in $\mathcal{N}_1$ and $\mathcal{P}_2$), etc. In general, the examples of the i-th class are contained in $\mathcal{P}_i$ and all $\mathcal{N}_j, j < i$, i.e., in i training sets. The ordered class binarization therefore generates $i \cdot \gamma_i \cdot E$ examples of class c_i. The only exceptions are the examples of the last class c_C, which appear in all $C-1$ training sets as part of the negative training sets $\mathcal{N}_i, i = 1 \ldots C-1$. For this class, we therefore only obtain $\gamma_c \cdot E \cdot (C-1)$ examples. □

For example, if all classes have the same number of examples (i.e., $\gamma_i = \frac{1}{C}$), the total number of training examples is

$$E \cdot \left(\sum_{i=1}^{C} i \cdot \frac{1}{C} - \frac{1}{C}\right) = E \cdot \frac{C \cdot (C+1)}{2 \cdot C} - \frac{E}{C} = E \cdot \frac{C+1}{2} - \frac{E}{C}.$$

Obviously, the order of the classes has a crucial influence on the final decision list. Multiple strategies are possible for selecting the order of classes. A good practice is to select the majority class as the C-th class, resulting in classification into the majority class whenever no other rule fires. This suggests to order classes according to increasing class frequency, starting with the smallest class and having the largest class as the prediction of the default rule. This is, in fact, the default strategy of RIPPER. However, it is not clear whether this approach is optimal. Alternatively, one could order the classes according to decreasing class frequency. A motivation for using this strategy is that it minimizes the total number of examples because in this case the largest γ_i come first and are thus multiplied with lower factors. Of course, RIPPER also allows to sort the classes in any user-provided order.[2]

[2]We here refer to William Cohen's original C-implementation of the algorithm. At the time of this writing, JRIP, the more accessible WEKA re-implementation of RIPPER, does not support these options.

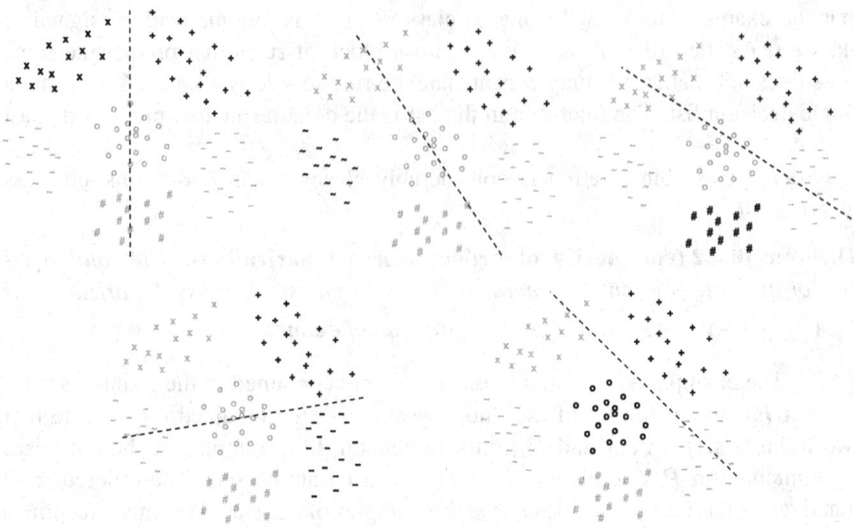

Fig. 10.6 Five of the 15 binary learning problems that are the result of a pairwise class binarization of the multiclass dataset of Fig. 10.2. Shown are all problems that pair the class + against one of the other classes. The grey examples in each training set are ignored

10.4.4 Pairwise Classification

This section discusses a more complex class binarization procedure, the *pairwise classifier*, also known as *1-against-1 classification* or *round robin learning*. The basic idea is simply to learn one classifier for each pair of classes. It has been used in the areas of statistics (Bradley & Terry, 1952; Friedman, 1996), neural networks (Knerr, Personnaz, & Dreyfus, 1990, 1992; Lu & Ito, 1999; Price, Knerr, Personnaz, & Dreyfus, 1995) support vector machines (Hastie & Tibshirani, 1998; Hsu & Lin, 2002; Kreßel, 1999; Schmidt & Gish, 1996), rule learning (Fürnkranz, 2002b; Hühn & Hüllermeier, 2009a) and others. We refer to (Fürnkranz) for a brief survey of the literature on this topic.

Definition 10.4.4 (pairwise classification). A *pairwise classifier* solves a C-class problem by transforming it into $C \cdot (C-1)/2$ two-class problems one for each pair of classes $<c_i, c_j>$, $i = 1 \ldots C-1$, $j = i+1 \ldots C$. The rule set $\mathcal{R}_{i,j}$ for problem $<c_i, c_j>$ is only trained on examples of classes c_i and c_j, examples of classes $c_k, k \neq i, j$ are ignored for this problem.

The rule set $\mathcal{R}_{i,j}$ is intended to separate the objects with label c_i from those having label c_j. At classification time, an example e is submitted to all rule sets $\mathcal{R}_{i,j}$, and their predictions $\mathcal{R}_{i,j}(e)$ are combined into an overall prediction. The application of pairwise classification to the 6-class problem shown in Fig. 10.2 is illustrated in Fig. 10.6. In this case, the pairwise classifier consists of 15 binary classifiers, one for each pair of classes. Figure 10.6 shows only the five classifiers that pair the class + against any of the other classes.

Comparing this to the one-against-all binarization shown in Fig. 10.4, it becomes apparent that in the pairwise classification, the base classifiers are trained on fewer examples and thus have more freedom for fitting a decision boundary between the two classes. In fact, in our example, all binary classification problems of the pairwise binarization could be solved with a simple linear discriminant, while neither the multiclass problem nor its unordered binarization have a linear solution. The phenomenon that pairwise decision boundaries can be considerably simpler than those originating from unordered binarization has also been observed in real-world domains. For example, Knerr et al. (1992) observed that the classes of a digit recognition task were pairwise linearly separable, while the corresponding one-against-all task was not. The results of Hsu and Lin (2002), who obtained a larger advantage of pairwise binarization over unordered binarization for support vector machines with a linear kernel than for support vector machines with a nonlinear kernel on several benchmark problems, could also be explained by simpler decision boundaries in the pairwise case.

A crucial point, of course, is how to decode the predictions of the pairwise classifiers to a final prediction. In the simplest case, each prediction $\mathcal{R}_{i,j}(e)$ is interpreted as a vote for either c_i or c_j, and the label with the highest number of votes is proposed as a final prediction. Note that some examples will be forced to be classified erroneously by some of the binary base classifiers because each classifier must label all examples as belonging to one of the two classes it was trained on. Consider, for example, the classifier $\mathcal{R}_{\mathsf{x},+}$ shown in the upper left of Fig. 10.6. It will arbitrarily assign all examples of classes o and # to either x or +, depending on which side of the decision boundary they are. In principle, such 'unqualified' votes may lead to an incorrect final classification. However, the votes of the five classifiers that contain examples of class o should be able to overrule the votes of the other ten classifiers, which pick one of their two constituent classes for each o example. If the class values are independent, it is unlikely that all classifiers would unanimously vote for a wrong class. However, the likelihood of such a situation could increase if there is a systematic correlation between the classes, such as, e.g., the examples of class ~, which are systematically classified as x, or those of class –, which are all classified as + by the base classifier in the upper left of Fig. 10.6. Note that such systematic correlations could appear, e.g., in problems with an ordinal or a hierarchical class structure, but are not as frequent as one would expect (Fürnkranz & Sima, 2010). In any case, if the five o classifiers unanimously vote for o, no other class can accumulate more than four votes because they lost their direct match against o.

Despite its simplicity, voting is the most popular and one of the most effective strategies for aggregating the predictions of pairwise classifiers. Particularly popular is weighted voting, which allows to split a base classifier's vote among the two classes, for example according to their estimated probabilities. Having queried all base rule sets, the class with the highest number of votes (sum of weighted votes) is eventually predicted. Further below, in Sect. 10.5.1, we will see that this simple approach also has a theoretical foundation.

At first sight, it seems that pairwise classification is very inefficient for a high number of class labels because one has to train a quadratic number of classifiers. However, a closer look reveals that this is often outweighed by a positive effect, namely that the individual problems are much smaller.

Theorem 10.4.3 (complexity of pairwise classification). *The total number of training examples generated by pairwise classification is* $(C - 1) \cdot E$.

Proof. Each of the E original training examples appears in the training sets of all binary classifiers where its class is paired against one of the $C - 1$ other classes. □

Thus, pairwise classification is more efficient in training than one-against-all classification. In fact, its advantage increases for expensive base classifiers with a superlinear time complexity, because they profit more if the training effort is distributed over a larger number of comparably smaller problems (Fürnkranz, 2002b).

A more interesting problem is the efficiency at prediction time. In principle, one has to query a quadratic number of classifiers in order to derive a final ranking of the classes. However, it is often not necessary to query all classifiers in order to determine the winning class. For example, if one class has received more votes than any other class can possibly achieve in their remaining evaluations, this class can be safely predicted without querying the remaining classifiers. The QWEIGHTED algorithm (Park & Fürnkranz, 2007) tries to enforce this situation by always focusing on the class that has lost the least amount of voting mass, thus reducing the prediction time to about $O(C \cdot \log C)$ without changing the predictions.

Even if training is quite efficient, and classification can be handled without the need to query all classifiers, we still have to store all $C \cdot (C - 1)/2$ binary classifiers because each classifier will be needed for some examples (unless some labels are never predicted). This is the key problem with pairwise classification for a large number of classes. For the particular case of rule learning, Sulzmann and Fürnkranz (2011) proposed an approach that uses *rule stacking*, a variant of stacking,[3] for compressing an ensemble of pairwise classifiers into a single set of rules.

Definition 10.4.4 assumes that the problem of discriminating class c_i from class c_j is identical to the problem of discriminating class c_j from class c_i. However, this is typically not the case for rule learning algorithms, where learning a rule set for one class and using the other as the default class will yield a different result as learning the rule set for the other class. We have already seen in Sect. 10.2 that rule learning algorithms often deal with binary classification by learning two rule sets, one for each class. The same idea can be applied for pairwise classification, resulting in the double round-robin classifier (Fürnkranz, 2002b).

Definition 10.4.5 (double round robin). A *double round robin* classifier solves a C-class problem by transforming it into $C \cdot (C - 1)$ two-class problems one for

[3]Stacking (Wolpert, 1992) denotes a family of techniques that use the predictions of a set of classifiers as inputs for a meta-level classifier that makes the final prediction.

each pair of classes $<c_i, c_j>$, $i, j = 1 \ldots C$, $j \neq i$, where the examples of class c_i are used as positive and the examples of class c_j as negative examples for training a rule set $\mathcal{R}_{i,j}$.

Thus, a double round robin has twice as many classifiers as a single round robin. This idea can also be straightforwardly combined with bagging,[4] yielding multiple classifiers for each pairwise comparison (Fürnkranz, 2003).

10.4.5 Error-Correcting Output Codes (ECOC)

Error-correcting codes (ECC) are a well-known topic in the field of coding and information theory (MacWilliams & Sloane, 1983). Messages, represented as bit sequences, can be distorted during the communication through a physical communication channel, i.e., a 0 may be falsely received as a 1, or vice versa. ECCs are able to detect and correct errors by enhancing the message sequence with redundant information bits, which are used for verification and correction on the receiver side. As we will see, this idea may also be exploited for error-correction in learning and can, in fact, be turned into a general framework for class binarization approaches.

Dietterich and Bakiri (1995) proposed *error correcting output codes (ECOC)* as an adaptation of ECCs to multiclass classification. They consider classifier predictions as information signals which ideally describe the correct class for a given instance. Due to external influences (such as, e.g., a too-small sample size) the signals emitted by classifiers are sometimes wrong, and such errors have to be detected and corrected. Formally, each class c_i $(i = 1 \ldots C)$ is associated with a so-called *code word* $\mathbf{cw}_i \in \{-1, 1\}^l$ of length l. In the context of ECOC, all relevant information is summarized in a so-called *coding matrix* $(m_{i,j}) = M \in \{-1, 1\}^{C \times l}$, whose ith row describes code word $\mathbf{cw}_i$, whereas the jth column represents the training protocol for the rule model $\mathcal{R}_j$. The classifier $\mathcal{R}_j$ is trained on all examples, where examples of classes for which the corresponding entry in the jth column is 1 are labeled as positive and examples for which the corresponding entry is -1 are labeled as negative. Thus, the coding matrix implicitly describes a class binarization scheme for the original multiclass problem.

As an example consider the following coding matrix, which encodes four classes with six classifiers:

$$M = \begin{pmatrix} 1 & 1 & 1 & -1 & -1 & -1 \\ 1 & -1 & -1 & 1 & 1 & -1 \\ -1 & -1 & -1 & 1 & -1 & 1 \\ -1 & -1 & 1 & -1 & 1 & 1 \end{pmatrix}$$

[4] Bagging (Breiman, 1996) is a popular ensemble technique which trains a set of classifiers, each on a sample of the training data that was generated by sampling uniformly and with replacement. The predictions of these classifiers are then combined, which often yields a better practical performance than using the predictions of a single classifier.

Each row corresponds to a class value. For example, the first class is encoded with the code vector $\mathbf{cw}_1 = (+1, +1, +1, -1, -1, -1)$. Each column corresponds to a binary classifier. Thus, the first classifier uses the examples of classes 1 and 2 as positive examples, and the examples of classes 3 and 4 as negative examples.

For the classification of a test instance e, all binary classifiers are evaluated. Their predictions form a *prediction vector* $\mathbf{p} = (\mathcal{R}_1(e), \mathcal{R}_2(e), \ldots, \mathcal{R}_l(e))$, which is then compared to the code words. The class c^* whose associated code word $\mathbf{cw}_{c^*}$ is closest to $\mathbf{p}$ according to some distance measure $d(.,.)$ is returned as the overall prediction. A popular choice for d is the Hamming distance, which simply returns the number of positions in which the code vector and the prediction vector differ, resulting in the well-known *Hamming decoding*.

Many different class binarization schemes may be encoded as code matrices, so that ECOCs may be viewed as a general framework for class binarization. For example, the coding matrix of one-against-all learning (Sect. 10.4.2) is a simple diagonal matrix, where all elements on the diagonal have the value $+1$, and all elements off the diagonal have the value -1.

The practical performance of ECOCs clearly depends on the choice of the coding matrix. In order to be able to maximize the error-correcting capabilities, the matrix should be chosen in a way that maximizes the difference between any pair of code words $\mathbf{cw}_i$ and $\mathbf{cw}_j$. In fact, a well-known theorem from coding theory states that if the minimal Hamming distance between two arbitrary code words is h, the ECC framework is capable of correcting up to $\lfloor \frac{h}{2} \rfloor$ bits. This is easy to see, since every code word $\mathbf{cw}$ has a $\lfloor \frac{h}{2} \rfloor$ neighborhood, for which every code in this neighborhood is nearer to $\mathbf{cw}$ than to any other code word. Thus, it is obvious that good error correction crucially depends on the choice of a suitable coding matrix.

Unfortunately, many of the results in coding theory are not fully applicable to the machine learning setting. For example, the above result assumes that the bit-wise error is independent, which does not necessarily hold in machine learning. Classifiers are learned with similar training examples and therefore their predictions tend to correlate. Thus, a good ECOC code also has to consider column distances, which may be taken as a rough measure for the independence of the involved classifiers. In any case, ECOCs have shown good performance on practical problems (e.g., Ghani, 2000; Kittler, Ghaderi, Windeatt, & Matas, 2003; Melvin, Ie, Weston, Noble, & Leslie, 2007). In particular, it has been shown that ECOCs can reduce both bias and variance of the underlying learning algorithm (Kong & Dietterich, 1995).

Since the introduction of ECOC, a considerable amount of research has been devoted to code design (see, e.g., Crammer & Singer 2002; Pimenta, Gama, & de Leon Ferreira de Carvalho 2008), but without reaching a clear conclusion. Popular choices include:

Random codes: A random code consist of randomly chosen code words of a prespecified length for each of the C classes.

Exhaustive codes: An exhaustive code for C classes has length $2^{C-1} - 1$, where each column encodes a possible partition of the classes into two nonempty sets.

k**-exhaustive codes:** A variant of exhaustive codes that always uses k class for the positive class and $C-k$ classes for the negative class, thus the code length is $\binom{k}{C}$ (1-exhaustive codes corresponds to one-against-all class binarization).

BCH codes: BCH codes (Bose & Ray Chaudhuri, 1960; Hocquenghem, 1959) produce near-optimal error correcting codes for a given minimal distance between code words. They are particularly well-studied in coding theory because they allow very fast decoding. They assume that the number of code words is a power of 2. Random selection of code words for the next largest power can be used for other values of C.

A brief survey and analysis of different coding techniques can be found in Windeatt and Ghaderi (2003). However, in many cases, the choice of the coding matrix depends on knowledge about the domain and the dataset. For example, the knowledge of an inherent hierarchy or order among the classes can be used to model classifiers which exploit this information (e.g., Cardoso & da Costa, 2007; Melvin et al., 2007). Pujol, Radeva, and Vitriá (2006) propose to generate a coding matrix, whose columns consist of the best discriminating classifiers on the considered dataset.

10.4.6 Ternary ECOCs

Conventional ECOCs as described in the previous section always use all classes and all training examples for training a binary classifier. Thus, class binarization schemes which use only parts of the data (such as pairwise classification) cannot be modeled in this framework.

For this reason, Allwein, Schapire, and Singer (2000) extended the ECOC approach to the ternary case, where code words are now of the form $\mathbf{cw}_i \in \{-1, 0, 1\}^l$. The additional code $m_{i,j} = 0$ denotes that examples of class c_i are ignored for training the base classifier $\mathcal{R}_j$. This extension increases the expressive power of ECOCs, so that now nearly all common multiclass binarization methods can be modeled. For example, pairwise classification (cf. Sect. 10.4.4) could not be modeled in the original framework, but can be modeled with ternary ECOCs. Its coding matrix has $l = C \cdot (C-1)/2$ columns, each consisting of exactly one positive value (+1), exactly one negative value (−1), and $C-2$ neutral values (0). Thus, the coding matrix of a pairwise classifier for a four-class problem looks as follows:

$$M = \begin{pmatrix} 1 & 1 & 1 & 0 & 0 & 0 \\ -1 & 0 & 0 & 1 & 1 & 0 \\ 0 & -1 & 0 & -1 & 0 & 1 \\ 0 & 0 & -1 & 0 & -1 & -1 \end{pmatrix}$$

Most coding strategies can be generalized to the ternary case. For example, exhaustive codes can be defined to cover all possible binary classifiers involving a given number of classes m (and ignoring the remaining $C-m$ classes). More formally, a m-exhaustive ternary code defines a ternary coding matrix M, for which every column j contains exactly m nonzero values, i.e., $\sum_{i=1}^{C} |m_{i,j}| = m$. Obviously, only columns with at least one positive (+1) and one negative (−1) class are useful. The following example shows a 3-exhaustive ternary code for a four-class problem.

$$M = \begin{pmatrix} 1 & 1 & -1 & 1 & 1 & -1 & 1 & 1 & -1 & 0 & 0 & 0 \\ 1 & -1 & 1 & 1 & -1 & 1 & 0 & 0 & 0 & 1 & 1 & -1 \\ -1 & 1 & 1 & 0 & 0 & 0 & 1 & -1 & 1 & 1 & -1 & 1 \\ 0 & 0 & 0 & -1 & 1 & 1 & -1 & 1 & 1 & -1 & 1 & 1 \end{pmatrix}$$

Note that 2-exhaustive ternary codes correspond to pairwise classification.

Similarly, most decoding techniques can be extended to the ternary case. For example, in the case of Hamming decoding, we need to determine the code word that is closest to the prediction vector. However, in the ternary case, the code word can contain 0-values, whereas the prediction vector is a set of binary predictions (either −1 or 1). A straightforward generalization of the Hamming distance to this scenario is that a 0-value in the code word will always increase the distance by $1/2$ (independent of the prediction).

Many alternative decoding strategies have been proposed in the literature. For example, Allwein et al. (2000) propose the use of general loss functions for cases where the base classifiers do not return binary predictions but instead estimated probability distributions over the two classes. Escalera, Pujol, and Radeva (2006) discuss the shortcomings of traditional Hamming distance for ternary ECOCs and present two novel decoding strategies, which should be more appropriate for dealing with the 0-value.

The training complexity of error-correcting output codes depends on the dimensions and the sparsity of the coding matrix, and the prediction complexity depends on the length of code vectors. However, just as with pairwise classification, often not all base classifiers have to be queried in order to reliably determine the winning class (Park & Fürnkranz, 2009).

10.5 Learning in Structured Output Spaces

So far, we have seen methods for adapting rule learning algorithms to multiclass problems where the goal is to assign one out of a list of possible classes to an instance to be classified. Many prediction problems are more complex in the sense that there either is an *inherent structure* in the set of possible values or that there is an *external structure* upon which the output needs to be mapped. Examples for inherent structures in the output space include *hierarchical classification*, where the class values have a tree structure, or *ordinal classification*, where the class values have a list structure. Example learning tasks whose goal is to predict class labels from an external structure include *multilabel classification*, where the task is to assign multiple class labels to a test instance, and *label ranking*, where the task is to sort the labels according to their relevance for a test instance.

This section presents approaches that allow rule learning algorithms to be applied to such problems. This will become possible by embedding the above-mentioned problem types into the general framework of *preference learning* (Fürnkranz & Hüllermeier, 2010a). For example, as we will discuss in more detail below, a classification problem can be reformulated as a preference learning problem by specifying that the true label of an example is preferred over all other possible labels. The task of the preference learner is to order all possible labels according to their predicted preference degree, and the most-preferred item can be predicted as a solution for the classification problem.

We start this section by generalizing pairwise classification to the setting of label ranking, as will be explained in Sect. 10.5.1. This can, in turn, be specialized to several generalized classification problems, including multilabel classification (Sect. 10.5.2) as well as ordinal and hierarchical classification (Sect. 10.5.3). Finally, Sect. 10.5.4 is devoted to the application of pairwise learning for the problems of object and instance ranking.

10.5.1 Label Ranking

In the label ranking scenario, the problem is to predict, for a given test example e, a ranking over all possible values of the class attribute $\mathcal{C} = \{c_1, c_2, \ldots, c_C\}$. In this context, the values c_i are also known as *labels*. The training information is available in the form of pairwise comparisons of the form

$$c_i \succ_e c_j$$

which means that for example e, the class label c_i is preferred to the class label c_j. When it is clear to which example the preferences belong, we will sometimes omit the subscript e and simply write $c_i \succ c_j$.

The training information consists of a set of instances for which (partial) knowledge about the associated preference relation is available. More precisely, each training instance e is associated with a subset of all pairwise preferences $c_i \succ_e c_j$, $1 \leq i, j \leq C$. The top of Fig. 10.7 shows a training set consisting of seven examples, each described in terms of three features $\mathbf{f}_1$, $\mathbf{f}_2$, and $\mathbf{f}_3$. Each training example is associated with some preferences over the set of possible labels $\mathcal{C} = \{a, b, c\}$. For example, for the second training instance, we know that $a \succ b$ and $c \succ b$, but we do not know whether a or c is the most preferred option. Thus, even though we assume the existence of an underlying ('true') ranking, we do not expect the training data to provide full information about that ranking. Besides, in order to increase the practical usefulness of the approach, we even allow for inconsistencies, such as pairwise preferences which are conflicting (cyclic) due to observation errors.

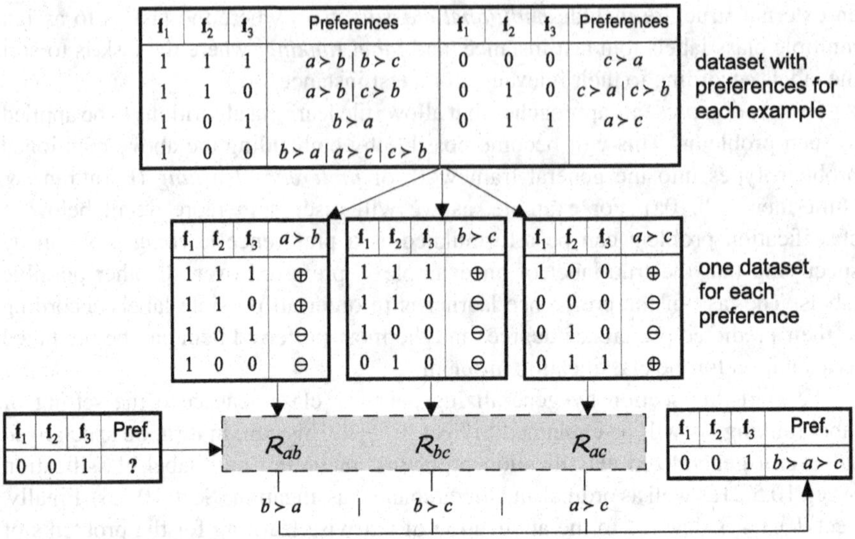

Fig. 10.7 Schematic illustration of learning by pairwise label ranking

Pairwise classification (cf. Sect. 10.4.4) can be extended to the above problem of learning from label preferences in a natural way (Fürnkranz & Hüllermeier, 2003; Hüllermeier, Fürnkranz, Cheng, & Brinker, 2008). To this end, all examples with associated preferences $c_i \succ c_j$ or $c_j \succ c_i$ are used as training examples for learning the rule set $\mathcal{R}_{i,j}$, where the example is labeled as positive if $c_i \succ c_j$ and as negative if $c_j \succ c_i$. Examples for which nothing is known about the preference between c_i and c_j are ignored. Thus, $\mathcal{R}_{i,j}$ is intended to learn the mapping that outputs $\oplus$ if $c_i \succ c_j$ and $\ominus$ if $c_j \succ c_i$.

Figure 10.7 illustrates the entire process. First, the original training set is transformed into three two-class training sets, one for each possible pair of labels, containing only those training examples for which the relation between these two labels is known. Then three rule sets $\mathcal{R}_{ab}$, $\mathcal{R}_{bc}$, and $\mathcal{R}_{ac}$ are trained. In our example, the rule sets could be defined as follows:

$$\mathcal{R}_{ab} : \oplus \leftarrow \mathbf{f}_2. \quad \mathcal{R}_{bc} : \oplus \leftarrow \mathbf{f}_3. \quad \mathcal{R}_{ac} : \oplus \leftarrow \mathbf{f}_1. \atop \qquad\qquad\qquad\qquad\qquad\qquad\qquad \oplus \leftarrow \mathbf{f}_3.$$

Given a new instance with an unknown preference structure (shown in the bottom left of Fig. 10.7), the predictions of these rule sets are then used to predict a ranking. In this case, $\mathcal{R}_{ab}$ predicts $\ominus$ because $\mathbf{f}_2$ is false, and $\mathcal{R}_{bc}$ and $\mathcal{R}_{ac}$ both predict $\oplus$ because $\mathbf{f}_3$ is true. Thus, the predicted preferences for this example are $b \succ a$, $b \succ c$, and $a \succ c$. Combining them naturally leads to the label ranking $b \succ a \succ c$.

However, combining the predicted preferences into a ranking is not always as trivial as in this example, because the predicted preferences may not be transitive. The voting and weighted voting strategies that we know from pairwise classification

(Sect. 10.4.4) already produce a ranking of all options and can be used for this task. Even though weighted voting may appear rather ad hoc at first sight, it does have a theoretical justification. For example, Hüllermeier et al. (2008) have shown that, assuming that the binary classifiers produce good class probability estimates for the pairwise comparisons, weighted voting minimizes the Spearman rank correlation with an assumed target ranking. Other standard loss functions on rankings, such as Kendall's tau or the position error, can also be minimized in expectation. On the other hand, there are also loss functions that cannot be minimized by the aggregation of pairwise probability estimates because of the loss of information caused by decomposing the original problem into a set of pairwise problems (Hüllermeier & Fürnkranz, 2010). The general problem of obtaining a probability distribution over all classes from the pairwise probabilities is known as *pairwise coupling* (Hastie & Tibshirani, 1998). Wu, Lin, and Weng (2004) provide a good overview of pairwise coupling techniques.

It has been observed by several authors (Dekel, Manning, & Singer, 2004; Fürnkranz & Hüllermeier, 2003; Har-Peled, Roth, & Zimak, 2002) that, in addition to classification, many learning problems, such as multilabel classification, ordered classification, or ranking may be formulated in terms of label preferences. These will be briefly tackled in the following sections.

10.5.2 Multilabel Classification

Multilabel classification refers to the task of learning a function that assigns to a test instance e not only a single label but a subset $\mathcal{P}_e$ of all possible labels. Thus, in contrast to multiclass learning, the class values c_i are not assumed to be mutually exclusive such that multiple labels may be associated with a single instance. The set of labels $\mathcal{P}_e$ are called *relevant* for the given instance, the set $\mathcal{N}_e = \mathcal{C} \setminus \mathcal{P}_e$ are the *irrelevant* labels.

The idea behind the reduction of multilabel classification to a label ranking problem is based on the observation that this information can be expressed equivalently in terms of a set of preferences that encode for each example that all its relevant labels are preferred over all irrelevant labels:

$$c_i \succ_e c_j \ \Leftrightarrow \ c_i \in \mathcal{P}_e \land c_j \in \mathcal{N}_e.$$

The resulting preferences (illustrated in Fig. 10.8a) can be used to train a label ranker, which is then able to predict a ranking over all possible labels of a new, unseen example.

The predicted ranking determines an order of the labels, with the idea that the upper part of the ranking consists of relevant labels, and the lower part of the ranking consists of irrelevant labels. What is still missing is a method for determining the exact split point for partitioning the ranking. Various techniques are possible at this place, such as the use of a fixed number of relevant labels for all examples or the

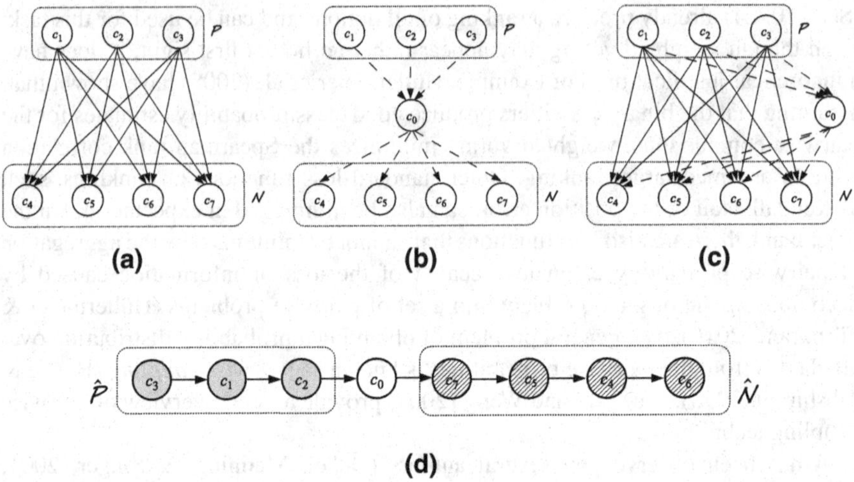

Fig. 10.8 Calibrated label ranking. (**a**) The set of preferences representing a multilabel classification problem. (**b**) Introducing a calibration label c_0 that separates $\mathcal{P}$ and $\mathcal{N}$. (**c**) The set of preferences representing a calibrated label ranking problem. (**d**) At prediction time, the calibration label c_0 indicates the split into labels that are predicted relevant ($\hat{\mathcal{P}}$) and labels that are predicted irrelevant ($\hat{\mathcal{N}}$)

use of a separate classifier to predict the number of labels. Fürnkranz, Hüllermeier, Loza Mencía, and Brinker (2008) proposed to incorporate the split-point into the learning process. This was achieved by introducing an artificial (neutral) label which is associated with the split-point and thus calibrates the ranking: for each training instance, this label is preferred to all irrelevant labels, but is less preferable than all relevant labels; see Fig. 10.8b. A *calibrated label ranker* trained on the enhanced set of preferences (Fig. 10.8c) is then able to predict a ranking of all labels, *including the artificial one*. The position of this label in the predicted ranking indicates where the ranking has to be split into the relevant and the irrelevant part. Experiments have shown that this approach outperforms standard methods for multilabel classification, despite a slight tendency to underestimate the number of relevant labels.

The crucial factor determining the efficiency of the approach is the average number d of relevant labels per training example (which is often small in comparison to the total number of labels). It can then be shown that the training complexity in this case is $O(d \cdot C \cdot E)$, by arguments similar to Theorem 10.4.3. It is also possible to speed up the prediction phase by avoiding to compute all possible comparisons (Loza Mencía, Park, & Fürnkranz, 2009).

10.5.3 Ordinal and Hierarchical Classification

Ordinal classification and *hierarchical classification* are problems in which the target label set has an inherent structure. In ordinal classification, this structure is a total order, whereas in hierarchical problems, the structure is a partial order in the form of a hierarchy, typically defined by various subconcept/superconcept relations (see Fig. 4.1).

In principle, one could use regular multiclass classification for solving these problems. However, such an approach does not take the relation between class labels into account. If low is the true class, for instance, then average is a better prediction than high, despite the fact that both are incorrect. An obvious idea is to express this type of information in the form of label preferences.

Within the label ranking framework, this can be done by associating a training instance with all pairwise preferences that can be *inferred* from the label structure. For example, if we know that an instance is of class low, we not only know that the label low is preferred to all other labels, but we can also infer that average would be a better classification than high. In other words, the set of training examples for which it is known that average $\succ$ high could be *enriched* by instances from the class low. Similarly, if we know that an object belongs to the class Slovenia, we can infer that the label Croatia would be preferred over China, because the former is geographically closer to the true class.

However, it seems that implementing the above-mentioned approach in a pairwise label ranker does not yield the expected improvements over a regular pairwise classifier. One of the reasons is presumably that enriching the training sets of the binary classifiers as described above results in more complex decision boundaries for the binary problems, which are then harder to learn (Fürnkranz & Sima, 2010). This should, however, not be interpreted as an entirely negative result because the pairwise classifier on its own already shows good performance on ordinal and hierarchical classification problems. For example, in Fürnkranz (2003), pairwise classification outperformed the ordinal classification technique proposed by Frank and Hall (2001). Also, Fürnkranz and Sima (2010) showed that enriching the pairwise classifier with hierarchical information leads to a classification that is essentially equivalent to the so-called Pachinko-machine classifier (Koller & Sahami, 1997), which contains one binary classifier for each internal node of the label hierarchy. It is, however, well-known that this classifier does not perform as well as the competing approaches.

10.5.4 Object and Instance Ranking

As discussed in the previous sections, label ranking is concerned with the ranking of all possible values of a class attribute. A different ranking task is to sort a set of examples according to a preference relation that can be observed for the training examples. Fürnkranz and Hüllermeier (2010a) discriminate between *object ranking*, where the training information specifies a binary preference relation between the data, and *instance ranking*, where the training information is given in the form of labeled training instances. Instance ranking can be further divided into *bipartite ranking*, where the training information consists of bipartition into positive and negative examples, and *multipartite ranking*, where the class attribute has an ordinal scale.

Bipartite ranking. A common problem is to generalize concept learning, where the task is to decide whether an example is positive or negative, to ranking, where the task is to predict the 'degree of positiveness' of the test example. This scenario is also known as *bipartite ranking*. The quality of the predicted ranking is typically evaluated by computing the area under ROC curve (AUC; see Sect. 3.2). Thus, many learning algorithms aim at maximizing the AUC instead of maximizing accuracy.

Fawcett (2001) first studied the ranking quality of conventional rule learning algorithms, in particular with respect to various conflict resolution strategies. The results showed that rule sets may have a similar ranking performance as, e.g., a naïve Bayes classifier, which tries to directly estimate the probability that a given example is positive.

In Sect. 8.2.3 we have already mentioned that the use of precision as a search heuristic leads to a local optimization of the ROC curve by selecting the condition or rule that promises the steepest ascent (Fürnkranz & Flach, 2005). This property was exploited in the ROCCER algorithm (Prati & Flach, 2005) which can be used to iteratively select a subset of rules that maximizes the AUC. The key idea is that in each iteration, a rule is inserted if it extends the ROC curve generated by the current rule set. A similar idea has been investigated by Boström (2007) with a particular focus on pruning rules so that they locally maximize the AUC.

To date, PRIE (Fawcett, 2008) is arguably the most advanced rule learning system that aims at maximizing the AUC. It incorporates multiclass classification by maintaining a separate ROC space for each individual class, and selecting the best rule over all ROC spaces. In each space, a set of rules is maintained, which are iteratively combined in order to maximize the AUC. As it is not possible to consider all possible combinations, heuristics are used to estimate which combination will lead to a good extension of the ROC curve. The key idea is to interpolate between the expected location of the combination when independence of the rules is assumed and an optimistic estimate resulting from a favorable dependence assumption. The experimental evaluation showed that PRIE compares favorably to several of the competing approaches mentioned above.

Multipartite ranking. Multipartite ranking proceeds from the setting of ordinal classification, where an instance e belongs to one among a finite set of classes $\mathcal{C}$, which have a natural order. In contrast to the classification setting, however, the goal is not to learn a classifier but learn a function that allows to rank a set of instances according to their presumed target value. Ideally, a ranking is produced in which instances from higher classes precede those from lower classes. Thus, multipartite ranking may also be considered as a generalization of bipartite ranking, the difference being that data is partitioned into multiple groups instead of only positive and negative. The target ranking can be evaluated using the *C-index* (Gönen & Heller, 2005), which estimates the probability that a randomly chosen pair of examples is ranked correctly according to their true target values. Obviously, the C-index is an immediate generalization of the area under ROC curve.

Like label ranking, the multipartite ranking problem can also be reduced to concept learning via a pairwise decomposition (Fürnkranz, Hüllermeier, & Vanderlooy, 2009). The key idea of the transformation is that instead of computing a score for each label and predicting the label with the maximum score, an overall score is computed for each example that reflects whether the examples tends to have a low or high target value. Examples can then be sorted according to that score. An alternative method, which performed better in an experimental evaluation, is to use a straightforward adaptation of the ordinal classification algorithm proposed by Frank and Hall (2001). Most recently, Quevedo, Montañés, Luaces, and del Coz (2010) improved the pairwise approach by organizing the pairwise classifiers in a directed acyclic graph. Contrary to Platt, Cristianini, and Shawe-Taylor (2000), the class order in the DAG is not arbitrary but is determined by the underlying ordinal scale of the class attribute. Bipartite rankings of all instances in each leaf node can then be concatenated or combined probabilistically.

Object ranking. Object ranking has been mostly studied in the context of statistical learning, where the goal is to learn a numerical utility function that underlies the target ranking. A prototypical example of that line of work is the RANK-SVM support vector machine for ranking tasks (Joachims, 2002, 2006).

In the realm of rule learning, not much work has been performed on this problem. A notable exception is Cohen, Schapire, and Singer (1999), where it is proposed to solve the problem by explicitly learning a binary preference predicate that compares two examples and predicts which one is preferable. Obviously, such a predicate can be learned by any concept learner, and in particular by a rule learning algorithm. This predicate is then used for computing a ranking over a set of test examples. Just as for label ranking, the NP-hard problem of finding an optimal aggregation of the possibly inconsistent predictions for the pairwise preferences is solved by a greedy algorithm quite similar to weighted voting.

10.6 Regression

Whereas classification algorithms predict a nominal target variable, *regression* algorithms are trained to predict a numerical target value. Thus, the head of a *regression rule* typically predicts a constant, numerical value. More elaborate techniques are able to learn rules that have statistical regression models in the head of the rule. In such cases, the model is evaluated for an example that is covered by the rule, and the resulting value is predicted for this example.

In general, rule-based regression has not been particularly well investigated. There are some published results as well as a few commercial solutions (e.g., CUBIST[5]), but none of them has had a significant impact. They differ from classification rule learning mainly by the different heuristics they use for evaluating the quality of a candidate rule or a refinement. Typically, discrimination of positive and negative examples is replaced by reduction of the variance around the predicted mean value of a rule. Stopping criteria are a key problem for such algorithms because refining a rule will lead to a reduction in variance, so that rules will typically be refined to the point where they have no variance because they only cover a single example. A simple approach for countering this is to request a minimum coverage for each rule. FORS (Karalič & Bratko, 1997) uses an MDL-based pruning technique.

A prototypical algorithm of this type is the R^2 system (Torgo, 1995). The algorithm starts by focusing on a yet uncovered region of the search space. Such a region can be identified as a negation of any of the previously used features in the current rule set. It then learns a conventional regression model for this region, and initializes the next regression rule with the negated feature in the body and the model in the conclusion. It then tries to add further conditions that improve a trade-off between the number of covered examples and the fit of the model.

More popular are techniques that first learn regression trees (Breiman, Friedman, Olshen, & Stone, 1984; Kramer, 1996; Quinlan, 1992) and straightforwardly convert the resulting trees into regression rules. A somewhat more elaborate technique is used by the M5RULES algorithm (Holmes, Hall, & Frank, 1999) which extracts a single rule from a learned regression tree, removes all covered examples, and subsequently learns a new tree and rule from the remaining examples until all the examples are covered. This is quite similar to the PART algorithm for learning classification rules (Frank & Witten, 1998).

Another popular technique is to deal with a continuous target attribute by reducing it to conventional classification. For example, Langford, Oliveira, and Zadrozny (2006) convert a regression problem into a set of binary classification problems, where each classifier essentially tests whether an example is above or below a certain threshold. Another common approach is to discretize the numeric values as a preprocessing step and apply conventional classification algorithms on the discretized target attribute. Research following this path can be found in (Torgo & Gama, 1997; Weiss & Indurkhya, 1995). The main problem here is that the performance of this technique strongly depends on the choice of the number of classes. The key idea of the approach of Janssen and Fürnkranz (2011) is to dynamically convert a regression problem into a classification problem by defining a region of positive examples around the mean of the current rule, and use classification learning heuristics to find good conditions that discriminate these positive examples from all others. A key difference to ϵ*-insensitive* loss functions, which form the basis of several support-vector regression algorithms (Smola & Schölkopf, 2004), is that the width of the insensitive region is dynamically adapted after each refinement step.

[5]http://www.rulequest.com/cubist-info.html

One of the key problems of regression rule sets is that the value that is predicted for a test example depends only on a single rule, which (at least in the case of constant predictions) results in a very large bias. This can be weakened by combining the predictions of many rules. Thus, state-of-the-art rule-based regression algorithms are typically based on ensemble techniques. For example, RULEFIT (Friedman & Popescu, 2008) performs a gradient descent optimization, allows the rules to overlap, and the final prediction is calculated by the sum of all predicted values of the covering rules. REGENDER (Dembczyński, Kotłowski, & Słowiński, 2008) provides a general framework that allows the use of forward stagewise additive modeling for minimizing various different loss functions with a rule-based learner.

10.7 Unsupervised Learning

Undoubtedly the most popular unsupervised rule learning technique is the discovery of association rules, which we briefly covered in Sect. 1.6.1. Apart from that, there have been a few attempts to develop rule-based clustering algorithms, i.e., algorithms that find a set of rules that identify homogeneous groups of examples that are similar to each other (*intraclass similarity*) but differ from examples in other groups (*interclass dissimilarity*; Fisher, 1987). The clustering problem with a strong focus on interpretable concepts is also known as *conceptual clustering* (Stepp & Michalski, 1986).

The advantages of rule-based clustering have been recognized early on. Michalski and Stepp (1983) introduced CLUSTER/2, an adaptation of the AQ rule learning methodology (Michalski, 1969, 1980) that replaced its evaluation criterion for supervised learning with a rule heuristic that measured the cohesiveness of the examples that are covered by a rule. However, while the classification system AQ produced many successor systems (essentially all algorithms of the separate-and-conquer family of rule learning algorithms), the impact of CLUSTER/2 was rather limited. There has been some later work on conceptual clustering, mostly in structural domains (Bisson, 1992; Blockeel, De Raedt, & Ramon, 1998; Cook & Holder, 1994; Pelleg & Moore, 2001; Stepp & Michalski, 1986) but there is no direct follow-up that brought the ideas introduced into the CLUSTER/2 system up to the current state-of-the-art of rule learning algorithms.

Recently, there has again been some work on bridging the gap between supervised and unsupervised rule learning. The key idea of *predictive clustering rules* is to learn rules that trade off the precision with respect to a target attribute and the compactness with respect to the input attributes. The key advantage of this approach is that it can easily be adapted to different prediction tasks. In particular, it is also well-suited for prediction problems with multiple target attributes (Zenko, 2007; Zenko, Džeroski, & Struyf, 2006). *Cluster grouping* (Zimmermann & De Raedt, 2009) provides a unifying framework for many descriptive rule learning algorithms (cf. Chap. 11) that incorporates clustering as a special case.

10.8 Conclusion

In this chapter, we have seen how we can tackle various prediction tasks with rule learning algorithms. In particular, we have considered multiclass classification, multilabel classification, ordinal and hierarchical classification, ranking, regression, and clustering. Many of these complex learning tasks can be tackled by reducing them to concept learning, which we have discussed in previous chapters.

Particularly versatile has been the framework of preference learning, in which several of the above-mentioned tasks can be framed. Learning by pairwise comparisons, a straightforward generalization of pairwise classification, aims at directly learning a binary preference predicate that allows the comparison of two class labels (in the case of instance ranking) or two examples (in the case of object ranking). Such comparisons may be viewed as basic building blocks upon which more complex decision making procedures can be based. Eventually, complex prediction problems can thus be reduced to the solution of a set of simpler binary classification problems, which makes this approach especially appealing from a machine learning point of view. Despite the high number of binary classifiers needed (quadratic in the number of class labels), the training time of the approach seems to be competing with alternative approaches. Storing and querying such a large number of classifiers, however, is an important problem requiring further research.

Chapter 11
Supervised Descriptive Rule Learning

This chapter presents subgroup discovery (SD) and some of the related supervised descriptive rule induction techniques, including contrast set mining (CSM) and emerging pattern mining (EPM). These descriptive rule learning techniques are presented in a unifying framework named *supervised descriptive rule learning*. All these techniques aim at discovering patterns in the form of rules induced from labeled data. This chapter contributes to the understanding of these techniques by presenting a unified terminology and by explaining the apparent differences between the learning tasks as variants of a unique supervised descriptive rule learning task. It also shows that various rule learning heuristics used in CSM, EPM, and SD algorithms all aim at optimizing a trade off between rule coverage and precision.

The chapter is organized as follows. After a brief introduction in Sect. 11.1, the main supervised descriptive rule learning approaches are presented: subgroup discovery in Sect. 11.2, contrast set mining in Sect. 11.3, and emerging pattern mining in Sect. 11.4. Section 11.5 is dedicated to unifying the terminology, definitions, and heuristics. An overview of other related supervised descriptive rule learning approaches is presented in Sect. 11.6, followed by a brief discussion of possibilities for visualizing subgroups (Sect. 11.7) and a brief summary which concludes the chapter (Sect. 11.8).

11.1 Introduction

Most rule learning techniques aim at discovering comprehensible predictive models. These rule learning techniques are referred to as *predictive rule learning* techniques, where models in the form of sets of rules, induced from class-labeled data, are used to predict the class value of previously unlabeled examples. In contrast, this

---

†Parts of this chapter are based on Kralj Novak, Lavrač, and Webb (2009).

J. Fürnkranz et al., *Foundations of Rule Learning*, Cognitive Technologies,
DOI 10.1007/978-3-540-75197-7_11, © Springer-Verlag Berlin Heidelberg 2012

chapter focuses on *descriptive rule learning*, where the aim is to find understandable patterns described by individual rules.

In a typical descriptive rule learning task, descriptive rules are induced from unlabeled data, as is the case in association rule learning. The approach to association rule learning from unlabeled data, which is well-covered in the data mining literature, is out of the scope of this book. On the other hand, a recent focus of descriptive rule learning is on the task of supervised descriptive rule learning from class-labeled data, where the aim is to construct descriptive patterns in the form of independent rules. Supervised descriptive rule learning tasks include *subgroup discovery* (Klösgen, 1996; Wrobel, 1997), *contrast set mining* (Bay & Pazzani, 2001) and *emerging pattern mining* (Dong & Li, 1999).

This chapter gives a survey of subgroup discovery, contrast set mining and emerging pattern mining in a unifying framework, named *supervised descriptive rule learning*. Typical applications of supervised descriptive rule learning include patient risk group detection in medicine, bioinformatics applications like finding sets of over-expressed genes for specific treatments in microarray data analysis, and identifying distinguishing features of different customer segments in customer relationship management. The main goal of these applications is to understand the underlying phenomena and not to classify new unlabeled instances. Some exemplary applications are described in more detail in Chap. 12.

Throughout this chapter, we will use examples that refer to the sample database of Table 1.1 consisting of 14 individuals who are described with four attributes: Education (with possible values primary school, secondary school, or university), MaritalStatus (with possible values single, married, or divorced), Sex (male or female), and HasChildren (yes or no), which encode rudimentary information about the sociodemographic background. We are also given a class attribute Approved with the possible values yes and no, which describes whether the person approves or disapproves of a certain issue. Since there is no need for expert knowledge to interpret the results, this dataset is appropriate for illustrating the results of supervised descriptive rule learning algorithms, whose task is to find interesting patterns describing individuals that are likely to approve or disapprove the issue, based on the four demographic characteristics.

In contrast to predictive rule learning algorithms, which would try to find a small rule set that collectively covers the entire example space (such as the rule set shown in Fig. 1.3), *descriptive rule learning* algorithms typically produce individual rules which try to highlight interesting relations that can be found in the database. Unsupervised techniques, such as the APRIORI association rule learner (Agrawal, Mannila, Srikant, Toivonen, & Verkamo, 1995), have no restriction on the attributes. In our example, they would treat the class no differently than any other attribute (cf., e.g., the rules shown in Fig. 1.5). Supervised techniques, on the other hand, always discover rules of the form $\mathrm{B} \rightarrow c$, which relate some of the attributes to the class attribute C. Figure 1.6 shows a few exemplary rules.

Exactly which rules will be induced by a supervised descriptive rule learning algorithm depends on the task definition, the selected algorithm, as well as the user-defined constraints concerning minimal rule support, precision, etc. In the following sections, the example set of Table 1.1 will be used to illustrate the outputs of a subgroup discovery algorithm, a contrast set mining algorithm, and an emerging pattern mining algorithm.

11.2 Subgroup Discovery

The task of subgroup discovery was defined by Klösgen (1996) and Wrobel (1997) as follows: *Given a population of individuals and a property of those individuals that we are interested in, find population subgroups that are statistically 'most interesting', e.g., are as large as possible and have the most unusual statistical (distributional) characteristics with respect to the property of interest.*

11.2.1 Subgroup Discovery Algorithms

Subgroup descriptions are conjunctions of features that are characteristic for a selected class of individuals (property of interest). A subgroup description B can be seen as the condition part of a rule $SubgroupDescription \rightarrow C$, where C is some property of interest. Obviously, subgroup discovery is a special case of the more general rule learning task.

Subgroup discovery research has evolved in several directions. On the one hand, exhaustive approaches guarantee the optimal solution given the optimization criterion. An early system that can use both exhaustive and heuristic discovery algorithms is EXPLORA by Klösgen (1996). Other algorithms for exhaustive subgroup discovery are the SD-MAP method by Atzmüller and Puppe (2006) and APRIORI-SD by Kavšek and Lavrač (2006). On the other hand, classification rule learners have been adapted to perform subgroup discovery with heuristic search techniques drawn from classification rule learning coupled with constraints appropriate for descriptive rules. Examples include algorithm SD by Gamberger and Lavrač (2002) and algorithm CN2-SD by Lavrač, Kavšek, Flach, and Todorovski (2004).

Relational subgroup discovery approaches have been proposed by Wrobel (1997, 2001) with algorithm MIDOS, by Klösgen and May (2002) with algorithm SUBGROUPMINER, which is designed for spatial data mining in relational space databases, and by Zelezný and Lavrač (2006) with the RSD algorithm for relational subgroup discovery. RSD uses a propositionalization approach to relational subgroup discovery, achieved through appropriately adapting rule learning and first-order feature construction. Other nonrelational subgroup discovery algorithms were developed, including an algorithm for exploiting background knowledge in subgroup discovery (Atzmüller, Puppe, & Buscher, 2005a), and an iterative genetic algorithm SDIGA by del Jesus, González, Herrera, and Mesonero (2007) implementing a fuzzy system for solving subgroup discovery tasks.

```
IF    MaritalStatus = married
THEN Approved = yes

IF    MaritalStatus = divorced
 AND HasChildren = no
THEN Approved = yes

IF    Sex = female
THEN Approved = yes

IF    Education = university
THEN Approved = yes
```

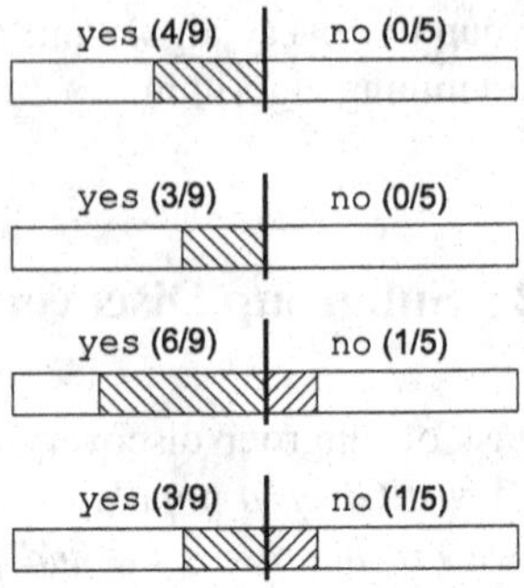

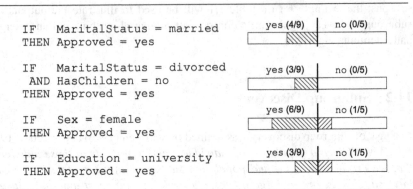

Fig. 11.1 Four subgroup descriptions induced from the data of Table 1.1. Note that in subgroup discovery, for a selected target class (yes) all the rules have just the selected target class label in rule consequents

Different heuristics have been used for subgroup discovery. By definition, the interestingness of a subgroup depends on its unusualness and size, therefore the rule quality evaluation heuristics needs to combine both factors. Weighted relative accuracy (*WRA*, see p. 143) is used by algorithms CN2-SD, APRIORI-SD, and RSD and—in a different formulation and in different variants—also by MIDOS and EXPLORA, the generalization quotient (*GQuotient*, see p. 148) is used by the SD algorithm. SUBGROUPMINER uses the classical binomial test to verify if the share of examples of the target class is significantly different in a subgroup.

Another important problem is the elimination of redundant subgroups. Algorithms CN2-SD, APRIORI-SD, SD and RSD use weighted covering (Lavrač et al., 2004) to achieve rule diversity (see Sect. 8.3). Algorithms EXPLORA and SUBGROUPMINER use an approach called subgroup suppression (Klösgen, 1996). A sample set of subgroup describing rules, induced by APRIORI-SD with parameters *Support* (see p. 141) set to 15 % (requiring at least two covered training examples per rule) and *Confidence* (*Precision*, see p. 144) set to 65 %, is shown in Fig. 11.1.

11.2.2 Selected Applications of Subgroup Discovery

Subgroup discovery was used in numerous real-life applications. The applications in medical domains include the analysis of coronary heart disease (Gamberger & Lavrač, 2002) and brain ischaemia data analysis (Kralj, Lavrač, Gamberger, & Krstačić, 2007a, 2007b; Lavrač, Kralj, Gamberger, & Krstačić, 2007), as well as profiling examiners for sonographic examinations (Atzmüller, Puppe, & Buscher, 2005b). Spatial subgroup mining applications include mining of census data (Klösgen, May, & Petch, 2003) and mining of vegetation data (May & Ragia, 2002).

There are also applications in other areas like marketing (del Jesus et al., 2007; Lavrač, Cestnik, Gamberger, & Flach, 2004) and analysis of manufacturing shop floor data (Jenkole, Kralj, Lavrač, & Sluga, 2007). Chapter 12 will present a few applications in more detail.

11.3 Contrast Set Mining

A *contrast set* was defined by Bay and Pazzani (2001) as a 'conjunction of attributes and values that differ meaningfully in their distributions across groups'. Thus, whereas subgroup mining targets primarily groups that are characteristic for a single class value, contrast set mining focuses on the discovery of rules that capture situations which have a strongly different probability of occurring in two example groups G_i and G_j.

11.3.1 Contrast Set Mining Algorithms

The STUCCO algorithm (Search and Testing for Understandable Consistent Contrasts) by Bay and Pazzani (2001) is based on the MAX-MINER rule discovery algorithm (Bayardo 1998). STUCCO discovers a set of contrast sets along with their support values for the different groups G_i in the head of the rule. STUCCO employs a number of pruning mechanisms. A potential contrast set is discarded if it fails a statistical test for independence with respect to the group variable G_i. It is also subjected to what Webb (2007) calls a test for productivity. Rule B → **C** is *productive* iff

$$\forall \mathrm{X} \subset \mathrm{B} : \mathit{Confidence}(\mathrm{X} \rightarrow \mathbf{C}) < \mathit{Confidence}(\mathrm{B} \rightarrow \mathbf{C}) \qquad (11.1)$$

Therefore a more specific contrast set must have higher confidence than any of its generalizations. Further tests for minimum counts and effect sizes may also be imposed.

It was shown by Webb, Butler, and Newlands (2003) that contrast set mining is a special case of the more general rule learning task. A contrast set can be interpreted as the antecedent of rule B → **C**, and group G_i for which it is characteristic—in contrast with some other group G_j—as the rule consequent, leading to rules of the form *ContrastSet* → G_i. A standard descriptive rule discovery algorithm, such as an association rule learner, can be used for the task if the consequent is restricted to a class variable **C** whose values $\mathbf{C} = \{G_1, \ldots, G_C\}$ denote group membership. In particular, Webb et al. showed that when STUCCO and the general-purpose descriptive rule learning system MAGNUM OPUS[1] were each run with

[1] http://www.giwebb.com/

their default settings, but the consequent restricted to the contrast variable in the case of MAGNUM OPUS, the contrasts found differed mainly as a consequence of differences in the statistical tests employed to screen the rules.

Hilderman and Peckham (2005) proposed a different approach to contrast set mining called CIGAR (ContrastIng Grouped Association Rules). CIGAR uses different statistical tests to STUCCO or MAGNUM OPUS for both independence and productivity and introduces a test for *minimum support*. Wong and Tseng (2005) have developed techniques for discovering contrasts that can include negations of terms in the contrast set.

STUCCO introduced a novel variant of the Bonferroni correction for multiple tests which applies ever more stringent critical values to the statistical tests employed as the number of conditions in a contrast set is increased. In comparison, the other techniques discussed below do not, by default, employ any form of correction for multiple comparisons, as result of which they have high risk of making *false discoveries* (Webb, 2007).

In general, contrast set mining approaches operate within the feature-based data view put forward in this book (Chap. 4). A few works attempt to directly deal with numerical data values. A data discretization method developed specifically for contrast set mining purposes is described by Bay (2000). This approach does not appear to have been further utilized by the contrast set mining community, except by Lin and Keogh (2006), who extended contrast set mining to time series and multimedia data analysis. They introduced a formal notion of a time series contrast set along with a fast algorithm to find time series contrast sets. An approach to quantitative contrast set mining without discretization in the preprocessing phase is proposed by Simeon and Hilderman (2007) with the algorithm GENQCSETS. In this approach, a slightly modified equal-width binning method is used.

Common to most contrast set mining approaches is that they generate all candidate contrast sets and later use statistical tests to identify the interesting ones. Figure 11.2 shows six descriptive rules constructed by the MAGNUM OPUS (Webb, 1995) rule learning algorithm applied in the contrast set mining setting. These rules were found using the default settings except that the critical value for the statistical test was relaxed to 0.25.

11.3.2 Selected Applications of Contrast Set Mining

The contrast set mining paradigm does not appear to have been pursued in many published applications. Webb et al. (2003) investigated its use with retail sales data. Wong and Tseng (2005) applied contrast set mining for designing customized insurance programs. Siu et al. (2005) have used contrast set mining to identify patterns in synchrotron X-ray data that distinguish tissue samples of different forms of cancerous tumors. Kralj et al. (2007b) have addressed a contrast set mining problem of distinguishing between two groups of brain ischaemia patients by transforming the contrast set mining task to a subgroup discovery task.

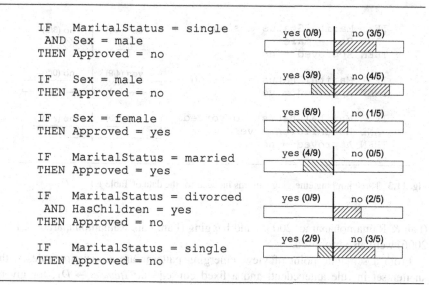

```
IF   MaritalStatus = single
 AND Sex = male
THEN Approved = no

IF   Sex = male
THEN Approved = no

IF   Sex = female
THEN Approved = yes

IF   MaritalStatus = married
THEN Approved = yes

IF   MaritalStatus = divorced
 AND HasChildren = yes
THEN Approved = no

IF   MaritalStatus = single
THEN Approved = no
```

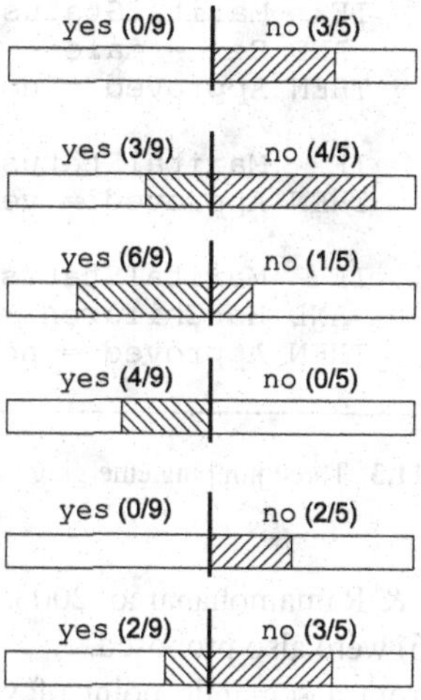

Fig. 11.2 Rules describing contrast sets in the data of Table 1.1, induced by MAGNUM OPUS

11.4 Emerging Pattern Mining

An *emerging pattern* was defined by Dong and Li (1999) as 'an itemset whose support increases significantly from one dataset to another.' Emerging patterns are said to 'capture emerging trends in time-stamped databases, or to capture differentiating characteristics between classes of data.'

11.4.1 Emerging Pattern Mining Algorithms

Efficient algorithms for mining emerging patterns were proposed by Dong and Li (1999), and Fan and Ramamohanarao (2003b). When first defined by Dong and Li, the purpose of emerging patterns was to capture emerging trends in time-stamped data, or useful contrasts between data classes. Subsequent emerging pattern research has largely focused on the use of the discovered patterns for classification purposes, for example, classification by emerging patterns (Dong, Zhang, Wong, & Li, 1999; Li, Dong, & Ramamohanarao, 2000) and classification by jumping emerging patterns[2] (Li, Dong, & Ramamohanarao, 2001). An advanced Bayesian approach

[2]Jumping emerging patterns are emerging patterns with support zero in one dataset and greater then zero in the other dataset.

```
IF   MaritalStatus = single
 AND Sex = male
THEN Approved = no

IF   MaritalStatus = married
THEN Approved = yes

IF   MaritalStatus = divorced
 AND HasChildren = yes
THEN Approved = no
```

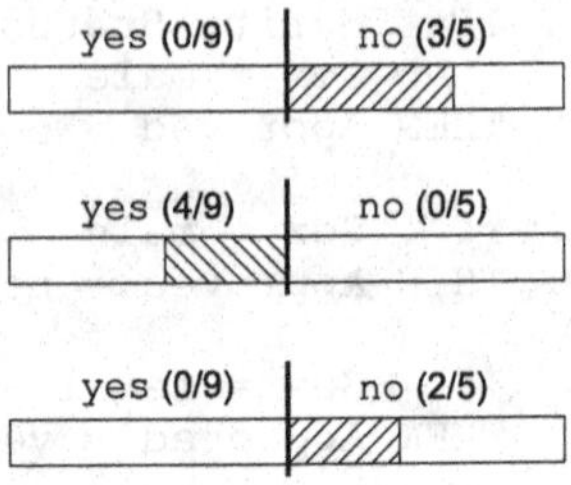

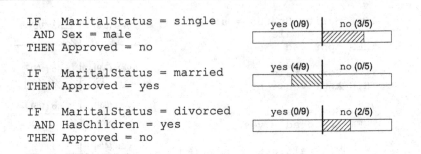

Fig. 11.3 Three jumping emerging patterns induced for the data of Table 1.1

(Fan & Ramamohanarao, 2003a) and bagging (Fan, Fan, Ramamohanarao, & Liu, 2006) were also proposed.

From a semantic point of view, emerging patterns are association rules with an itemset in rule antecedent, and a fixed consequent: *ItemSet* $\rightarrow D_1$, for given dataset D_1 being compared to another dataset D_2. The measure of quality of emerging patterns is the *growth rate*, the ratio of the two supports (cf. also Sect. 11.5.3).

Some researchers have argued that finding all the emerging patterns above a minimum growth rate generates too many patterns to be analyzed by a domain expert. Fan and Ramamohanarao (2003b) have worked on selecting the interesting emerging patterns, while Soulet, Crémilleux, and Rioult (2004) have proposed condensed representations of emerging patterns.

Boulesteix, Tutz, and Strimmer (2003) introduced a CART-based approach to discover emerging patterns in microarray data. The method is based on growing decision trees from which the emerging patterns are extracted. It combines pattern search with a statistical procedure based on Fisher's exact test to assess the significance of each emerging pattern. Subsequently, sample classification based on the inferred emerging patterns is performed using maximum-likelihood linear discriminant analysis.

Figure 11.3 shows all the jumping emerging patterns found for the data in Table 1.1 when using a minimum support of 2. These were discovered using the MAGNUM OPUS software, limiting the consequent to the variable `Approved`, setting minimum confidence to 1.0, and setting minimum support to 2.

11.4.2 Selected Applications of Emerging Patterns

Emerging patterns have been mainly applied to the field of bioinformatics, more specifically to microarray data analysis. Li, Liu, Downing, Yeoh, and Wong (2003) present an interpretable classifier based on simple rules that is competitive to the state of the art black-box classifiers on the acute lymphoblastic leukemia (ALL)

Table 11.1 Table of synonyms from different communities, showing the compatibility of terms

Contrast set mining	Emerging pattern mining	Subgroup discovery	Rule learning
Contrast Set	Itemset	Subgroup description	Rule Body
Groups $G_1, \ldots G_n$	Datasets D_1 and D_2	Class/Property C	Class variable **C** concept c_i
Examples in groups $G_1, \ldots G_n$	Transactions in datasets D_1 and D_2	Examples of C and $\overline{C}$	Examples of $c_1 \ldots c_n$
Examples for which the contrast set is true	Transactions containing the itemset	Subgroup of instances	Covered examples
Support of contrast set on G_i	Support of EP in dataset D_1	True positive rate	True positive rate
Support of contrast set on G_j	Support of EP in dataset D_2	False positive rate	False positive rate

microarray dataset. Li and Wong (2002b) have focused on finding groups of genes by emerging patterns and applied it to the ALL/AML dataset and the colon tumor dataset. Song, Kimb, and Kima (2001) used emerging patterns together with unexpected change and the added/perished rule to mine customer behavior.

11.5 A Unifying View

This section presents a unifying framework for contrast set mining, emerging pattern mining, and subgroup discovery as the main representatives of supervised descriptive rule discovery algorithms. This is achieved by unifying the terminology, the task definitions, and the rule learning heuristics used in supervised descriptive rule learning.

11.5.1 Unifying the Terminology

Contrast set mining (CSM), emerging pattern mining (EPM), and subgroup discovery (SD) were developed in different communities, each developing their own terminology that needs to be clarified before proceeding. Below we show that terms used in different communities are compatible and can be translated to the standard rule learning terminology through a term dictionary proposed in Table 11.1. More specifically, this table provides a dictionary of equivalent terms in a unifying terminology of concept learning (considering class c_i as the concept to be learned from the positive examples of this concept, and the negative examples formed of examples of all other classes).

Table 11.2 Table of task definitions from different communities, showing the compatibility of task definitions in terms of output rules

Contrast set mining	Emerging pattern mining	Subgroup discovery	Rule learning
Given examples in G_i *vs.* G_j from $G_1, \dots G_n$	Given transactions in D_1 and D_2 from D_1 and D_2	Given examples in C from C and $\overline{C}$	Given examples in c_i from $c_1 \dots c_C$
Find $ContrastSet_{ik} \rightarrow G_i$ $ContrastSet_{jl} \rightarrow G_j$	Find $ItemSet_{1_k} \rightarrow D_1$ $ItemSet_{2_l} \rightarrow D_2$	Find $SubgrDescr_k \rightarrow C$	Find $\{RuleCond_{ik} \rightarrow c_i\}$

11.5.2 Unifying the Task Definitions

Having established a unifying view of the terminology, the next step is to provide a unifying view of the different task definitions.

CSM (Bay & Pazzani, 2001) Given a set of user defined groups G_1, G_2, …, G_n of data instances, a contrast set is a pattern that best discriminates the instances of different user-defined groups. A special case of contrast set mining considers only two contrasting groups (G_1 and G_2). In this case, we wish to find characteristics of one group discriminating it from the other and vice versa.

EPM (Dong & Li, 1999) For two datasets D_1 and D_2 that are characterized with the same set of features, emerging pattern mining aims at discovering patterns whose support increases significantly from one dataset to another.

SD (Wrobel, 1997) Given a property of interest C, and the population of examples of C and $\overline{C}$, the subgroup discovery task aims at finding population subgroups of instances of interest which are as large as possible and have the most unusual statistical (distributional) characteristics with respect to the property of interest C.

The definitions of contrast set mining, emerging pattern mining, and subgroup discovery appear different: contrast set mining searches for discriminating characteristics of groups called contrast sets, emerging pattern mining aims at discovering itemsets whose support increases significantly from one dataset to another, while subgroup discovery searches for subgroup descriptions. By using the dictionary from Table 11.1 we can see that the goals of these three mining tasks are very similar, it is primarily the terminology that differs.

To show the compatibility of task definitions, meaning that the tasks can be translated into other tasks without substantially changing the learning goal, we propose a unifying table of task definitions (Table 11.2). This table allows us to see that emerging pattern mining task $EPM(D_1, D_2)$ is equivalent to $CSM(G_i, G_j)$. It is also easy to show that a two-group contrast set mining task $CSM(G_i, G_j)$ can be directly translated into the following two subgroup discovery tasks: $SD(G_i)$ for $C = G_i$ and $\overline{C} = G_j$, and $SD(G_j)$ for $C = G_j$ and $\overline{C} = G_i$.

Having proved that the subgroup discovery task is compatible with a two-group contrast set mining task, it is by induction compatible with a general contrast set mining task, as shown below.

$CSM(G_1, \dots G_n)$
for i=2 to n **do**
for j=1, j$\neq$ i to n-1 **do**
$SD(C = G_i\ vs.\ \overline{C} = G_j)$

Note that in Table 11.2, the column 'Rule Learning' again corresponds to a concept learning setting where the task is to discriminate class c_i from all other classes. Single rules can be learned with the LEARNONERULE algorithm (Fig. 2.8) or, more generally, with the FINDBESTRULE algorithm (Fig. 6.1). Sets of rules can be learned with the WEIGHTEDCOVERING algorithm (Fig. 8.4), which ensures the necessary diversity of the found rules.

While the primary tasks are very closely related, each of the three communities has concentrated on different sets of issues around this task. The contrast set discovery community has paid the greatest attention to the statistical issues of multiple comparisons that, if not addressed, can result in a high risk of false discoveries. The emerging patterns community has investigated how supervised descriptive rules can be used for classification. The contrast set and emerging pattern communities have primarily addressed only categorical data, whereas the subgroup discovery community has also considered numeric and relational data. The subgroup discovery community has also explored techniques for discovering small numbers of supervised descriptive rules with high coverage of available examples.

11.5.3 Unifying the Rule Learning Heuristics

The aim of this section is to provide a unifying view on rule learning heuristics used in different communities. To this end we first investigate the rule quality measures.

CSM Contrast set mining aims at discovering contrast sets that best discriminate the instances of different user-defined groups. The support of contrast set B with respect to group G_i, *Support*(B $\rightarrow G_i$), is the percentage of examples in G_i for which the contrast set is true (cf. p. 141). A derived goal of contrast set mining, proposed by Bay and Pazzani (2001), is to find contrast sets whose support differs meaningfully across two groups G_i and G_j, or, equivalently, whose support difference is maximal.

$$\mathit{SuppDiff}(\mathrm{B}, G_j, G_i) = |\mathit{Support}(\mathrm{B} \rightarrow G_i) - \mathit{Support}(\mathrm{B} \rightarrow G_j)|$$

Thus, Bay and Pazzani (2001) define the goal of contrast set mining as to find contrast sets B for which $\mathit{SuppDiff}(\mathrm{B}, G_i, G_j) \geq \delta$ for some user-defined parameter δ.

Table 11.3 Table of relationships between the pairs of heuristics, and their equivalents in classification rule learning

Contrast set mining	Emerging pattern mining	Subgroup discovery	Rule learning
SuppDiff (B, G_i, G_j)		*WRA*$(\mathrm{B} \rightarrow C)$	Piatetski-Shapiro leverage
	GrowthRate (B, D_1, D_2)	*GQuotient*$(\mathrm{B} \rightarrow C)$	Odds ratio for $g = 0$ accuracy/precision for $g = p$

EPM Emerging pattern mining aims at discovering itemsets whose support increases significantly from one dataset to another (Dong & Li, 1999). Suppose we are given an ordered pair of datasets D_1 and D_2. The *growth rate* of an itemset X from D_1 to D_2 is defined as follows:

$$\boxed{GrowthRate(\mathrm{B}, D_1, D_2) = \frac{Support(\mathrm{B} \rightarrow D_1)}{Support(\mathrm{B} \rightarrow D_2)}}$$

Thus, the growth rate computes the ratio of the supports of the pattern B in the two groups D_1 and D_2, which is equivalent to the ratio of the two confidences, because both rules have the same rule body B. The growth ratio determines, for example, that a pattern with a 10% support in one dataset and 1% in the other is better than a pattern with support 70% in one dataset and 10% in the other (as $\frac{10}{1} > \frac{70}{10}$). In order to prevent division by 0 if $Support(\mathrm{B} \rightarrow D_2) = 0$, the above definition is extended as follows: if also $Support(\mathrm{B} \rightarrow D_1) = 0$ then $GrowthRate(\mathrm{B}, D_1, D_2) = 0$, otherwise $GrowthRate(\mathrm{B}, D_1, D_2) = \infty$.

SD Subgroup discovery aims at finding population subgroups that are as large as possible and have the most unusual statistical (distributional) characteristics with respect to the property of interest (Wrobel, 1997). There were several heuristics developed and used in the subgroup discovery community. Since they follow from the task definition, they try to maximize subgroup size and the distribution difference at the same time. Examples of such heuristics are the weighted relative accuracy (WRA) and the generalization quotient ($GQuotient$).

Kralj Novak et al. (2009) have shown that the heuristics used in CSM, EPM, and SD are compatible, and derive the relationships between pairs of heuristics shown in Table 11.3.

11.5.4 Comparison of Rule Selection Mechanisms

Having established a unifying view on the terminology, definitions, and rule learning heuristics, the last step is to analyze rule selection mechanisms used by different algorithms. The motivation for rule selection can be either to find only significant rules or to avoid overlapping rules (too many too similar rules), or to avoid showing redundant rules to the end users. Note that rule selection is not always necessary and that depending on the goal, redundant rules can be valuable (e.g., classification by aggregating emerging patterns by Dong et al., 1999). Two approaches are commonly used: statistic tests and the (weighted) covering approach. In this section, we compare these two approaches.

Webb et al. (2003) have first proved that contrast set mining is a special case of the more general rule discovery task, and that it can be performed by standard rule discovery algorithms such as MAGNUM OPUS. The experimental study of the paper, which compared MAGNUM OPUS to STUCCO, highlighted the importance of appropriate rule filtering, where the evidence seemed to suggest that STUCCO filters out some promising rules, whereas MAGNUM OPUS admits some spurious rules.

STUCCO (see Bay and Pazzani (2001) for more details) uses several mechanisms for rule pruning. Statistical significance pruning removes contrast sets that, while significant and large, derive these properties only due to being specializations of more general contrast sets: any specialization is pruned that has a similar support to its parent or that fails an χ^2 test of independence with respect to its parent.

The emphasis of recent work on MAGNUM OPUS has been on developing statistical tests that are robust in the context of the large search spaces explored in many rule discovery applications (Webb, 2007). These include tests for independence between the antecedent and consequent, and tests to assess whether specializations have significantly higher confidence than their generalizations.

For subgroup discovery, Lavrač et al. (2004) proposed the *weighted covering* approach with the aim of ensuring the diversity of rules induced in different iterations of the algorithm. As described in detail in Sect. 8.3, in each iteration, after selecting the best rule, the weights of positive examples are decreased according to the number of rules covering each positive example. For selecting the best rule in consequent iterations, the SD algorithm (Gamberger & Lavrač, 2002) uses a weighted variant of the generalization quotient, while the CN2-SD (Lavrač et al., 2004) and APRIORI-SD (Kavšek & Lavrač, 2006) algorithms use weighted relative accuracy modified with example weights.

Unlike in the sections on the terminology, task definitions, and rule learning heuristics, the comparison of rule pruning mechanisms described in this section does not result in a unified view; although the goals of rule pruning may be the same, the pruning mechanisms used in different subareas of supervised descriptive rule discovery are—as shown above—very different. It is still an open question which approach, if any, is preferable.

11.6 Other Approaches to Supervised Descriptive Rule Learning

Research in some closely related areas of rule learning, performed independently from the above described approaches, is outlined below.

11.6.1 Change Mining

The paper by Liu, Hsu, and Ma (2001) on *fundamental rule changes* proposes a technique to identify the set of fundamental changes in two given datasets collected from two time periods. The proposed approach first generates rules and in the second phase it identifies changes (rules) that cannot be explained by the presence of other changes (rules). This is achieved by applying the statistical χ^2 test for homogeneity of support and confidence. This differs from contrast set discovery through its consideration of rules for each group, rather than itemsets. A change in the frequency of just one itemset between groups may affect many association rules, potentially all rules that have the itemset as either an antecedent or consequent.

Liu, Hsu, Han, and Xia (2000a) and Wang, Zhou, Fu, & Yu (2003) present techniques that identify differences in the decision trees and classification rules, respectively, found on two different data sets.

11.6.2 Mining Closed Sets from Labeled Data

Closed sets have proved successful in the context of compact data representation for association rule learning. A closed set may be regarded as a rule that cannot be made more specific without decreasing its support. However, their use is mainly descriptive, dealing only with unlabeled data. It was recently shown that when considering labeled data, closed sets can be adapted for classification and discrimination purposes by conveniently contrasting covering properties on positive and negative examples (Garriga, Kralj, & Lavrač, 2006). The approach was successfully applied in potato microarray data analysis to a real-life problem of distinguishing between virus sensitive and resistant transgenic potato lines (Kralj et al., 2006).

11.6.3 Exception Rule Mining

Exception rule mining considers a problem of finding a set of rule pairs, each of which consists of an exception rule (which describes a regularity for fewer objects) associated with a strong rule (description of a regularity for numerous objects with few counterexamples). An example of such a rule pair is *'using a seat belt is safe'* (strong rule) and *'using a seat belt is inappropriate for an infant'* (exception rule). While the goal of exception rule mining is also to find descriptive rules from labeled data, in contrast with other rule discovery approaches described in this paper, the goal of exception rule mining is to find 'weak' rules—surprising rules that are an exception to the general belief represented by the background knowledge.

Suzuki (2006) and Daly and Taniar (2005), summarizing the research in exception rule mining, reveal that the key concerns addressed by this body of research include interestingness measures, reliability evaluation, practical application, parameter reduction, and knowledge representation, as well as providing fast algorithms for solving the problem.

11.6.4 Descriptive Induction of Regression Rules

Supervised descriptive rule discovery seeks to discover sets of conditions that are related to deviations in the class distribution, where the class is a qualitative variable. A related body of research seeks to discover sets of conditions that are related to deviations in a target quantitative variable. Such techniques include *bump hunting* (Friedman & Fisher, 1999), *quantitative association rules* (Aumann & Lindell, 1999), and *impact rules* (Webb, 2001).

11.7 Subgroup Visualization

Visualization is an important practical issue in supervised descriptive rule learning. The goal is to present the quality of induced rules and the relevance of the results for the application domain. Several methods for rule visualization were developed by Wettschereck (2002), Wrobel (2001), Gamberger, Lavrač, and Wettschereck (2002), Kralj, Lavrač, and Zupan (2005), and Atzmüller and Puppe (2005). In this section, they are illustrated using the coronary heart disease dataset that will be presented in Sect. 12.4. In particular, we compare different visualizations of the five subgroups **a1**, **a2**, **b1**, **b2**, and **c1** shown in Table 12.9 on p. 284. However, it is not necessary to look up the rules to understand the presented visualization techniques. Note that although these techniques were all initially developed for subgroup discovery applications, they can all be used to present any type of class labeled rules.

11.7.1 Visualization by Pie Charts

Slices of pie charts are the most common way of visualizing parts of a whole. They are widely used and understood. Subgroup visualization by pie chart, proposed by Wettschereck (2002), consists of a two-level pie for each subgroup. The base pie represents the distribution of individuals in terms of the property of interest of the entire example set, whereas the inner pie contrasts this information with their distribution in a specific subgroup. The size of the inner pie encodes the size of this subgroup (cf. Fig. 11.4).

The main weakness of this visualization is that the representation of the relative size of subgroups by the radius of the circle is somewhat misleading because the surface of the circle increases with the square of its radius. For example, a subgroup

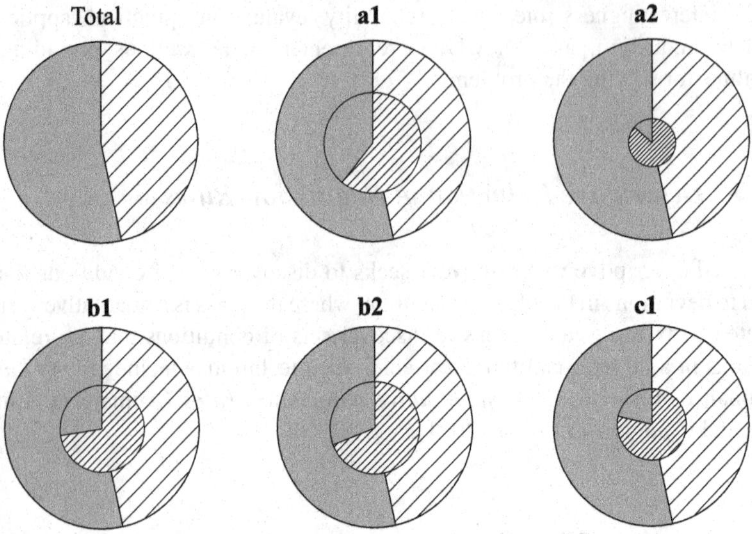

Fig. 11.4 Subgroup visualization by pie charts

that covers 20 % of examples is represented by a circle that covers only 4 % of the whole surface, while a subgroup that covers 50 % of examples is represented by a circle that covers 25 % of the whole surface. In terms of usefulness, this visualization is not very handy since—in order to compare subgroups—one would need to compare sizes of circles, which is difficult to estimate for humans. On the other hand, this visualization could be straightforwardly adapted to multiclass problems.

11.7.2 Visualization by Box Plots

In subgroup visualization by box plots, introduced by Wrobel (2001), each subgroup is represented by one box plot (all examples are also considered as one subgroup and are displayed in the top box). Each box shows the entire set of examples; the diagonally striped area on the left represents the positive examples and the white area on the right-hand side of the box represents the negative examples. The gray area within each box indicates the respective subgroup. The overlap of the gray area with the hatched area shows the overlap of the group with the positive examples. Hence, the further to the left the gray area extends the better. The less the gray area extends to the right of the hatched area, the more specific a subgroup is (less overlap with the subjects of the negative class). Finally, the location of the box along the x-axis indicates the relative share of the target class within each subgroup: the more to the right a box is placed, the higher the share of the target value within this subgroup. The vertical line (in Fig. 11.5 at value 46.6 %) indicates the default

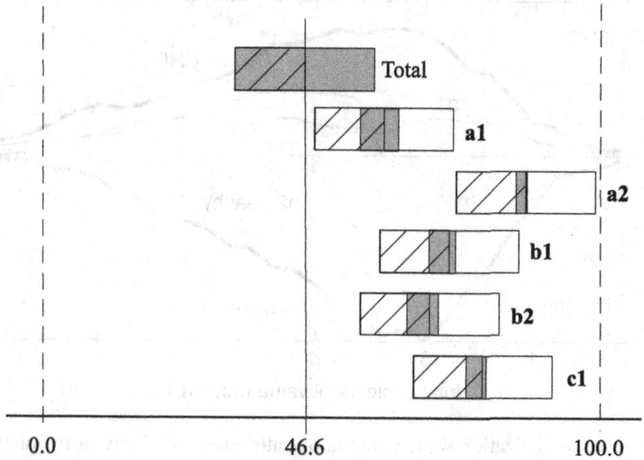

Fig. 11.5 Subgroup visualization by box plots

accuracy, i.e., the number of positive examples in the entire population. An example box plot visualization of our five sample subgroups is shown in Fig. 11.5.

On the negative side, the intuitiveness of this visualization is relatively poor since an extensive explanation is necessary for understanding it. This visualization is not very attractive since most of the image is white; the gray area (the part of the image that really represents the subgroups) is a relatively tiny part of the entire image. On the positive side, the subgroups are arranged by their confidence. It is also easier to contrast the sizes of subgroups in this representation than in a pie chart. However, it would be difficult to extend this visualization to multiclass problems.

11.7.3 Visualizing Subgroup Distribution with a Continuous Attribute

The distribution of examples with regards to a continuous attribute, introduced by Gamberger and Lavrač (2002) and Gamberger et al. (2002), was used in the analysis of several medical domains. It is the only subgroup visualization method that offers an insight into the visualized subgroups. The approach assumes the existence of at least one numeric (or ordered discrete) attribute of an expert's interest for subgroup analysis. The selected attribute is plotted on the x-axis of the diagram. The y-axis represents the target variable, or more precisely, the number of instances belonging to target property C (shown on the positive part of the y-axis) or not belonging to C (shown on the negative part of the y-axis) for the values of the attribute on the x-axis. It must be noted that both directions of the y-axis are used to indicate the number of instances. The entire dataset and two subgroups **a1** and **b2** are visualized by their distribution over a continuous attribute in Fig. 11.6.

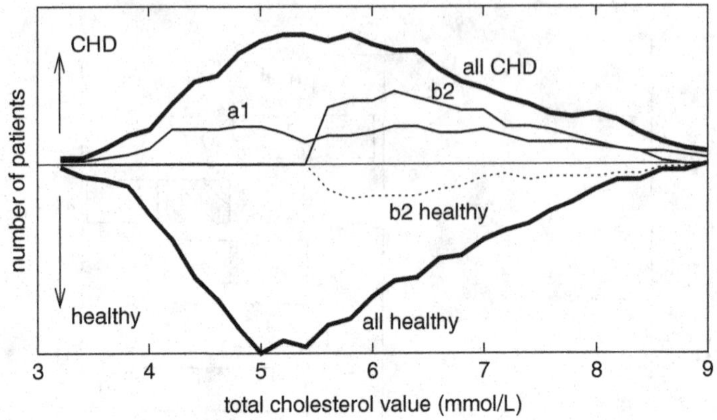

Fig. 11.6 Subgroup visualization w.r.t. a continuous attribute. For clarity of the picture, only the positive side of subgroup **a1** is depicted

For computing this graph, a sliding window is moved over the x-axis and the number of positive examples in this window are counted and plotted. Note that this approach is problematic when the attribute from the x-axis appears in the subgroup description because values that are not covered by the subgroup and should thus have a 0-value on the y-axis may still have values $\neq 0$ because the sliding window may include some covered examples for neighboring values.

This visualization method is very intuitive since it does not need much explanation. It is attractive and very useful for the end user since it offers an insight into the contents of examples. However, the displayed data are based on sampling in a window around the target value, so that the correctness is not guaranteed. It is impossible to generalize this visualization to multiclass problems.

11.7.4 Bar Charts Visualization

The visualization by bar charts was introduced by Kralj et al. (2005). In this visualization, the purpose of the first line is to visualize the distribution of the entire example set. The area on the right represents the positive examples, and the area on the left represents the negative examples of the target class. Each following line represents one subgroup. The positive and the negative examples of each subgroup are drawn below the positive and the negative examples of the entire example set. Subgroups are sorted by the relative share of positive examples (precision). The five subgroups visualized by bar charts are shown in Fig. 11.7.

This visualization method allows a simple comparison between subgroups and is therefore useful. It is straightforward to understand and can be extended to multiclass problems. It does not display the contents of data, though. In comparison with the visualization of subgroups with regards to a continuous attribute, which

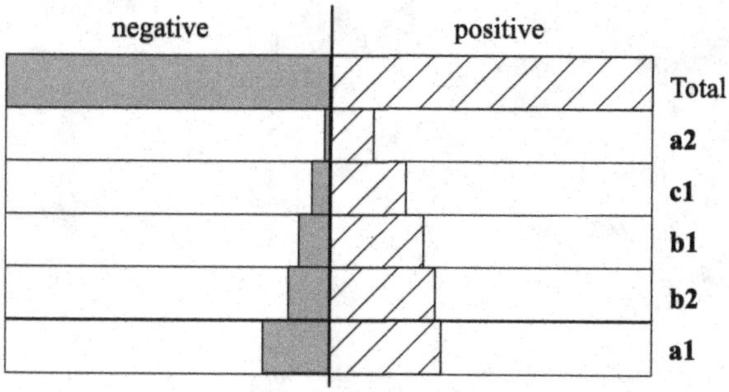

Fig. 11.7 Subgroup visualization by bar charts

is the only visualization that directly shows the contents of the data but requires a continuous or ordered discrete attribute in the data, the advantage of the bar chart visualization is that it combines the good properties of the pie chart and the box plot visualization. For this reason, we had already used it for visualizing the class distribution of the examples covered by a rule throughout this book.

11.8 Conclusion

Patterns in the form of rules are intuitive, simple, and easy for end users to understand. Therefore, it is not surprising that members of different communities have independently addressed supervised descriptive rule learning, each of them solving similar problems in similar ways and developing vocabularies according to the conventions of their respective research communities.

This chapter sheds light on previous work in this area by providing a systematic comparison of the terminology, definitions, goals, algorithms and heuristics of contrast set mining (CSM), emerging pattern mining (EPM), and subgroup discovery (SD) in a unifying framework called supervised descriptive rule learning. We have also shown that the heuristics used in CSM and EPM can be translated into two well-known heuristics used in SD, both aiming at trading off between coverage and distributional difference.

Chapter 12
Selected Applications

Rule learning is among the oldest machine learning techniques and it has been used in numerous applications. When optimal prediction quality of induced models is the primary goal then rule learning is not necessarily the best choice. For such applications one may often expect better results from complex models constructed by more modern machine learning approaches like support vector machines or random forests. The key quality of rule learning has been—and still is—its inherent simplicity and the human understandability of the induced models and patterns. Research results clearly demonstrate that rule learning, together with decision tree learning, is an essential part of the knowledge discovery process.

A thorough overview of applications of rule learning would be beyond the scope of this book. After a brief discussion of the strengths of rule learning in applications (Sect. 12.1), we will present illustrative examples of applications selected from our own work. We will start with an application in a demographic domain, in which a predictive rule learner was used to discover differences between population groups (Sect. 12.2). Thereafter, we discuss applications of descriptive rule learning in several biomedical domains. These applications are meant to illustrate and evaluate relevant parts of the rule learning process like feature construction and feature selection (Sect. 12.3), as well as to demonstrate the applicability and usefulness of rule learning for expert-guided data analysis in a concrete real-world medical task (Sect. 12.4). Finally, Sect. 12.5 presents an approach to systematic insightful analysis of collected data based on the spectrum of sensitivity levels of induced rules. Presentation of each application ends with a discussion on the lessons learned.

12.1 Introduction

The fact that the result of modeling is presented in the form of rules that can be directly integrated into the knowledge bases of expert systems has been important for knowledge acquisition applications. A nice overview by Langley and Simon (1995) describes about a dozen success stories in which rule learning enabled

J. Fürnkranz et al., *Foundations of Rule Learning*, Cognitive Technologies,
DOI 10.1007/978-3-540-75197-7_12,

construction of effective and useful expert systems. Included applications are for credit decisions by loan companies, diagnosis of mechanical devices, automatic classification of celestial objects, and monitoring quality of rolling emulsions. In these types of applications the usefulness of rule learning is not evaluated by prediction quality of each of the induced rules but by the effectiveness of the complete expert system that is constructed from the acquired knowledge.

In intelligent data analysis tasks, the understandability of the induced models enables us to identify inherent relationships in the data. Of relevance are not only the constructed models but also the detected outliers, the recognition of most relevant attributes, and, last but not least, the property that rules consist of features and that features for numerical attributes are of the form 'attribute value is greater than or less than some fixed value'. The boundary values of such features are determined automatically from the available data as an ideal decision point for the given classification task. Its value itself may present very useful information about the collected data, especially if it is compared with existing domain knowledge or computed values in other, related data sets.

Thus, rule learning emerges as a versatile tool for complex data analysis tasks. Although high prediction quality of constructed rules is important because from such rules we can expect that they reflect relevant relationships in the data, it is often more important to include existing domain knowledge into the induction process and to be able to construct a broad spectrum of models for detecting concepts hidden in the data. For the latter purpose, parametric rule evaluation measures discussed in Sect. 7.3.5 are very useful as well as the weighted covering approaches for constructing sets of rules described in Sect. 8.3. Together they form the subgroup discovery methodology of Gamberger and Lavrač (2002), which is presented and discussed in Sect. 11.2. It has been successfully applied in the biological and medical domains presented later in this chapter. First, we will show an application of predictive rule learning to the analysis of differences in the life courses of different population groups.

12.2 Demographic Domain: Dealing with Sequential Data

Our first application addresses a demographic domain, in which predictive rule learner RIPPER was used to construct rules with the aim of discovering differences between population groups in Italy and Austria, concerning their habits in the period of transition to adulthood and starting of independent living.

The transition to adulthood is one of the areas where present-day European countries exhibit a high behavioral heterogeneity. In some countries events characterizing the transition to adulthood are experienced at an early age, and at much later ages in others. The sequence of these events is also very different, even in neighboring countries.

In the following, we recapitulate the key results of Billari, Fürnkranz, and Prskawetz (2006), where the differences between the life courses in Italy and Austria were analyzed. The choice of these two countries is justified by the different patterns of transition to adulthood they exhibit—this provides us with a clear benchmark. In Austria, the duration of education is quite standardized, and the vocational training system allows for a potentially smooth transition from school to work. Furthermore, leaving home occurs to a great extent before marriage, and there is a traditionally high share of births outside of cohabiting (married or unmarried) unions. In Italy, the duration of formal education and the entry into the labor market are experienced in a rather heterogeneous way. Leaving home occurs at a late age—the latest age observed among Western countries for which data are available. Leaving home is highly synchronized with marriage, and it is not common to leave home before finishing education. Finally, childbearing outside of marriage is still less common than in other European countries.

In this application we address a two class learning problem through the construction of sets of rules, one for each class: Italy and Austria. To this end, we use the RIPPER classification rule algorithm, which is standardly used for predictive learning tasks. However, in this application we use it in a supervised descriptive rule learning setting, as an alternative to subgroup discovery algorithms.

12.2.1 Data

The data for the analysis originate from the Austrian and Italian *Fertility and Family Surveys (FFS)*, which were conducted between December 1995 and May 1996 in Austria and between December 1995 and January 1996 in Italy. After some preprocessing of the data, such as a homogenization of the age range in both samples and the removal of records with missing or incorrect values, the final dataset contained 11,107 individuals, 5,325 of which were of Austrian and 5,782 of Italian origin.

In the FFS, retrospective histories of partnerships, births, employment, and education (in a more or less complete fashion) were collected on a monthly time scale. Thus, each example essentially corresponds to a time series of events that occurred in the person's life course. The key problem that one has to solve in order to make conventional data mining algorithms applicable to this type of data is how to encode the time-sequential nature of the data in a single data table that can be analyzed with such algorithms. Billari et al. (2006) proposed to encode the information as is shown in Table 12.1. Four general descriptors are related to sex, age, and birth cohort (with two potentially different categorizations for cohorts). Binary variables are employed to indicate whether each of the six events used to characterize the transition to adulthood has occurred up to the time of the interview (*quantum*). If an event has occurred, the corresponding *timing* variable contains the

Table 12.1 Variables used in the experiments of Billari et al. (2006)

General Descriptors	
Sex	Female, male
Birth cohort (5 years)	1946–1950, 1951–1955, 1956–1960, 1961–1965, 1966–1970, 1971–1975
Birth cohort (10 years)	1946–1955, 1956–1965, 1966–1975g
Age	Age at interview in years
Quantum	
Education finished?	Yes, no
Had job?	Yes, no
Left home?	Yes, no
Formed union?	Yes, no
Married?	Yes, no
Had child?	Yes, no
Timing	
Education	Age at end of education
First job	Age at first job
Left home	Age at leaving home
Union	Age at first union
Marriage	Age at first marriage
Children	Age at the birth of first child
Age is measured in years.	
If the event has not yet occurred, the interview date is used.	
Sequencing	
Education / job	<, >, =, n.o.
Education / left home	<, >, =, n.o.
Education / union	<, >, =, n.o.
Education / marriage	<, >, =, n.o.
Education / children	<, >, =, n.o.
First job / left home	<, >, =, n.o.
First job / union	<, >, =, n.o.
First job / marriage	<, >, =, n.o.
First job / children	<, >, =, n.o.
Left home / union	<, >, =, n.o.
Left home / marriage	<, >, =, n.o.
Left home / children	<, >, =, n.o.
Union / marriage	<, >, =, n.o.
Union / children	<, >, =, n.o.
Marriage / children	<, >, =, n.o.

For each possible combination of timing variables, their relative order is computed, or the value n.o. *is used if* both *events have not yet (i.e., before the interview date) occurred.*

age at which the person experienced the event (computed in years between the birth date and the date at which the event occurred). Finally, the *sequencing* information is encoded in the form of pairwise comparisons between the dates at which two events occurred. If both events have not occurred, this is encoded with the designated value n.o..

Table 12.2 Error rates (in %) and average size (no. of conditions) RIPPER on different problem representations (estimated by tenfold cross-validations)

Feature Set	Error	Size
Only general descriptors	**46.83**	15.7
Quantum	**33.99**	38.9
Timing	**19.15**	154.4
Sequencing	**18.40**	48.2
Quantum and Timing	**18.62**	188.8
Quantum and Sequencing	**17.94**	42.1
Timing and Sequencing	**15.18**	116.4
All features	**14.94**	99.4

12.2.2 Results

Decision tree and rule learning algorithms were applied to the dataset in order to detect the key features which distinguish between Austrians and Italians with respect to the timing, sequencing, and quantum of events in the transition to adulthood. We focus here on the results of the RIPPER rule learner, which were of equal accuracy but produced much simpler rule sets than the decision tree learner.

In order to determine the relative importance of the quantum, timing, and sequencing of events that characterize the transition to adulthood, different subsets of the available features were tested. Each line of Table 12.2 shows the achieved performance of one particular attribute subset. The first column describes which subset is used, the second and third columns show the error rate and the average size of the resulting rule set measured by the total number of conditions, respectively.

The first line shows the results from using only the four general descriptors shown at the top of Table 12.1 and none of the quantum, timing, or sequencing variables. Obviously, this information is insufficient, because the obtained error rate is only slightly better than the default error of 47.94 %, which is the error rate of a classifier that uniformly predicts that all examples belong to the majority class. This value is primarily shown as a benchmark, in order to demonstrate that the relevant information for satisfactory classification performance is captured by the other variables.

The next three lines show the results from using each attribute subset independently, i.e., using only quantum, only timing, or only sequencing information. Among these three, sequencing proves to be most important. Using only sequencing information, both learning algorithms are able to discriminate the life courses of Italians and Austrians with an error rate of about 18 %. Quantum information—in the simple encoding that shows only the occurrence or nonoccurrence of an event—seems to be the least important. These results are confirmed by the next three lines, which show the performance of each pair of sets of variables. The pair quantum & timing—the only one that does not use sequencing information—produces the worst results, while the timing & sequencing pair, which does not use quantum

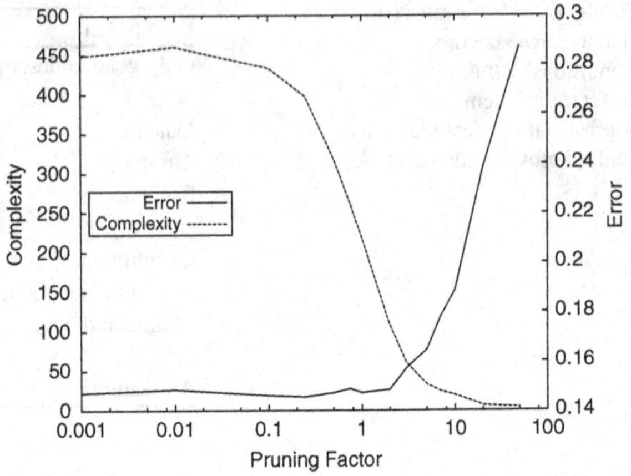

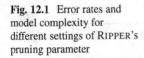

Fig. 12.1 Error rates and model complexity for different settings of RIPPER's pruning parameter

information, performs best. It should be noted, however, that quantum information is still of importance, as can be seen from the last line, which shows the results obtained using all variables shown in Table 12.1. Adding quantum to the timing and sequencing information further reduces the error rate (although this decrease is not statistically significant) and also results in simpler models.

However, error rate was not the sole concern in this application. It was equally important to generate understandable rules. The rule model that uses all features still has about 100 conditions, which is hardly comprehensible. Hence, the use of RIPPER's `-S` parameter was systematically explored, which controls the strength of RIPPER's MDL-based pruning by multiplying the encoding cost of the rule set with the provided factor (cf. Sect. 9.4.3). Larger factors lead to heavier pruning. Figure 12.1 shows the error and complexity curves for various settings of this parameter. One can see that at values close to 0, overfitting occurs. The best performance can be noticed around a value 0.1 (the default value). Larger values lead to a worse predictive performance, but yield considerably simpler theories, which are much easier to understand.

Figure 12.2 shows an example rule set, which was learned in a setting that favored simplicity over accuracy in order to optimize comprehensibility. Its error rate is about 16.5 %, as opposed to 14.94 % of the best rule set which has approximately 100 conditions. From this rule set one can, e.g., see that whereas 2,851 Italians left home when they marry, this is only the case for 592 Austrians. On the other hand, leaving home precedes marriage for 3,476 Austrians versus 976 Italians.

12.2.3 Lessons Learned

In this application we addressed a two-class learning problem, using the RIPPER classification rule learner for finding differences between two demographic groups.

```
IF   LeftHome = Marriage
THEN ITA

IF   LeftHome = n.o.
 AND Union = n.o.
 AND Child = n.o.
THEN ITA

IF   Union = Marriage
 AND Education <= 14
THEN ITA

IF   Union = Marriage
 AND Education >= 22
THEN ITA

IF   LeftHome < Marriage
THEN AUT

IF   LeftHome > Marriage
 AND HasLeftHome = yes
 AND Education >= 15
THEN AUT

IF   Education >= 15
 AND FirstJob <= 18
 AND Education <= 18
 AND Union <= 21
THEN AUT

DEFAULT AUT
```

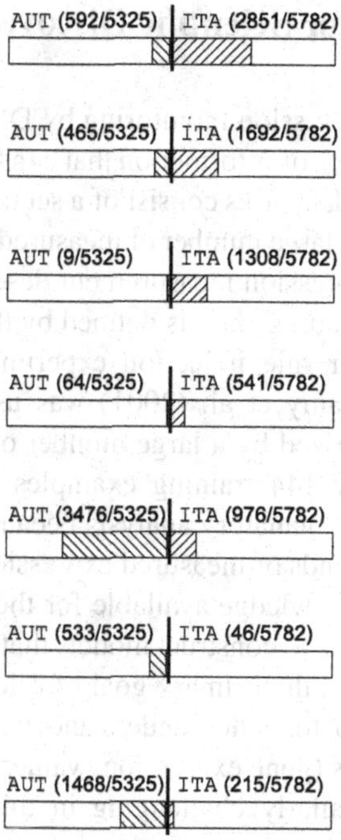

Fig. 12.2 A rule set describing the data on the transition to adulthood in Austria and Italy

Note that RIPPER is normally used for predictive learning in a supervised descriptive rule learning setting, but here it was used as an alternative to subgroup discovery algorithms. Although we have not performed a comparative analysis of the RIPPER results and the results of subgroup discovery algorithms designed to address this task, we see that with appropriate tuning of RIPPER parameters, the result is adequate for the application purpose. More importantly, this section aims to illustrate the difficult problem of analyzing sequential data, which has been solved in the data preparation phase by representing the timing variables with binary features that capture the relative order of each pairwise combination of the original attributes.

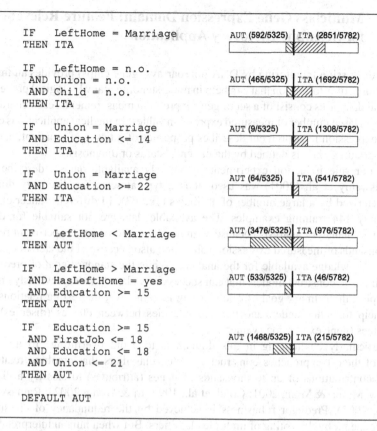

12.3 Multiclass Gene Expression Domain: Feature Relevancy for Scientific Discovery Applications

Gene expression monitoring by DNA microarrays (gene chips) provides an important source of information that can help in understanding many biological processes. Typical databases consist of a set of gene expression measurements, each characterized by a large number of measured expression values. In medical applications every gene expression measurement describes properties of a tissue of a patient, and the corresponding class is defined by the patient's status or diagnosis.

In our rule induction experiments a publicly available dataset described by Ramaswamy et al. (2001) was used. It is a typical scientific discovery domain characterized by a large number of attributes (16,063), 14 different cancer classes and only 144 training examples. The available data are not suitable for direct human explanatory analysis because a single DNA micro array experiment results in thousands of measured expression values and also because of the lack of existing expert knowledge available for the analysis. While the standard goal of predictive learning is to construct models that can successfully classify new, previously unseen examples, the primary goal of rule learning is to uncover interesting relations that can help to better understand the dependencies between classes (diseases) and attributes (gene expressions values).

Particularly challenging in this domain is to avoid overfitting. To this end, state-of-the-art approaches construct complex classifiers that combine relatively weak contributions of up to thousands of genes (attributes) to classify a disease (Chow, Moler, & Mian, 2001; Golub et al., 1999; Li & Wong, 2002a; Ramaswamy et al., 2001). Predictor robustness is achieved by the redundancy of classifiers, realized, e.g., by the voting of multiple classifiers. But when human interpretability of results is necessary such an approach is not appropriate.

In order to prevent overfitting in the rule construction process it is most of all necessary to ensure that only really relevant features may enter the rule construction process. Checking feature relevancy, as described in Sect. 4.4, is decisive for this task. In this section we present experimental results of the detection and elimination of irrelevant features for such complex scientific domains. In some of these experiments random attributes have intentionally been inserted in order to demonstrate that the approach is really able to detect irrelevant features. The rules generated from the selected relevant features, their prediction quality evaluated on an independent set of 54 examples, and expert evaluation demonstrate the effectiveness of the process.

In gene expression domains it is possible to choose between two different attribute types. The first type is *signal intensity*, represented by continuous values obtained as direct output of gene expression scanners. The second type is *presence call*, which is computed from measured signal intensity values by the Affymetrix GENECHIP software. The presence call form has discrete values A (absent), P (present), and M (marginal). Rule learning and feature relevancy are applicable both for signal intensity and/or the presence call attribute presentation types.

Typically, signal intensity values are used (Li & Wong, 2002a) because they impose less restrictions on the classifier construction process and because the results do not depend on the GENECHIP software presence call computation. For descriptive induction we prefer the latter approach based on presence call values. The reason is that features presented by conditions like `Gene = P` is true (meaning that signal intensity of gene is high and reliably measured) or `Gene = A` is true (meaning that the signal intensity of gene is low and reliably measured) are very natural for human interpretation, and that the approach can help to avoid overfitting, as the feature space is very strongly restricted to only three different values. Moreover, it can be further directed towards generation of only relevant features by specific treatment of these values.

12.3.1 Handling of Marginal Values as Unknown Values

From the three discrete values for the presence call attribute, in total six different features of the form `Gene = A` and `Gene <> P` can be constructed. But the values A, M, and P are a formalization of values low, medium, and high and when the danger of overfitting is high, a sound approach is to process marginal values so that they can neither build independent features nor that they can support low and high values. The suggested approach is to treat marginal values as if they were unknown. This approach actually implements the approach used for handling imprecise attribute values discussed in Sect. 4.6, here adapted for nominal attributes.

In this approach, attributes with three discrete values (A, M, and P) are transformed to attributes with two discrete values (A and P) and potentially many unknown values. The consequence is that no features of the form `Gene = M` and `Gene <> M` may be generated and that features `Gene = A` and `Gene <> P` will always have the same coverage properties. The result is that only two features per attribute (`Gene = A` and `Gene = P`) have to be generated, which represents a significant additional reduction of the feature space.

By eliminating marginal values as distinct nominal values we also want that they are not able to increase the relevance of other nominal values, i.e., that examples with value M should not increase the relevance of any feature of type `Gene = A` or `Gene = P`. This can be achieved if feature values for examples with attribute value M are in positive examples always set to value false, while in negative examples they are always set to value true. The approach was described in Sect. 4.5 as the pessimistic value strategy for handling unknown attribute values. The process is illustrated in Table 12.3. The table presents five positive and four negative examples for one of the target classes in the gene expression domain. Only features generated from presence call values for three attributes (genes) X, Y, and Z are presented. Feature values obtained for those examples with a corresponding attribute value 'marginal' are presented in bold.

Table 12.3 Transformation of presence call attribute values A, M, and P into feature values True and false. Transformations of marginal values M are presented in *bold*

Examples	Class	Attributes			Features					
Ex.	Cl.	X	Y	Z	X = A	X = P	Y = A	Y = P	Z = A	Z = P
p_1	⊕	A	A	A	True	False	True	False	True	False
p_2	⊕	P	P	A	False	True	False	True	True	False
p_3	⊕	A	A	P	True	False	True	False	False	True
p_4	⊕	P	P	A	False	True	False	True	True	False
p_5	⊕	M	A	A	**False**	**False**	True	False	True	False
n_1	⊖	A	P	P	True	False	False	True	False	True
n_2	⊖	P	P	P	False	True	False	True	False	True
n_3	⊖	M	M	A	**True**	**True**	**True**	**True**	True	False
n_4	⊖	P	P	A	False	True	False	True	True	False

Table 12.4 Reduced set of relevant attributes and relevant features selected for the examples set presented in Table 12.3

Examples	Class	Attributes		Features		
Ex.	Cl.	Y	Z	Y = A	Z = A	Z = P
p_1	⊕	A	A	True	True	False
p_2	⊕	P	A	False	True	False
p_3	⊕	A	P	True	False	True
p_4	⊕	P	A	False	True	False
p_5	⊕	A	A	True	True	False
n_1	⊖	P	P	False	False	True
n_2	⊖	P	P	False	False	True
n_3	⊖	M	A	True	True	False
n_4	⊖	P	A	False	True	False

The result of the described process is that for each attribute only two features are generated and the example covering quality of these features is determined only by example values that are for sure either low (absent) or high (present). This set of features presents a starting point for the process of detecting and eliminating irrelevant features based on constrained and relative irrelevancy of features described in Sect. 4.4. As the covering quality has been very strictly defined, there is a high chance that features may be detected as irrelevant. By Theorem 4.4.2, feature Y = P in Table 12.3 is totally irrelevant because it has value true for all negative examples. Additionally, following Definition 4.4.5, feature X = A is relatively irrelevant because of feature Y = A, and feature X = P is relatively irrelevant because of feature Z = A. Consequently, both features generated for gene X can be eliminated as irrelevant and the complete attribute X is eliminated from the rule construction process. The reduced set of attributes and features is shown in Table 12.4.

Table 12.5 This table shows the mean numbers of generated features for the lymphoma, leukemia, and CNS domains. Presented are the total number of features (All), the number of features after the elimination of totally irrelevant features (Total), the number of features after the elimination of constrained irrelevant features (Constr.), and the number of features after the elimination of constrained and relatively irrelevant features (Relative). These values are shown for the following training sets: the real training set with 16,063 genes (with 32,126 gene expression activity values, constructed as `Gene = A` and `Gene = P`), a randomly generated set with 16,063 genes, and a set with 32,126 genes which is a combination of 16,063 real and 16,063 random attributes

	Tasks	All	Total	Constr.	Relative
Task 1	Real domain with 16,063 att.	32,126	23,500	9,628	4,445
Task 2	Randomly generated domain with 16,063 att.	32,126	27,500	16,722	16,722
Task 3	Combination of 16,063 real and 16,063 randomly generated attributes	64,252	51,000	26,350	15,712

12.3.2 Experiments with Feature Relevancy

In order to demonstrate the importance of feature relevancy for overfitting prevention, a series of experiments—including experiments with intentionally added random attributes—was performed on three concept learning tasks. Their results are presented in Table 12.5. These experiments concentrate on tasks for a subset of the cancer classes for which there is a sufficient number of available examples—lymphoma, leukemia, and CNS—for which the achieved prediction quality on an independent test set (see Table 12.7) was high. Concept learning tasks were constructed in a way that examples of the target cancer class are used as positive examples while all others are defined as the negative class.

Experiments were performed as follows. In task 1 the starting point was the original domain with 16,063 attributes; in task 2 we experimented with two randomly generated domains with 16,063 and 32,126 examples, respectively; while in task 3 we joined the original domain with 16,063 examples with the randomly generated domain with 16,063 examples. For constrained relevance, default values $min_{\hat{P}} = P/2$ and $min_{\bar{N}} = \sqrt{N}$ were used.

Results for Task 1. In the real domain with 16,063 attributes both concepts of constrained and relative relevancy were very effective in reducing the number of features. About 60 % of all features were detected as constrained irrelevant while relative irrelevancy was even more effective as it managed to eliminate up to 75 % of all the features. Their combination resulted in the elimination of 75–85 % of all the features. These results are presented in the first row of Table 12.5. The set of all features in these experiments was generated so that for each gene (attribute) two features were generated (`Gene = A` and `Gene = P`), followed by eliminating totally irrelevant features (with $\hat{P} = 0$ or $\bar{N} = 0$), which substantially reduced the total number of features.

Results for Task 2. An artificial domain with 16,063 completely randomly generated attribute values was constructed, and the same experiments were repeated on this artificial domain as for the real gene expression domain. The results (repeated with five different randomly generated attribute sets and presented as their mean values) were significantly different: there were only about 40 % of constrained irrelevant features and practically no relatively irrelevant features eliminated. The results are presented in the second row of Table 12.5. Comparing the results for the real and for the randomly generated domain, especially large differences can be noticed in the performance of the relative relevancy filter. This is the consequence of the fact that in the real domain there are some features that are indeed relevant; they cover many target class examples and a few non-target class examples and in this way they make many other features relatively irrelevant. The results prove the importance of relative relevancy for domains in which strong and relevant dependencies between classes and attribute values exist.

Results for Task 3. The experiments with feature relevancy continued with another domain with 32,126 attributes, generated as the combination of two previous domains with 16,063 attributes each: the real and the randomly generated domain. The results are presented in the last row of Table 12.5. After the elimination of constrained irrelevant features the number of features is equal to the sum of features that remained in the two independent domains with 16,063 attributes. In contrast, relative relevancy was much more effective. Besides eliminating many features from the real attribute part it was now possible to eliminate also a significant part of features based on randomly generated attributes.

Interpretation. These results indicate that the elimination of features is very effective in real-life domains (Task 1) and also in the artificial test environment when random data is added to the original dataset (Task 3). This proves the effectiveness of the feature elimination techniques as discussed further in the lessons learned section.

Figure 12.3 illustrates the results presented in Table 12.5 with one added domain with 32,126 randomly generated attributes for which feature filtering performs similar as in Task 2.

12.3.3 Prediction Quality of Induced Rules

The described gene expression multiclass domain with 14 classes was transformed into 14 concept learning tasks so that each cancer type is once the target class while all other cancer types are treated as non-target class examples. For each concept learning task the generation of features and elimination of irrelevant features were performed as described in the previous section. After that, rules were induced using the subgroup discovery rule learning algorithm SD with its parameter g value set to 5 for all 14 concept learning tasks. The rules obtained for the three largest cancer classes are shown in Table 12.6.

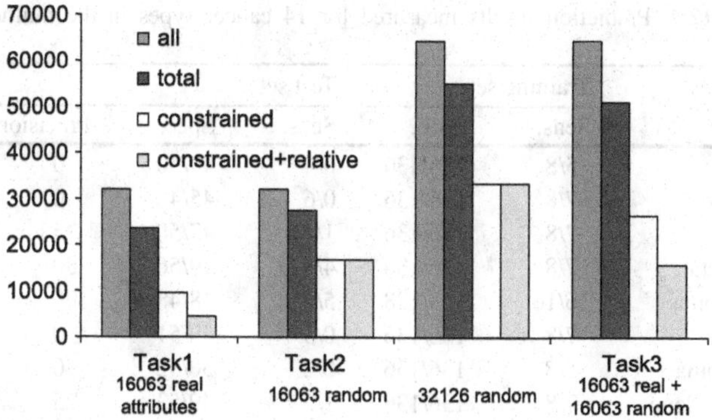

Fig. 12.3 Mean numbers of features for the three domains (lymphoma, leukemia, and CNS) for the following training sets: real training set with 16,063 attributes of gene expression activity values, a randomly generated set with 16,063 attributes, a randomly generated set with 32,126 attributes, and a set which is a combination of 16,063 real and 16,063 random attributes

Table 12.6 Rules induced for the multiclass cancer domain for cancer types with 16 (lymphoma and CNS) and 24 (leukemia) target class samples

```
IF    CD20_receptor = expressed
 AND Phosphatidylinositol_3_kinase_regulatory_alpha_subunit
                                               = NOT expressed
THEN lymphoma

IF    KIAA0128_gene = expressed
 AND Prostaglandin_d2_synthase_gene = NOT expressed
THEN leukemia

IF    Fetus_brain_mRNA_for_membrane_glycoprotein_M6 = expressed
 AND CRMP1_collapsin_response_mediator_protein_1 = expressed
THEN CNS
```

For these three diseases a very good prediction quality, tested on the independent test set with 54 examples was achieved. The sensitivity values are between 66 % and 83 %, while the specificity value is always excellent and equal or almost equal to 100 %, for all the three rules. The result indicates that no significant training data overfitting has occurred. Expert analysis of these rules confirmed that the rules are appropriate for human interpretation and that reasonable expert evaluation is possible for all genes included in these rules. Details of the expert evaluation can be found in Gamberger, Lavrač, Zelezny, and Tolar (2004).

But rule learning has not been as successful for the complete domain. The summary of the prediction results for the training and the test sets for all 14 classes is given in Table 12.7. In addition to the sensitivity and the specificity results for the training set and the available independent test set, the precision value is computed for the test set.

Table 12.7 Prediction results measured for 14 cancer types in the multiclass domain

	Training set		Test set		
Cancer	Sens.	Spec.	Sens.	Spec.	Precision (%)
breast	5/8	136/136	0/4	49/50	0
prostate	7/8	136/136	0/6	45/48	0
lung	7/8	136/136	1/4	47/50	25
colorectal	7/8	136/136	4/4	49/50	80
lymphoma	16/16	128/128	5/6	48/48	100
bladder	7/8	136/136	0/3	49/51	0
melanoma	5/8	136/136	0/2	50/52	0
uterus_adeno	7/8	136/136	1/2	49/52	25
leukemia	23/24	120/120	4/6	47/48	80
renal	7/8	136/136	0/3	48/51	0
pancreas	7/8	136/136	0/3	45/51	0
ovary	7/8	136/136	0/4	47/50	0
mesothelioma	7/8	136/136	3/3	51/51	100
CNS	16/16	128/128	3/4	50/50	100

From Table 12.7 an interesting and important relationship between prediction results on the test set and the number of target class examples in the training set can be noticed. The obtained prediction quality on the test set is very low for many classes, significantly lower than those reported in Ramaswamy et al. (2001). For 7 out of 14 classes the measured precision is 0 %. However, there are very large differences among the results for various classes (diseases). It can be noticed that the precision on the test set higher than 50 % was obtained for only 5 out of 14 classes. Notice that there are only three classes (lymphoma, leukemia, and CNS) with more than eight training samples and for all of them the induced rules have high precision on the test set, while for only 2 out of 11 classes with eitht training cases (colorectal and mesothelioma) a high precision has been achieved. The classification properties corresponding to classes with 16 and 24 target class examples are comparable to the performances reported for these classes in Ramaswamy et al. (2001), yet are achieved by predictors much simpler than in the mentioned work. The results are summarized in Fig. 12.4.

12.3.4 Lessons Learned

The results indicate that there is a certain threshold on the number of available training examples below which the rule learning approach is not appropriate because it cannot prevent overfitting despite the techniques designed for this purpose. However, it seems that for only slightly larger training sets it can effectively detect relevant relationships. This conclusion is very positive because we can expect significantly larger gene expression databases to become available in the future.

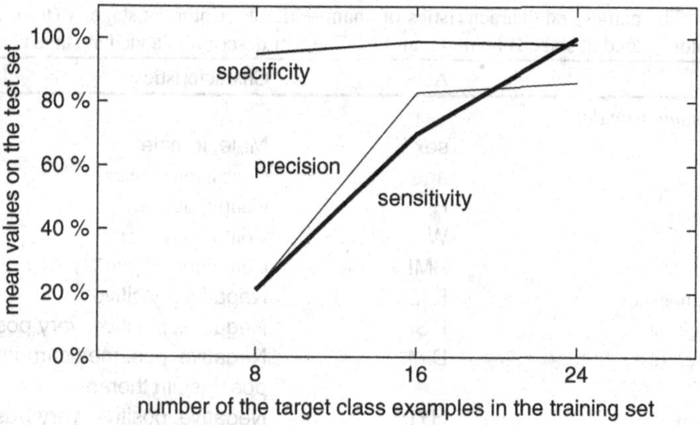

Fig. 12.4 Mean values of sensitivity, specificity, and precision on the independent test set for rules constructed in concept learning tasks with various numbers of examples in the positive class for the multiclass gene expression domain

From the experiments with feature relevancy it is obvious that the elimination of features is very effective in real-life domains. It is important that in domains which are combinations of real and random attributes the proposed feature filtering methodology is also very effective: in Task 3 less features remained after feature elimination (15,712 features) than in Task 2 (16,722 features) although the number of attributes in Task 3 was two times larger. This proves that the presented methodology, especially relative relevancy, can be very effective in avoiding overfitting by reducing the hypothesis search space through the elimination of non-significant dependencies between attribute values and classes. This property is important because it can be assumed that among 16,063 real attributes there are many which are irrelevant with respect to the target class.

12.4 Coronary Heart Disease Domain: Expert-Guided Data Analysis

Atherosclerotic coronary heart disease (CHD) is one of the world's most frequent causes of mortality and an important problem in medical practice. Today's CHD prevention relies on two significantly different concepts: general education of the population about known risk factors, especially lifestyle factors, and early detection of patients at risk. The diagnostic task is not difficult in cases with significantly pathological test values (especially, for example, left ventricular hypertrophy, increased LDL cholesterol, decreased HDL cholesterol, hypertension, and intolerance to glucose). However, the problem of disease prevention is to decide in cases with slightly abnormal values and in cases when combinations of different risk factors occur.

Table 12.8 The names and characteristics of anamnestic descriptors at stage A (*top*), laboratory test descriptors added at stage B (*center*), and ECG at rest descriptors added at stage C (*bottom*)

Descriptor	Abbr.	Characteristics
Stage A: Anamnestic data		
Sex	sex	Male, female
Age	age	Continuous (years)
Height	H	Continuous (m)
Weight	W	Continuous (kg)
Body mass index	BMI	Continuous (kg m^{-2})
Family anamnesis	F.A.	Negative, positive,
Present smoking	P.S.	Negative, positive, very positive
Diabetes mellitus	D.M.	Negative, pos. medicament therapy, pos. insulin therapy
Hypertension	HYP	Negative, positive, very positive
Stress	STR	Negative, positive, very positive
Stage B: Laboratory tests		
Total cholesterol	T.CH.	Continuous (mmol L^{-1})
Trygliceride	TR	Continuous (mmol L^{-1})
High-density lipoprotein	HDL/CH	Continuous (mmol L^{-1})
Low-density lipoprotein	LDL/CH	Continuous (mmol L^{-1})
Uric acid	U.A.	Continuous (μ mol L^{-1})
Fibrinogen	FIB	Continuous (g L^{-1})
Stage C: ECG at rest		
Heart rate	HR	Continuous (beats min^{-1})
ST segment depression	ECGst	Negative, positive 1 mm, positive $\geq$2 mm
Serious arrhythmias	ECGrhyt	Negative, positive
Conduction disorders	ECGcd	Negative, positive
Left ventricular hypertrophy	ECGhlv	Negative, positive

The database representing typical medical practice in CHD diagnosis was collected at the *Institute for Cardiovascular Prevention and Rehabilitation*, Zagreb, Croatia. The domain descriptor set, shown in Table 12.8, includes anamnestic parameters (ten attributes), parameters describing laboratory test results (six attributes), and ECG at rest (five attributes). CHD diagnosis is typically based on exercise test data, echocardiography results, vectorcardiography results, and long-term continuous ECG recording data. Some of the descriptors included in the data set may be collected only in specialized medical institutions and they are not appropriate for use in building patient risk models. But their collection is relevant for the correct classification of the patients that are used for building such models.

In this study, only patients with complete data were included in the data set, resulting in a database with 238 patient records: 111 CHD patients (positive cases), and 127 persons without CHD (negative cases). The database is in no respect a good epidemiological CHD database reflecting actual CHD occurrence in a general population, since about 50 % of gathered patient records represent CHD patients and this percentage is unrealistic in the general population. But the database is very valuable since it includes records of different types of the disease. Moreover, the included negative cases (patients who do not have CHD) are not randomly selected persons but individuals with some subjective problems or those considered by general practitioners as potential CHD patients, and hence sent for further investigations to the Institute. The consequences of the biased data set are twofold:

- The data set is very good for modeling purposes but model statistics measured on this data can be expected to be significantly different from those that can be obtained in other populations (e.g., the general population).
- Among the patients that are recorded in the data set, there are many patients who have had the disease for a long time and typically are subject to treatment which reduces important risk factors. Moreover, most patients have already changed their lifestyles concerning smoking and nutrition habits. On the other side, negative cases are people that do not have CHD but may be ill because of some other heart-related disease. Consequently, their test values may be different from typical values expected for a healthy person. All statistical approaches on collected attribute values can be significantly influenced by this bias.

Patient screening is performed in general practice in three different stages. Stage A are anamnestic data (top part of Table 12.8), stage B consists of stage A with added laboratory test results (center of Table 12.8), while stage C additionally includes also ECG at rest data (bottom of Table 12.8). The primary goal is the construction of at least one interesting pattern that can be used in early disease detection for each stage A, B, and C, respectively.

12.4.1 Results of CHD Patient Risk Group Detection

The process of expert-guided data analysis was performed as follows. For every data stage A, B, and C, 15–20 rules were constructed using the SD subgroup discovery algorithm with different covering properties. This was achieved by using the generalization quotient *GQuotient* as defined in Sect. 7.3.5 with values of the parameter g in the range 0.5 to 100 (values $0.5, 1, 2, 4, 6, \ldots$). Rule construction was performed in the subgroup discovery context by first performing relevancy tests of included features and their combinations as described in Sect. 4.4. For each g parameter value up to three different rules covering different subsets of positive examples were constructed. The inspection of the constructed rules by a medical expert triggered further experiments. Typical suggestions by the expert were to limit the number of features in the rule body and to avoid the generation of rules whose features would involve expensive and/or unreliable laboratory tests.

In the iterative process of rule generation and selection, the expert has selected the five most interesting rules describing CHD risk groups of patients. Table 12.9 shows the induced rules accompanied with the information on the g value with which the rule was constructed and true positive ($\hat{\pi}$) and false positive ($\hat{\nu}$) rates

Table 12.9 Expert-selected rules for different data stages. Rule **a1** is for male patients, rule **a2** for female patients, while rules **b1**, **b2**, and **c1** are for both male and female patients. The g column presents parameter values for which the rules were constructed, while the two last columns contain true positive rates and false positive rates measured on the training set

	Expert-selected rules	g	$\hat{\pi}$ (%)	$\hat{\upsilon}$ (%)
a1	`IF` sex = male `AND` positive family history `AND` age over 46 year `THEN` `CHD`	14	34.2	20.4
a2	`IF` sex = female `AND` body mass index over 25 kg m^{-2} `AND` age over 63 years `THEN` `CHD`	8	13.5	1.6
b1	`IF` total cholesterol over 6.1 mmol L^{-1} `AND` age over 53 years `AND` body mass index below 30 kg m^{-2} `THEN` `CHD`	10	28.8	9.4
b2	`IF` total cholesterol over 5.6 mmol L^{-1} `AND` fibrinogen over 3.7 g L^{-1} `AND` body mass index below 30 kg m^{-2} `THEN` `CHD`	12	32.4	12.6
c1	`IF` left ventricular hypertrophy `THEN` `CHD`	10	23.4	5.5

measured on the complete available data set. The described iterative process was relatively straightforward for data at stages B and C, but it turned out that medical history data on its own (stage A data) is not informative enough for inducing rules; it failed to fulfill the expert's subjective criteria of interestingness. Only after engineering the domain by separating male and female patients were interesting subgroups **a1** and **a2** discovered.

Separately for each data stage A, B, and C, we have investigated which of the induced rules are the best in terms of the $\hat{\pi}/\hat{\upsilon}$ trade-off in the coverage space, i.e., which of them are used to define the convex hull in this space. At stage B, for instance, seven rules (marked by +) are on the convex hull shown in Fig. 12.5. Two of these rules, **x1** and **x2**, are listed in Table 12.10. It is important to notice that both expert-selected rules **b1** and **b2** are according to the χ^2 test significant at the level of 99.9 % but they are not among the ones lying on the convex hull in Fig. 12.5. The reasons for selecting exactly these two rules at stage B are their simplicity (consisting of three features only), their generality (covering relatively many positive cases), and the fact that the used features are, from the medical point of view, inexpensive and reliable tests. Moreover, the two rules **b1** and **b2** were deemed interesting by the expert.

Rules **b1** and **b2** are interesting because of the feature body mass index below 30 kg m^{-2}, which is intuitively in contradiction with the expert knowledge

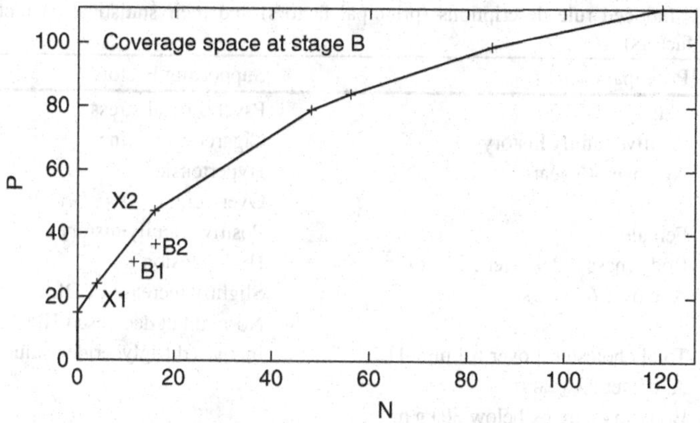

Fig. 12.5 The coverage space presenting the convex hull of rules induced at data stage B. Labels **b1** and **b2** denote positions of rules selected by the medical expert, while **x1** and **x2** are two of the seven rules forming the coverage convex hull

Table 12.10 Two of the best induced rules, **x1** and **x2**, induced for stage B using the g values 4 and 6, respectively. Their positions in the coverage space are marked in Fig. 12.5

	Best induced rules		g	$\hat{\pi}$ (%)	$\hat{\nu}$ (%)
x1	IF	age over 61 years	4	21.6	3.1
	AND	tryglicerides below 1.85 mmol L^{-1}			
	AND	high den. lipopr. below 1.25 mmol L^{-1}			
	THEN	CHD			
x2	IF	body mass index over 25	6	42.3	12.6
	AND	high den. lipopr. below 1.25 mmol L^{-1}			
	AND	uric acid below 360 mmol L^{-1}			
	AND	glucose below 7 mmol L^{-1}			
	AND	fibrinogen over 3.7 g L^{-1}			
	THEN	CHD			

that both increased body weight as well as increased total cholesterol values are CHD risk factors. It is known that increased body weight typically results in increased total cholesterol values while subgroups **b1** and **b2** actually point out the importance of increased total cholesterol when it is not caused by obesity as a relevant disease risk factor.

12.4.2 Statistical Characterization of Detected Risk Groups

Supporting factors are attributes that have significantly different distributions between the subpopulation of positive examples described by a rule and the

Table 12.11 Induced rule descriptions (principal factors) and their statistical characterizations (supporting factors)

	Principal Factors	Supporting Factors
a1	Male Positive family history Age over 46 year	Psychosocial stress Cigarette smoking Hypertension Overweight
a2	Female Body mass index over 25 kg m^{-2} Age over 63 years	Positive family history Hypertension Slightly increased LDL cholesterol Normal but decreased HDL cholesterol
b1	Total cholesterol over 6.1 mmol L^{-1} Age over 53 years Body mass index below 30 kg m^{-2}	Increased triglycerides value
b2	Total cholesterol over 5.6 mmol L^{-1} Fibrinogen over 3.7 mmol L^{-1} Body mass index below 30 kg m^{-2}	Positive family history
c1	Left ventricular hypertrophy	Positive family history Hypertension Diabetes mellitus

complete population of negative cases. These factors are important as they can help to confirm that a patient is a member of a subpopulation, also providing a better description of a typical member of that subpopulation.

Identification of supporting factors is useful for establishing a complete context describing identified patterns, patient risk groups in this application, and for increasing the user confidence when classifying new cases by induced models. Statistical differences in distributions are tested using the χ^2 test with 95 % confidence level ($p = 0.05$). For this purpose numerical attributes were partitioned in up to 30 intervals so that in every interval there are at least five instances.

In the expert-guided data analysis framework, the role of statistical analysis is to detect potentially relevant supporting factors, whereas the decision whether they will be actually used is left to the expert. The decisive issues are how reliable they are and how easily they can be measured in practice. In Table 12.11, expert-selected supporting factors are listed next to the individual CHD risk groups, each described by a list of principal factors.

Detected and selected supporting factors in combination with expert interpretation of computed mean values for the patient subpopulations enable the formulation of a textual description of relevant CHD risk groups. Existing medical knowledge can be used in this process to stress the relevance of some principal and supporting factors. The description for rule **a1**, which covers male patients with positive family history and an age over 46 years, illustrates how the final result might look.

The main supporting characteristic is psychosocial stress, but important are also cigarette smoking, hypertension, and overweight. Both principal risk factors for this model are nonmodifiable. Positive family history is a known important risk factor and requires careful screening of other risk factors. This model stresses the importance of a positive family history especially for the male population. The selected age margin in the second factor is rather low but it is in accordance with the existing medical experience. This low limit is good for prevention and early disease diagnosis although typical patients in this model are significantly older. The model describes well existing medical knowledge about the CHD risk, but its applicability is rather low because of the high false positive rate as seen in Table 12.9.

12.4.3 Lessons Learned

For early detection of potential CHD patients it would be important to construct patterns representing CHD risk groups from a large data collection obtained from general practice screening. As such a dataset was not available, we tried to overcome the deficiencies of the available relative small and biased data set by integrating expert knowledge and experience into the data mining process. The result is an active mining approach in which rule learning is used as a tool supporting the medical expert and the knowledge engineer in the interactive and iterative search for relevant results.

Semiautomated active mining may at first glance seem inappropriate for knowledge discovery. This application shows that, on the contrary, such an approach may be very productive and useful, overcoming the deficiencies of automated knowledge discovery in the cases when it is impossible to gather a sufficiently large unbiased data collection from general practice screening. The results are achieved by active involvement of the expert in all steps of the discovery process. Although the rule learning process is data driven, the main characteristic of the methodology is that the obtained results significantly reflect the existing knowledge and experience of medical experts.

The selection of relevant rules is based on objective prediction quality of rules (true positive and false positive rates) but also on subjective properties like the understandability, unexpectedness, and actionability of induced models (Silberschatz & Tuzhilin, 1995), which depend on the attributes used in the conditions of induced rules and their combinations. In this application the medical expert preferred short rules that did not contain attributes based on expensive and/or unreliable medical tests.

Besides that, the expert's role was recognized as indispensable in:

1. Extensive evaluation of explicitly detected noise in the data set. Iterative correction of data collection and example classification errors ensured high quality of the data entering the rule induction process.
2. Partitioning the CHD risk group problem into three data stages A–C that was based on the expert's understanding of the typical diagnostic process.

3. The suggestion to try to build separate models for male and female patients at stage A.

All these issues have turned out to be important for the success of the data analysis process and formation of operational predictive models.

12.5 Brain Ischaemia Domain: Insightful Data Analysis

The application described in Sect. 12.4 is an approach in which rule learning was used as a tool in an expert-centric knowledge discovery process. In this section we use a brain ischaemia domain to illustrate the applicability of rule learning for systematic analysis of collected data. The underlying idea is to organize and systematize the data analysis process based on the complete spectrum of the sensitivity levels of induced rules. Additionally, we illustrate human reasoning based on the inspection of induced rules and the intellectual effort necessary to convert the rules into useful domain knowledge.

The brain ischaemia database consists of records of patients who were treated at the Intensive Care Unit of the *Department of Neurology, University Hospital Center* in Zagreb, Croatia, in year 2003. In total, 300 patients are included in the database: 209 with the computed tomography (CT) confirmed diagnosis of brain attack (stroke), and 91 patients who entered the same hospital department with adequate neurological symptoms and disorders, but who were diagnosed (based on the outcomes of neurological tests and CT) as patients with transition ischaemic brain attack (TIA, 33 patients), reversible ischaemic neurological deficit (RIND, 12 patients), and serious headache or cervical spine syndrome (46 patients). The main goal of these data analysis experiments is to discover regularities that characterize brain stroke patients.

Patients are described with 26 different descriptors representing anamnestic data, physical examination data, laboratory test data, ECG data, and information about previous hospital therapies. Descriptors and their abbreviations used in the rules are listed in Table 12.12. The classification of patients is based on physical examination confirmed by the CT test. All the patients in the control group have normal brain CT in contrast with the positive CT test result for patients with a confirmed stroke.

It should be noted that the target class are the patients with brain stroke, and the control group does not consist of healthy persons but of patients with suspected serious neurological symptoms and disorders. In this sense, the available database is particularly appropriate for studying the specific characteristics and subtle differences that distinguish patients with strokes. The detected relationships can be accepted as the actual characteristics for these patients. However, the computed evaluation measures—including probability, specificity, and sensitivity of induced rules—only reflect characteristics specific to the available data, not necessarily holding for the general population or other medical institutions.

Table 12.12 The descriptors in the brain ischaemia domain. The last column provides reference values representing the ranges typically accepted as normal in the medical practice

Descriptor	Abbr.	Characteristics
Sex	sex	m, f
Age	age	Continuous (years)
Family anamnesis	fhis	Positive, negative
Present smoking	smok	Yes, no
Stress	str	Yes, no
Alcohol consumption	alcoh	Yes, no
Body mass index	bmi	Continuous (kg m^{-2}) ref. value 18.5–25
Systolic blood pressure	sys	Continuous (mmHg) normal value < 139 mmHg
Diastolic blood pressure	dya	Continuous (mmHg) normal value < 89 mmHg
Uric acid	ua	Continuous (μ mol L^{-1}) ref. value for men <412 ref. value for women <380
Fibrinogen	fibr	Continuous (g L^{-1}) ref. value 2.0–3.7
Glucose	gluc	Continuous (mmol L^{-1}) ref. value 3.6–5.8
Total cholesterol	chol	Continuous (mmol L^{-1}) ref. value 3.6–5.0
Trygliceride	tryg	Continuous (mmol L^{-1}) ref. value 0.9–1.7
Heart rate	ecgfr	Continuous ref. value 60–100 beats/min
Platelets	plat	Continuous ref. value 150000–400000
Protrombin time	pt	Continuous ref. value without th. 0.7–1.2 with anticoagulant th. 0.25–0.4
Atrial fibrillation	af	Yes, no
Left ventricular hypertr.	ecghlv	Yes, no
Fundus ocular	fo	Discrete value 0–4
Aspirin therapy	asp	Yes, no
Anticoagulant therapy	acoag	Yes, no
Antihypertensive therapy	ahyp	Yes, no
Antiarrhytmic therapy	aarrh	Yes, no
Statins	stat	Yes, no
Hypoglycemic therapy	hypo	None, oral, insulin

12.5.1 *Scanning the Sensitivity Space*

Insightful data analysis in this application is defined as a systematic approach to supervised learning with the goal to detect as many possible relevant properties that describe the target (positive) class cases (brain stroke patients) in contrast to cases in the non-target (negative or control) class (TIA, RIND, and similar patients). This means that for this process examples of two classes have to be available for the analysis. Sometimes the decision about what is the target class is not simple and the complete data analysis process can have a few task definitions with different choices of target and non-target classes. For example, in the same brain ischaemia domain the target class could also be patients with stroke receiving some treatment, and the non-target class being stroke patients not taking the treatment. In this setting, the process of data analysis is far from completely automatic. Moreover, the process should be sometimes repeated for different subpopulations with specific properties, like sex or age range, or with different subsets of descriptors. In this application we demonstrate only the process performed for the complete database with stroke patients selected as the target class. The same approach may be repeated for differently defined problems, potentially leading to other relevant results. Selection and definition of subproblems that have to be analyzed more thoroughly depends on medical expert review.

Having determined the analysis task, defined by the selection of the target and non-target class examples, the analysis process continues by systematically inducing rules at different generalization levels. By applying the SD subgroup discovery rule learning algorithm using the generalization quotient with parameter g (Sect. 7.3.5), a total of 15 rules, presented in Table 12.13, were induced. For each of the five selected values of parameter g there are three rules. By selecting a low parameter value the rule learning process tends to construct very specific rules with relatively low sensitivity. With the increase of the value of the parameter the sensitivity of rules typically improves at the cost of decreased specificity. The sensitivity and the specificity values for each rule are given in columns 3 and 4, respectively. The last column indicates the overlap between the current rule and one/two rules induced previously for the same g-value. The overlap value is defined as the number of positive cases that are covered both by the current rule and the previously generated rule(s) divided by the number of positive cases covered by either the current rule or the previously generated rule(s), whichever is smaller. Low overlap values mean a relative independence of rules.

From Table 12.13 it can be seen that the same rule (e.g., `ahyp = yes`) may be induced with different generalization parameter values. This is possible because the rule induction process has been independently repeated for each parameter value. The order of rules in each group is the order selected by the weighted covering algorithm presented in Fig. 8.4, which takes into account the covering relations between the current rule and other rules previously selected for the same g-value.

Table 12.13 Bodies of the rules for stroke patients induced for g-values 5, 10, 20, 50, and 100, together with their *Sensitivity* ($\hat{\pi}$) and *Specificity* ($\bar{\nu}$) values measured on the available data set as well as their overlap with previously induced rule(s) in the same g-value group

Ref.	Rule	$\hat{\pi}$ (%)	$\bar{\nu}$ (%)	Overlap (%)
generalization parameter g=5				
g5a	`fibr > 4.55 AND str = no`	25	100	–
g5b	`fibr > 4.45 AND age > 64.00`	41	100	94
g5c	`af = yes AND ahyp = yes`	28	95	36
generalization parameter g=10				
g10a	`fibr > 4.45 AND age > 64.00`	41	100	–
g10b	`af = yes AND ahyp = yes`	28	95	34
g10c	`str = no AND alcoh = yes`	28	95	67
generalization parameter g=20				
g20a	`fibr > 4.55`	46	97	–
g20b	`ahyp = yes AND fibr > 3.35`	65	73	71
g20c	`sys > 153.00 AND age >57.00 AND asp = no`	45	88	80
generalization parameter g=50				
g50a	`ahyp = yes`	74	54	–
g50b	`fibr > 3.35 AND age > 58.00`	79	63	76
g50c	`age > 52.00 AND asp = no`	64	63	96
generalization parameter g=100				
g100a	`age > 52.00`	96	20	–
g100b	`dya > 75.00`	98	8	98
g100c	`ahyp = yes`	74	54	100

12.5.2 *Rule Analysis for Different Sensitivity Levels*

The interpretation of induced rules starts by independent interpretations of each individual rule. There is no prior preference for either more specific or more sensitive rules. But typically more sensitive rules covering many cases tend to be shorter. This fact can be verified also from Table 12.13: very sensitive subgroup descriptions are described by a small number of conditions and because of that they are easier to be analyzed by the domain experts.

Analysis of sensitive rules. Highly sensitive rules, like those induced with parameter $g = 100$, describe general characteristics of the target class. In the given domain we can observe that stroke is characteristic for a middle-aged and elderly population (`age > 52.00`), that people with stroke typically have normal or increased diastolic blood pressure (`dya > 75.00`), and that they have already detected hypertension problems and take some therapy (anti-hypertension therapy `ahyp = yes`). We also see that the selected boundary values are relatively low (52 years for the age and 75 *mmHg* for the diastolic pressure) which is due to the fact that the rules should satisfy a large number of cases. This is the reason why the rules are not applicable as decision rules but they provide useful descriptive information about the target class.

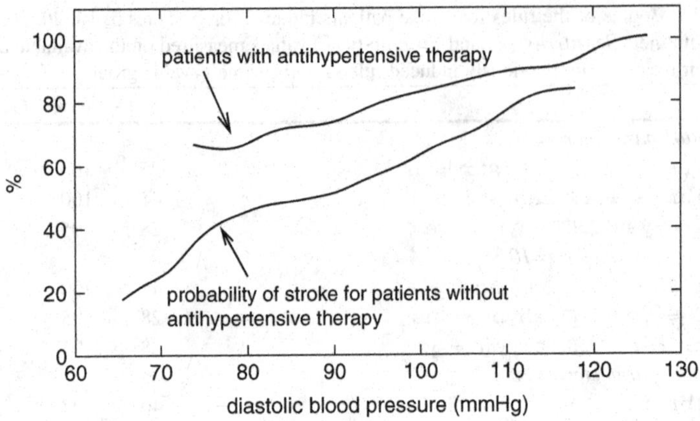

Fig. 12.6 Probability of brain stroke, estimated by the proportion of patients with brain attack (stroke) relative to the total number of patients in the hospital department, shown dependent of diastolic blood pressure values, presented separately for patients with and without antihypertensive therapy

Expert interpretation of each individual rule is essential for the generation of useful knowledge. For example, the interpretation of rules like (`age > 52.00`) or (`dya > 75.00`) is straightforward. In contrast, the interpretation of the rule (`ahyp = yes`) could lead to the conclusion that antihypertensive therapy itself is dangerous for the incidence of stroke. A much more appropriate interpretation is that hypertension is dangerous, therefore people with detected hypertension problems, characterized by the fact that they already take antihypertensive therapy, have a greater probability of having a stroke. Indirectly, this rule also means that we have little chance to recognize the danger of high blood pressure, as suggested by rule g100b from Table 12.13, directly from their measured values. The reason is that many seriously ill patients have these values artificially low due to a previously prescribed therapy.

This is a good example of expert reasoning stimulated by an induced rule. In this situation we may try to answer the question how the probability of stroke with respect to the transitory ischaemia cases changes with increasing diastolic blood pressure. From the induced rule we have learned that we should compare only patients without anti-hypertension therapy. The result is presented in Fig. 12.6. It can be noticed that the probability of stroke grows with the increase of diastolic blood pressure. The same dependency can be drawn also for patients with the therapy. The differences between the two curves are large and from them some interesting conclusions can be derived. The first is that antihypertensive therapy helps in reducing the risk of stroke: this can be concluded from the fact that the probability of stroke is decreasing with the decrease of diastolic blood pressure also for the patients with the therapy. But it is also true that for diastolic blood pressure up to $100\,mmHg$ the probability of stroke is higher for patients with recognized hypertension problems than for other patients. The interpretation is that also in cases

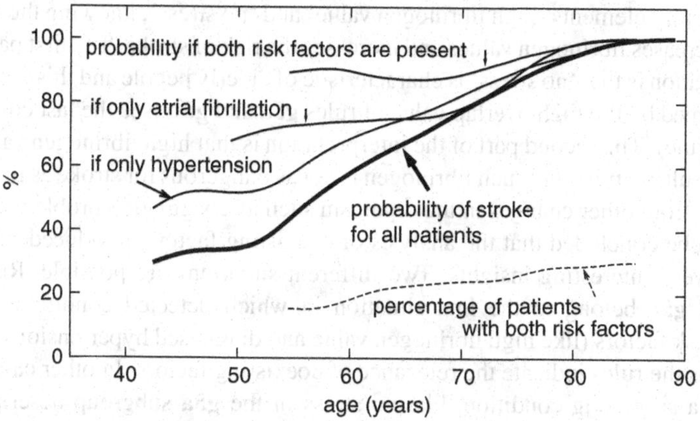

Fig. 12.7 Probability of stroke, estimated by the proportion of stroke patients, shown dependent of patients' age presented for all patients in the available hospital population (*thick line*), probability of stroke for persons with hypertension problems, with atrial fibrillation problems, and with both hypertension and atrial fibrillation problems (*thin solid lines*). The percentage of patients with both risk factors is about 20–25 % of the given hospital population (*dashed line*). The curves are drawn only for the range with sufficiently large numbers of patients in the database

when successful treatment of hypertension is possible, the risk of stroke still remains relatively high and it is higher than for patients without hypertension problems.

Analysis of specific rules. As noticed earlier, very sensitive rules are appropriate for extracting general properties of the target class. In contrast, very specific rules induced by generalization parameter values 5 or 10 are reliable classification rules for the target class. For example, rule **g5c** (`af = yes AND ahyp = yes`) well reflects the existing expert knowledge that hypertension and atrial fibrillation are important risk factors for stroke. The rule is relevant because it emphasizes the importance of the combination of these two risk factors, which is not a generally known fact. The relevance of the detected association is illustrated in Fig. 12.7. It shows that the probability of stroke is at least 85 % in the age range 55–80 years for persons with both risk factors measured on the available hospital population. We cannot estimate this probability on the general population but we can assume that it is even larger. The observation might be important for prevention purposes in general medical practice, especially because both factors can be easily detected.

Two other rules induced for the g-value equal 5 contain conditions based on the fibrinogen values of about 4.5 or more (reference values for negative fibrinogen finding are in the range 2.0–3.7 $\mathrm{g\,L^{-1}}$). These rules demonstrate the importance of high fibrinogen values for brain stroke patients. In the first rule the second necessary condition is the absence of stress, while in the second rule the second condition is age over 64 years. The interpretation of the second rule is relatively easy, leading to the conclusion that fibrinogen above 4.5 is itself very dangerous, which is confirmed also by rule g20a, being especially dangerous for elderly people. The interpretation of rule (`fibr > 4.55 AND str = no`) is not so easy because it includes contradictory elements ‘high fibrinogen value’ and ‘no stress’, knowing the fact that stress increases fibrinogen values and increases the risk of stroke. The first part of the interpretation is that ‘no stress’ is characteristic of elderly people and this conclusion is confirmed by the high overlap value of rules **g5a** and **g5b** (see the last column for the **g5b** rule). The second part of the interpretation is that high fibrinogen values can be the result of stress and such fibrinogen is not as dangerous for stroke as fibrinogen resulting from other changes in the organism such as coagulation problems.

It can be concluded that the analysis of coexisting factors in induced rules may lead to very interesting insights. Two different situations are possible. Rules like **g5b** and **g5c** belong to the first situation in which detected conditions present known risk factors (like high fibrinogen value and diagnosed hypertension). In such situations the rules indicate the relevance of coexisting factors. In other cases when there is a surprising condition, like no stress in the **g5a** subgroup description of brain ischaemia, the interpretation should necessarily be based on existing expert knowledge. The rule does not suggest the conclusion that ‘no stress’ is dangerous; instead, the conclusion is that increased fibrinogen is dangerous, *especially* when it is detected for patients that are not under stress. If there is a patient with increased fibrinogen value and the patient is under stress, it is possible to understand the reason for the increased fibrinogen value. In these circumstances the doctor will suggest the patient to avoid stressful situations. On the other hand, the situation when the patient has increased fibrinogen value without being exposed to stress is very different. In this case the fibrinogen value is increased without a known reason and, according to rule **g5a**, this may be a dangerous condition.

A very similar situation has been reported also for rules **b1** and **b2** in Table 12.9 when—in the coronary heart disease domain—the rules connected increased total cholesterol values with body mass index *below* 30. Again we had the situation that high body mass index and increased total cholesterol are known risk factors for the disease. The appropriate interpretation is that increased total cholesterol value is dangerous, *especially* if detected for patients without significant over-weight problems.

Analysis of moderately sensitive and specific rules. From the rules induced with generalization parameter values 10–50 it can be noticed that conditions on age and fibrinogen values repeat often, confirming already made conclusions about their importance. Generally, rules obtained in the middle range of parameter *g* may be analyzed in the same way as very sensitive or very specific rules. Potentially interesting are (`ahyp = yes AND fibr > 3.35`) or another rule (`age > 52.00 AND asp = no`). The latter rule stimulated the analysis presented in Fig. 12.8, which provides an excellent motivation for patients to accept prevention based on aspirin therapy. From the figure it can be easily noticed that the inductive learning approach correctly recognized the importance of the therapy for persons older than 52 years.

In addition, the moderately sensitive and specific rules are relevant also for the selection of appropriate boundary values for numeric descriptors included into rule conditions. Examples are age over 57 or 58 years, fibrinogen over 3.3, and systolic

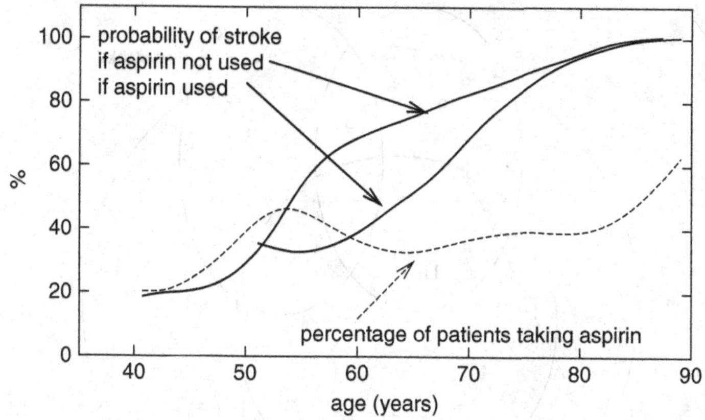

Fig. 12.8 The probability of brain stroke, estimated by the proportion of stroke patients, shown dependent of patient age presented for patients taking aspirin as the prevention therapy, and the probability of stroke for patients without this therapy. The percentage of patients with the aspirin therapy is presented by a dashed line

blood pressure over 153. These values, if significantly different from generally accepted reference values, could be reviewed as possible new decision points in medical decision-making practice. Even more importantly, it means that different boundary points can be suggested in combination with different conditions. This is in contradiction with existing medical practice, which tends to define unique reference values irrespective of the disease that has to be described and irrespective of other patient characteristics. In the case of fibrinogen, reference values above 3.7 are treated as positive while rules induced for the stroke domain suggest 4.55 as a standalone decision point, 4.45 in combination with age over 64 years, and 3.35 in combination with hypertension or age over 58 years for very sensitive detection of stroke.

Analysis of rule groups. Besides the possibility to analyze each rule separately, combinations of co-occurring rules can suggest other interpretations. In this respect it is useful to look at the overlap values of rules. A good example is a group of three rules induced for g-value 10. These rules have low overlap values, meaning that they describe relatively diverse subpopulations of the target class. Their analysis enables a global understanding of the hospital population in the *Intensive Care Unit* of the *Neurology Department*. Results of the analysis are presented in Fig. 12.9.

The figure graphically and numerically illustrates the importance of each subpopulation and its overlap with other groups. The textual description is also important, reflecting the results of basic statistical analysis (mean values of age and fibrinogen, as well as sex distribution) for the subpopulation described by the rule. Following this is a short list of supporting factors that additionally characterize the subpopulation. The results show that the induced rules describe three relatively different subpopulations of stroke patients.

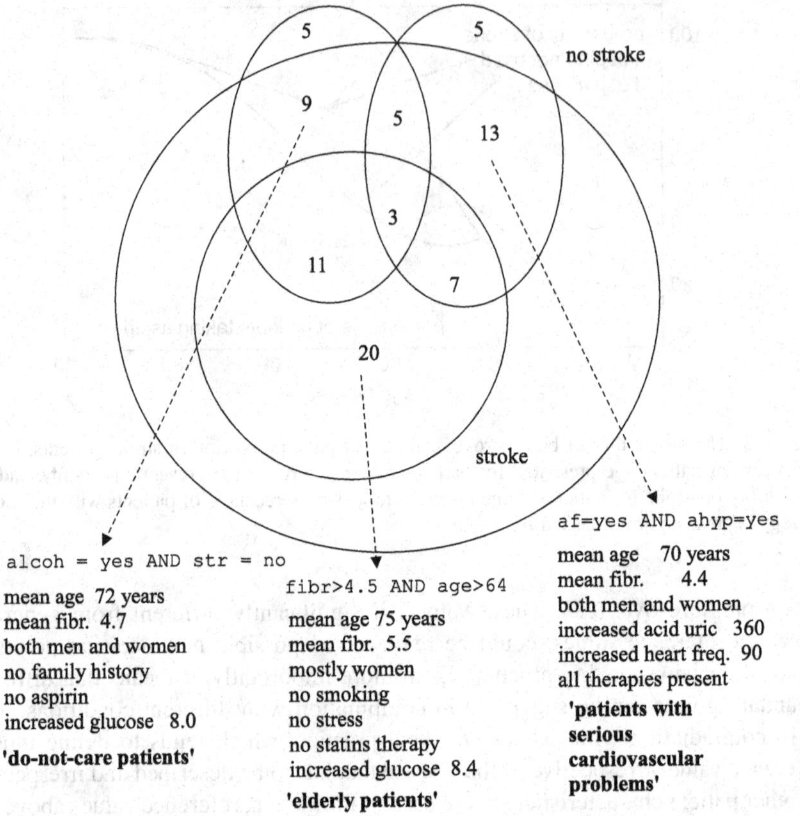

Fig. 12.9 Comparative study of three important subpopulations of stroke patients detected by rules induced with generalization parameter $g = 10$. The large circle presents the stroke patients, negative cases are outside the large circle. Small circles present three detected subpopulations. One of them includes only positive cases, while the other two also include a small portion of negative cases. The numbers present the percentages of patients that satisfy the conditions of one, two, or all three rules. In total, 68 % of positive cases are covered by at least one model. The definitions of patient groups (*in bold*) are followed by a list of most relevant properties that characterize the patient group (the supporting factors). The list ends with the concept name given to the group by the expert (*in bold*)

The largest subgroup can be called *elderly patients*; it is characterized by extremely high fibrinogen values (mean value 5.5) and increased glucose values (mean value 8.4). In most cases these are women (about 70 %) who do not smoke, do not suffer from stress, and do not have problems with lipoproteins (no statins therapy). Very different is the subpopulation that can be called *patients with serious cardiovascular problems* characterized by diagnosed hypertension and atrial fibrillation. It is a mixed male–female population. Its main characteristic is that these patients typically receive many different therapies but still have increased—but inside the allowable range—heart rate frequency (about 90) and uric acid (about 360). In between these two populations—in terms of age—is a subpopulation that can be called *do-not-care patients* characterized by alcohol consumption and no stress. It is a mixed male–female population characterized by the increased glucose values of laboratory tests, which one would not expect to find among the stroke patients because of their negative family history. Their do-not-care attitude is visible also from not taking aspirin as the prevention therapy.

12.5.3 Lessons Learned

The suggested inclusion of decision points into induced rule sets presents an important part of insightful data analysis results. These values are the result of an unbiased search for optimal decision functions incorporated in the rule learning process. In this respect these values represent a result that is difficult to achieve by classical statistical approaches. Their importance is in the fact that the detected values nicely integrate the properties of collected data into values that can be easily compared to generally acceptable reference values or results obtained on other databases. In this way, major discrepancies may also indicate diagnostic, methodological, or organizational problems inherent where the data has been collected.

In respect to the expert reasoning presented in this application it can be noticed that medical interpretations follow from a relatively simple visualization of existing statistical properties of the collected data. It is significant that the selection of properties that will be analyzed (like diastolic blood pressure) and conditions of the analysis (antihypertensive therapy present or absent) are suggested by medical reasoning based on induced rules. Obviously the same set of rules may stimulate other types of analyses, like the probability of stroke with respect to age with different levels of blood pressure as a parameter. What will be actually analyzed and how the analysis will be performed depends on the medical expert's preferences.

Rules with 'surprising' conditions are especially interesting because they may open different hypotheses, sometimes stimulating further research. For example, in the case of stroke we can speculate that various subtypes of fibrinogen exist: one type as a result of stress which is not very dangerous for stroke and the other subtype which is more dangerous but with unknown causes. The other possible speculation may be that increased fibrinogen is not dangerous by itself, but is dangerous because of some other unknown, possibly co-occurring phenomenon that is difficult to detect or measure. And stress is the exception in the sense that it results in increased fibrinogen values but without dangerous co-occurring phenomena.

Analysis of groups of rules is the final stage of insightful data analysis. Both the rules and their supporting factors are inputs for nontrivial medical expert reasoning. The problem is that this process cannot be formalized because it strongly depends on human experience. In successful cases the process results in better expert understanding of the basic properties of detected subpopulations and the recognition of their value by comparison with previous medical experience. The final point of this reasoning process is when experts are able to give names to the detected subpopulations and when these names or metaphors start to be used as concepts that describe existing medical practice or as novel knowledge that can be communicated throughout the medical community.

12.6 Conclusion

The presented applications demonstrate that rule learning is a powerful tool for various data analysis and knowledge discovery tasks. Appropriate data preparation, systematic prevention of overfitting, and collaboration with domain experts in the analysis of obtained results have been recognized as decisive ingredients that led to highly relevant results. Each of them has been presented on a separate domain but they may be combined in many different ways. They can be applied regardless if descriptive or predictive rule learning methods are used.

In the described applications the quality of obtained results is evaluated by the relevance of insight they have enabled. To this end, the last application is particularly interesting as it demonstrates that the construction of rules of various generality levels may be used for systematic exploration of the space of all possible descriptions. Most significant results for human interpretation are combinations of features occurring in the same rule, sets of rules describing disjoint subpopulations, and discriminatory values used in features constructed for numerical attributes. We have also demonstrated that the domain expert's engagement in the selection of attributes that will enter the induction process and in the selection of most interesting rules that will be evaluated represents a simple but effective way for integrating existing human knowledge into the knowledge discovery process.

In knowledge discovery tasks the complexity of induced rules is intentionally limited to a small number of features. The motivation for this practice is the requested generality of the resulting concepts and the necessity for human interpretation of the rules. An inherent problem is that such rules include only the most relevant features and that less relevant but also potentially interesting properties may not be detected in the rule learning process. The problem is solved by statistical evaluation of subpopulations defined by the resulting rules. The application in the coronary heart disease domain confirmed that generated supporting factors may be very useful additional input for domain expert evaluation.

Finally, the experiments and evaluation of the results in the multiclass gene expression domain demonstrate that reasonable predictive quality can be obtained by rule learning algorithms also in domains with a huge number of attributes and a very modest number of available examples. The result is important because it demonstrates the effectiveness of rule induction for predictive tasks and its applicability in challenging scientific discovery tasks. Due to the recent development of descriptive rule learning methods it may be expected that the later type of rule learning applications will be even more relevant in the future.

References

Adamo, J.-M. (2000). *Data mining for association rules and sequential patterns: Sequential and parallel algorithms*. New York: Springer. /15/

Adé, H., De Raedt, L., & Bruynooghe, M. (1995). Declarative bias for specific-to-general ILP systems. *Machine Learning, 20*(1–2), 119–154. Special Issue on Bias Evaluation and Selection. /133/

Agrawal, R., Mannila, H., Srikant, R., Toivonen, H., & Verkamo, A. I. (1995). Fast discovery of association rules. In U. M. Fayyad, G. Piatetsky-Shapiro, P. Smyth, & R. Uthurusamy (Eds.), *Advances in knowledge discovery and data mining* (pp. 307–328). Menlo Park, CA: AAAI. /14, 15, 121, 248/

Agrawal, R., & Srikant, R. (2000). Privacy-preserving data mining. In W. Chen, J. F. Naughton, & P. A. Bernstein (Eds.), *Proceedings of the 2000 ACM SIGMOD International Conference on Management of Data (SIGMOD-2000)*, Dallas, TX (pp. 439–450). New York: ACM. /17/

Aha, D. W., Kibler, D., & Albert, M. K. (1991). Instance-based learning algorithms. *Machine Learning, 6*, 37–66. /2/

Ali, K. M., & Pazzani, M. J. (1993). HYDRA: A noise-tolerant relational concept learning algorithm. In R. Bajcsy (Ed.), *Proceedings of the 13th Joint International Conference on Artificial Intelligence (IJCAI-93)*, Chambéry, France (pp. 1064–1071). San Mateo, CA: Morgan Kaufmann. /150, 220/

Allwein, E. L., Schapire, R. E., & Singer, Y. (2000). Reducing multiclass to binary: A unifying approach for margin classifiers. *Journal of Machine Learning Research, 1*, 113–141. /235, 236/

An, A., & Cercone, N. (1998). ELEM2: A learning system for more accurate classifications. In R. E. Mercer & E. Neufeld (Eds.), *Proceedings of the 12th Biennial Conference of the Canadian Society for Computational Studies of Intelligence* (pp. 426–441). Berlin, Germany/New York: Springer. /155/

Atzmüller, M., & Puppe, F. (2005). Semi-automatic visual subgroup mining using VIKAMINE. *Journal of Universal Computer Science, 11*(11), 1752–1765. Special Issue on Visual Data Mining. /261/

Atzmüller, M., & Puppe, F. (2006). SD-Map – A fast algorithm for exhaustive subgroup discovery. In *Proceedings of the 10th European Conference on Principles and Practice of Knowledge Discovery in Databases (PKDD-06)*, Berlin, Germany (pp. 6–17). Berlin, Germany: Springer. /249/

Atzmüller, M., Puppe, F., & Buscher, H.-P. (2005a). Exploiting background knowledge for knowledge-intensive subgroup discovery. In *Proceedings of the 19th International Joint Conference on Artificial Intelligence (IJCAI-05)*, Edinburgh, UK (pp. 647–652). San Francisco: Morgan Kaufmann. /249/

J. Fürnkranz et al., *Foundations of Rule Learning*, Cognitive Technologies,
DOI 10.1007/978-3-540-75197-7,

Atzmüller, M., Puppe, F., & Buscher, H.-P. (2005b). Profiling examiners using intelligent subgroup mining. In *Proceedings of the 10th Workshop on Intelligent Data Analysis in Medicine and Pharmacology (IDAMAP-05)* (pp. 46–51) Aberdeen: AIME. /250/

Aumann, Y., & Lindell, Y. (1999). A statistical theory for quantitative association rules. In *Proceedings of the 5th ACM SIGKDD International Conference on Knowledge Discovery and Data Mining (KDD-99)*, San Diego, CA (pp. 261–270). New York: ACM. /261/

Azevedo, P. J., & Jorge, A. J. (2010). Ensembles of jittered association rule classifiers. *Data Mining and Knowledge Discovery, 12*(5), 421–453. Special Issue on Global Modeling using Local Patterns. /185/

Badea, L. (2001). A refinement operator for theories. In C. Rouveirol, & M. Sebag (Eds.), *Proceedings of the 11th International Conference on Inductive Logic Programming (ILP-01)*, Strasbourg, France (pp. 1–14). Berlin, Germany/New York: Springer. /183/

Bay, S. D. (2000). Multivariate discretization of continuous variables for set mining. In *Proceedings of the 6th ACM SIGKDD International Conference on Knowledge Discovery and Data Mining (KDD-2000)*, Boston (pp. 315–319). New York: ACM. /252/

Bay, S. D., & Pazzani, M. J. (2001). Detecting group differences: Mining contrast sets. *Data Mining and Knowledge Discovery, 5*(3), 213–246. /248, 251, 256, 257, 259/

Bayardo, R. J., Jr. (1997). Brute-force mining of high-confidence classification rules. In *Proceedings of the 3rd International Conference on Knowledge Discovery and Data Mining (KDD-97)* (pp. 123–126). Menlo Park, CA: AAAI. /54, 122, 185/

Bayardo, R. J., Jr. (1998). Efficiently mining long patterns from databases. In *Proceedings of the 1998 ACM SIGMOD International Conference on Management of Data (SIGMOD-98)*, Seattle, WA (pp. 85–93). New York: ACM /251/

Bayardo, R. J., Jr., & Agrawal, R. (1999). Mining the most interesting rules. In *Proceedings of the 5th ACM SIGKDD International Conference on Knowledge Discovery and Data Mining (KDD-97)*, Newport Beach, CA (pp. 145–154). New York: ACM. /167/

Bennett, K. P., Buja, A., Freund, W. S. Y., Schapire, R. E., Friedman, J., & Hastie, T. et al. (2008). Responses to Mease and Wyner (2008). *Journal of Machine Learning Research, 9*, 157–194. /2/

Bergadano, F., Giordana, A., & Saitta, L. (1988). Automated concept acquisition in noisy environments. *IEEE Transactions on Pattern Analysis and Machine Intelligence, 10*, 555–578. /120/

Bergadano, F., Matwin, S., Michalski, R. S., & Zhang, J. (1992). Learning two-tiered descriptions of flexible concepts: The POSEIDON system. *Machine Learning, 8*, 5–43. /48, 68, 200, 202/

Berthold, M. R., Cebron, N., Dill, F., Gabriel, T. R., Kötter, T., & Meinl, T., et al. (2009). KNIME—The Konstanz information miner. Version 2.0 and beyond. *SIGKDD Explorations, 11*, 26–31. /5/

Billari, F. C., Fürnkranz, J., & Prskawetz, A. (2006). Timing, sequencing, and quantum of life course events: A machine learning approach. *European Journal of Population,* 22(1), 37–65. /269, 270/

Bishop, C. M. (1995). *Neural networks for pattern recognition*. Oxford, UK: Clarendon. /2/

Bisson, G. (1992). Conceptual clustering in a first order logic representation. In B. Neumann (Ed.), *Proceedings of the 10th European Conference on Artificial Intelligence (ECAI-92)*, Vienna (pp. 458–462). Chichester, UK/New York: Wiley. /245/

Blaszczynski, J., Stefanowski, J., & Zajac, M. (2009). Ensembles of abstaining classifiers based on rule sets. In J. Rauch, Z. W. Ras, P. Berka, & T. Elomaa (Eds.), *Proceedings of the 18th International Symposium on Foundations of Intelligent Systems (ISMIS-09)*, Prague, Czech Republic (pp. 382–391). Berlin, Germany: Springer. /223/

Blockeel, H., & De Raedt, L. (1998). Top-down induction of first-order logical decision trees. *Artificial Intelligence, 101*(1–2), 285–297. /13/

Blockeel, H., De Raedt, L., & Ramon, J. (1998). Top-down induction of clustering trees. In J. Shavlik (Ed.), *Proceedings of the 15th International Conference on Machine Learning*, Madison, WI (pp. 55–63). San Francisco: Morgan Kaufmann. /245/

Blockeel, H., & Vanschoren, J. (2007). Experiment databases: Towards an improved experimental methodology in machine learning. In J. N. Kok, J. Koronacki, R. L. de Mántaras, S. Matwin, D. Mladenic, & A. Skowron (Eds.), *Proceedings of the 11th European Conference on Principles and Practice of Knowledge Discovery in Databases (PKDD-07)*, Warsaw, Poland (pp. 6–17). Berlin, Germany/New York: Springer. */48/*

Bose, R. C., & Ray Chaudhuri, D. K. (1960). On a class of error correcting binary group codes. *Information and Control, 3*(1), 68–79. */235/*

Boström, H. (1995). Covering vs. divide-and-conquer for top-down induction of logic programs. In *Proceedings of the 14th International Joint Conference on Artificial Intelligence (IJCAI-95)*, Montréal, QC (pp. 1194–1200). San Mateo, CA: Morgan Kaufmann. */12/*

Boström, H. (2004). Pruning and exclusion criteria for un-ordered incremental reduced error pruning. In J. Fürnkranz (Ed.), *Proceedings of the ECML/PKDD Workshop on Advances in Inductive Rule Learning*, Pisa, Italy (pp. 17–29). */156, 208/*

Boström, H. (2007). Maximizing the area under the ROC curve with decision lists and rule sets. In *Proceedings of the 7th SIAM International Conference on Data Mining (SDM-07)*, Minneapolis, MN (pp. 27–34). Philadelphia: SIAM. */242/*

Botta, M., & Giordana, A. (1993). SMART+: A multi-strategy learning tool. In R. Bajcsy (Ed.), *Proceedings of the 13th Joint International Conference on Artificial Intelligence (IJCAI-93)*, Chambéry, France (pp. 937–944). San Mateo, CA: Morgan Kaufmann. */162, 163/*

Botta, M., Giordana, A., & Saitta, L. (1992). Comparison of search strategies in learning relations. In B. Neumann (Ed.), *Proceedings of the 10th European Conference on Artificial Intelligence (ECAI-92)*, Vienna (pp. 451–455). Chichester, UK/New York: Wiley. */120, 163/*

Boulesteix, A.-L., Tutz, G., & Strimmer, K. (2003). A CART-based approach to discover emerging patterns in microarray data. *Bioinformatics, 19*(18), 2465–2472. */254/*

Bradley, R. A., & Terry, M. E. (1952). The rank analysis of incomplete block designs—I. The method of paired comparisons. *Biometrika, 39*, 324–345. */230/*

Bratko, I. (1990). *Prolog programming for artificial intelligence* (2nd ed.). Wokingham, UK: Addison-Wesley. */104/*

Bratko, I. (1999). Refining complete hypotheses in ILP. In S. Džeroski & P. Flach (Eds.), *Proceedings of the 9th International Workshop on Inductive Logic Programming (ILP-99)*, Bled, Slovenia (pp. 44–55). Berlin, Germany/New York: Springer. */183/*

Bratko, I., & Muggleton, S. H. (1995). Applications of inductive logic programming. *Communications of the ACM, 38*(11), 65–70. */52/*

Breiman, L. (1996). Bagging predictors. *Machine Learning, 24*(2), 123–140. */2, 233/*

Breiman, L. (2001a). Random forests. *Machine Learning, 45*(1), 5–32. */2/*

Breiman, L. (2001b). Statistical modeling: The two cultures. *Statistical Science, 16*(3), 199–231. With comments by D. R. Cox, B. Efron, B. Hoadley, and E. Parzen, and a rejoinder by the author. */3/*

Breiman, L., Friedman, J. H., Olshen, R., & Stone, C. (1984). *Classification and regression trees*. Pacific Grove, CA: Wadsworth & Brooks. */2, 3, 147, 199, 200, 211, 244/*

Brin, S., Motwani, R., & Silverstein, C. (1997). Beyond market baskets: Generalizing association rules to correlations. In *Proceedings of the ACM SIGMOD International Conference on Management of Data*, Toronto, ON (pp. 265–276). New York: ACM. */154/*

Bringmann, B., Nijssen, S., & Zimmermann, A. (2009). Pattern-based classification: A unifying perspective. In A. Knobbe & J. Fürnkranz (Eds.), *From Local Patterns to Global Models: Proceedings of the ECML/PKDD-09 Workshop (LeGo-09)*, Bled, Slovenia (pp. 36–50). */54, 185/*

Bruha, I., & Franek, F. (1996). Comparison of various routines for unknown attribute value processing: The covering paradigm. *International Journal of Pattern Recognition and Artificial Intelligence, 10*(8), 939–955. */87/*

Brunk, C. A., & Pazzani, M. J. (1991). An investigation of noise-tolerant relational concept learning algorithms. In *Proceedings of the 8th International Workshop on Machine Learning (ML-91)*, Evanston, IL (pp. 389–393). San Mateo, CA: Morgan Kaufmann. */202, 203/*

Buntine, W., & Niblett, T. (1992). A further comparison of splitting rules for decision-tree induction. *Machine Learning, 8*, 75–85. */136/*

Cai, Y., Cercone, N., & Han, J. (1991). Attribute-oriented induction in relational databases. In G. Piatetsky-Shapiro & W. J. Frawley (Eds.), *Knowledge discovery in databases* (pp. 213–228). Menlo Park, CA: MIT. /70/

Cameron-Jones, R. M. (1996). The complexity of batch approaches to reduced error rule set induction. In N. Foo & R. Goebel (Eds.), *Proceedings of the 4th Pacific Rim International Conference on Artificial Intelligence (PRICAI-96)*, Cairns, QLD (pp. 348–359). Berlin, Germany/New York: Springer. /202, 205, 209/

Cameron-Jones, R. M., & Quinlan, J. R. (1993). Avoiding pitfalls when learning recursive theories. In R. Bajcsy (Ed.), *Proceedings of the 13th International Joint Conference on Artificial Intelligence (IJCAI-93)*, Chambéry, France (pp. 1050–1057). San Mateo, CA: Morgan Kaufmann. /111/

Cardoso, J. S., & da Costa, J. F. P. (2007). Learning to classify ordinal data: The data replication method. *Journal of Machine Learning Research, 8*, 1393–1429. /235/

Cendrowska, J. (1987). PRISM: An algorithm for inducing modular rules. *International Journal of Man-Machine Studies, 27*, 349–370. /13, 23, 35, 49, 146, 164/

Ceri, S., Gottlob, G., & Tanca, L. (1989). What you always wanted to know about datalog (and never dared to ask). *IEEE Transactions on Knowledge and Data Engineering, 1*(1), 146–166. /108, 109/

Ceri, S., Gottlob, G., & Tanca, L. (1990). *Logic programming and databases* (Surveys in computer science). Berlin, Germany: Springer. /108, 109/

Cestnik, B. (1990). Estimating probabilities: A crucial task in machine learning. In L. Aiello (Ed.), *Proceedings of the 9th European Conference on Artificial Intelligence (ECAI-90)*, Stockholm (pp. 147–150). London: Pitman. /149/

Chapman, P., Clinton, J., Kerber, R., Khabaza, T., Reinartz, T., & Shearer, C., et al. (2000). Crisp-Dm 1.0: Step-by-step data mining guide. SPSS. Available from http://www.the-modeling-agency.com/CRISP-DM.pdf. /4/

Chen, S. F., & Goodman, J. T. (1998). *An empirical study of smoothing techniques for language modeling* (Tech. Rep. TR-10-98). Cambridge, MA: Computer Science Group, Harvard University. /215/

Chow, M., Moler, J., & Mian, S. (2001). Identifying marker genes in transcription profiling data using a mixture of feature relevance experts. *Physiological Genomics, 3*(5), 99–111. /274/

Clark, P., & Boswell, R. (1991). Rule induction with CN2: Some recent improvements. In *Proceedings of the 5th European Working Session on Learning (EWSL-91)*, Porto, Portugal (pp. 151–163). Berlin, Germany: Springer. /10, 21, 50, 146, 150, 169, 199, 228/

Clark, P., & Niblett, T. (1987). Induction in noisy domains. In I. Bratko & N. Lavrač (Eds.), *Progress in Machine Learning*. Wilmslow, UK: Sigma Press. /48/

Clark, P., & Niblett, T. (1989). The CN2 induction algorithm. *Machine Learning, 3*(4), 261–283. /1, 21, 43, 44, 48, 49, 50, 67, 68, 118, 146, 196/

Cloete, I., & Van Zyl, J. (2006). Fuzzy rule induction in a set covering framework. *IEEE Transactions on Fuzzy Systems, 14*(1), 93–110. /92/

Cohen, W. W. (1993). Efficient pruning methods for separate-and-conquer rule learning systems. In *Proceedings of the 13th International Joint Conference on Artificial Intelligence (IJCAI-93)*, Chambéry, France (pp. 988–994). San Mateo, CA: Morgan Kaufmann. /202, 203, 204, 205, 209, 211/

Cohen, W. W. (1995). Fast effective rule induction. In A. Prieditis & S. Russell (Eds.), *Proceedings of the 12th International Conference on Machine Learning (ML-95)*, Lake Tahoe, CA (pp. 115–123). San Francisco: Morgan Kaufmann. /21, 45, 51, 52, 67, 68, 145, 162, 189, 198, 210, 211, 213/

Cohen, W. W. (1996). Learning trees and rules with set-valued features. In *Proceedings of the 13th National Conference on Artificial Intelligene (AAAI-96)* (pp. 709–716). Menlo Park, CA: AAAI. /69/

Cohen, W. W., Schapire, R. E., & Singer, Y. (1999). Learning to order things. *Journal of Artificial Intelligence Research, 10*, 243–270. /243/

Cohen, W. W., & Singer, Y. (1999). A simple, fast, and effective rule learner. In *Proceedings of the 16th National Conference on Artificial Intelligence (AAAI-99)* (pp. 335–342). Menlo Park, CA: AAAI/MIT. /178, 186/

Cook, D. J., & Holder, L. B. (1994). Substructure discovery using minimum description length and background knowledge. *Journal of Artificial Intelligence Research, 1*, 231–255. /245/

Cootes, A. P., Muggleton, S. H., & Sternberg, M. J. (2003). The automatic discovery of structural principles describing protein fold space. *Journal of Molecular Biology, 330*(4), 527–532. /52/

Crammer, K., & Singer, Y. (2002). On the learnability and design of output codes for multiclass problems. *Machine Learning, 47*(2–3), 201–233. /234/

Daly, O., & Taniar, D. (2005). Exception rules in data mining. In M. Khosrow-Pour (Ed.), *Encyclopedia of information science and technology* (Vol. II, pp. 1144–1148). Hershey, PA: Idea Group. /260/

Dasarathy, B. V. (Ed.). (1991). *Nearest neighbor (NN) norms: NN pattern classification techniques*. Los Alamitos, CA: IEEE. /2/

Davis, J., Burnside, E., Castro Dutra, I. d., Page, D., & Santos Costa, V. (2004). Using Bayesian classifiers to combine rules. In *Proceedings of the 3rd SIGKDD Workshop on Multi-Relational Data Mining (MRDM-04)*, Seattle, WA. /222/

Dehaspe, L., & Toivonen, H. (2001). Discovery of relational association rules. In S. Džeroski & N. Lavrač (Eds.), *Relational data mining* (pp. 189–212). Berlin, Germany/New York: Springer. /96/

Dekel, O., Manning, C. D., & Singer, Y. (2004). Log-linear models for label ranking. In S. Thrun, L. K. Saul, & B. Schölkopf (Eds.), *Advances in neural information processing systems (NIPS-03)* (pp. 497–504). Cambridge, MA: MIT. /239/

del Jesus, M. J., González, P., Herrera, F., & Mesonero, M. (2007). Evolutionary fuzzy rule induction process for subgroup discovery: A case study in marketing. *IEEE Transactions on Fuzzy Systems, 15*(4), 578–592. /249, 251/

Dembczyński, K., Kotłowski, W., & Słowiński, R. (2008). Solving regression by learning an ensemble of decision rules. In L. Rutkowski, R. Tadeusiewicz, L. A. Zadeh, & J. M. Zurada (Eds.), *Proceedings of the 9th International Conference on Artificial Intelligence and Soft Computing (ICAISC-08)*, Zakopane, Poland (pp. 533–544). Berlin, Germany/New York: Springer. /215, 245/

Dembczyński, K., Kotłowski, W., & Słowiński, R. (2010). ENDER – A statistical framework for boosting decision rules. *Data Mining and Knowledge Discovery, 12*(5), 385–420. Special Issue on Global Modeling using Local Patterns. /178, 185/

Demšar, J. (2006). Statistical comparisons of classifiers over multiple data sets. *Journal of Machine Learning Research, 7*, 1–30. /90/

Demšar, J., Zupan, B., & Leban, G. (2004). Orange: From experimental machine learning to interactive data mining. White Paper, Faculty of Computer and Information Science, University of Ljubljana. Available from http://orange.biolab.si/. /5/

De Raedt, L. (1992). *Interactive theory revision: An inductive logic programming approach*. London/San Diego, CA/Sydney, NSW: Academic. /133/

De Raedt, L. (Ed.). (1995). *Advances in inductive logic programming* (Frontiers in artificial intelligence and applications, Vol. 32). Amsterdam/Washington, DC: IOS Press. /96/

De Raedt, L. (1996). Induction in logic. In R. Michalski & J. Wnek (Eds.), *Proceedings of the 3rd International Workshop on Multistrategy Learning (MSL-96)* (pp. 29–38). Fairfax, VA: Machine Learning and Inference Laboratory, George Mason University. /174/

De Raedt, L. (2008). *Logical and relational learning*. Berlin, Germany: Springer. /2, 129/

De Raedt, L., & Dehaspe, L. (1997). Clausal discovery. *Machine Learning, 26*(2/3), 99–146. Special Issue on *Inductive Logic Programming*. /96, 174/

De Raedt, L., & Van Laer, W. (1995). Inductive constraint logic. In K. Jantke, T. Shinohara, & Zeugmann, T. (Eds.), *Proceedings of the 5th Workshop on Algorithmic Learning Theory (ALT-95)*, Fukuoka, Japan (pp. 80–94). Berlin, Germany/New York: Springer. /50, 174/

Dietterich, T. G. (2000). Ensemble methods in machine learning. In J. Kittler & F. Roli (Eds.), *Proceedings of the 1st International Workshop on Multiple Classifier Systems*, Cagliari, Italy (pp. 1–15). Berlin, Germany/New York: Springer. */180/*

Dietterich, T. G., & Bakiri, G. (1995). Solving multiclass learning problems via error-correcting output codes. *Journal of Artificial Intelligence Research, 2*, 263–286. */233/*

Domingos, P. (1996a). Linear-time rule induction. In *Proceedings of the 2nd International Conference on Knowledge Discovery and Data Mining (KDD-96)* pp. 96–101. Menlo Park, CA: AAAI. */182/*

Domingos, P. (1996b). Unifying instance-based and rule-based induction. *Machine Learning, 24*, 141–168. */181, 182/*

Domingos, P. (1999). The role of Occam's Razor in knowledge discovery. *Data Mining and Knowledge Discovery, 3*(4), 409–425. */9/*

Dong, G., & Li, J. (1999). Efficient mining of emerging patterns: Discovering trends and differences. In *Proceedings of the 5th ACM SIGKDD International Conference on Knowledge Discovery and Data Mining (KDD-99)*, San Diego, CA (pp. 43–52). New York: ACM */248, 253, 256, 258/*

Dong, G., Zhang, X., Wong, L., & Li, J. (1999). CAEP: Classification by aggregating emerging patterns. In *Proceedings of the 2nd International Conference on Discovery Science (DS-99)*, Tokyo, Japan (pp. 30–42). Berlin, Germany/New York: Springer. */253, 259/*

Dubois, D., & Prade, H. (1980). *Fuzzy sets and systems*. New York: Academic. */92/*

Duda, R. O., Hart, P. E., & Stork, D. G. (2000). *Pattern classification* (2nd ed.). New York: Wiley. */2/*

Džeroski, S., & Bratko, I. (1992). Handling noise in inductive logic programming. In S. H. Muggleton & K. Furukawa (Eds.), *Proceedings of the 2nd International Workshop on Inductive Logic Programming (ILP-92)* (pp. 109–125). No. TM-1182 in ICOT Technical Memorandum, Institute for New Generation Computer Technology, Tokyo, Japan. */50, 118, 150, 196/*

Džeroski, S., Cestnik, B., & Petrovski, I. (1993). Using the m-estimate in rule induction. *Journal of Computing and Information Technology, 1*, 37–46. */45/*

Džeroski, S., & Lavrač, N. (Eds.). (2001). *Relational data mining: Inductive logic programming for knowledge discovery in databases*. Berlin, Germany/New York: Springer. */7, 95, 96, 129/*

Džeroski, S., Schulze-Kremer, S., Heidtke, K. R., Siems, K., Wettschereck, D., & Blockeel, H. (1998). Diterpene structure elucidation from 13CNMR spectra with inductive logic programming. *Applied Artificial Intelligence, 12*(5), 363–383. Special Issue on First-Order Knowledge Discovery in Databases. */120/*

Egan, J. P. (1975). *Signal detection theory and ROC analysis*. New York: Academic Press */58/*

Eineborg, M., & Boström, H. (2001). Classifying uncovered examples by rule stretching. In C. Rouveirol & M. Sebag (Eds.), *Proceedings of the Eleventh International Conference on Inductive Logic Programming (ILP-01)*, Strasbourg, France (pp. 41–50). Berlin, Germany/New York: Springer. */223/*

Elmasri, R., & Navathe, S. B. (2006). *Fundamentals of database systems* (5th ed.). Boston: Addison-Wesley. */107/*

Escalera, S., Pujol, O., & Radeva, P. (2006). Decoding of ternary error correcting output codes. In J. F. M. Trinidad, J. A. Carrasco-Ochoa, & J. Kittler (Eds.), *Proceedings of the 11th Iberoamerican Congress in Pattern Recognition (CIARP-06)*, Cancun, Mexico (pp. 753–763). Berlin, Germany/Heidelberg, Germany/New York: Springer. */236/*

Esposito, F., Malerba, D., & Semeraro, G. (1993). Decision tree pruning as a search in the state space. In P. Brazdil (Ed.), *Proceedings of the 6th European Conference on Machine Learning (ECML-93)*, Vienna (pp. 165–184). Berlin, Germany/New York: Springer. */200/*

Esposito, F., Semeraro, G., Fanizzi, N., & Ferilli, S. (2000). Multistrategy theory revision: Induction and abduction in INTHELEX. *Machine Learning, 38*(1–2), 133–156. */183/*

Everitt, B., & Hothorn, T. (2006). *A handbook of statistical analyses using R*. Boca Raton, FL: Chapman & Hall/CRC. */5/*

Fan, H., Fan, M., Ramamohanarao, K., & Liu, M. (2006). Further improving emerging pattern based classifiers via bagging. In *Proceedings of the 10th Pacific-Asia conference on Knowledge*

Discovery and Data Mining (PAKDD-06), Singapore (pp. 91–96). Berlin, Germany/Heidelberg, Germany/New York: Springer. /254/

Fan, H., & Ramamohanarao, K. (2003a). A Bayesian approach to use emerging patterns for classification. In *Proceedings of the 14th Australasian Database Conference (ADC-03)*, Adelaide, SA (pp. 39–48). Darlinghurst, NSW: Australian Computer Society /254/

Fan, H., & Ramamohanarao, K. (2003b). Efficiently mining interesting emerging patterns. In *Proceeding of the 4th International Conference on Web-Age Information Management (WAIM-03)*, Chengdu, China (pp. 189–201). Berlin, Germany/New York: Springer. /253, 254/

Fawcett, T. E. (2001). Using rule sets to maximize ROC performance. In *Proceedings of the IEEE International Conference on Data Mining (ICDM-01)*, San Jose, CA (pp. 131–138). Los Alamitos, CA: IEEE. /242/

Fawcett, T. E. (2006). An introduction to ROC analysis. *Pattern Recognition Letters, 27*(8), 861–874. /60/

Fawcett, T. E. (2008). PRIE: A system for generating rulelists to maximize ROC performance. *Data Mining and Knowledge Discovery, 17*(2), 207–224. /242/

Fawcett, T., & Niculescu-Mizil, A. (2007). PAV and the ROC convex hull. *Machine Learning, 68*(1), 97–106. /215/

Fayyad, U. M., & Irani, K. B. (1992). On the handling of continuous-valued attributes in decision tree generation. *Machine Learning, 8*(2), 87–102. /82/

Fayyad, U. M., Piatetsky-Shapiro, G., & Smyth, P. (1996). From data mining to knowledge discovery in databases. *AI Magazine, 17*(3), 37–54. /4/

Fayyad, U. M., Piatetsky-Shapiro, G., Smyth, P., & Uthurusamy, R. (Eds.). (1995). *Advances in knowledge discovery and data mining*. Menlo Park, CA: AAAI. /2/

Fensel, D., & Wiese, M. (1993). Refinement of rule sets with JoJo. In P. Brazdil (Ed.), *Proceedings of the 6th European Conference on Machine Learning (ECML-93)*, Vienna (pp. 378–383). Berlin, Germany/New York: Springer. /128/

Fensel, D., & Wiese, M. (1994). From JoJo to Frog: Extending a bi-directional strategy to a more flexible three-directional search. In C. Globig & K.-D. Althoff (eds.) *Beiträge zum 7. Fachgruppentreffen Maschinelles Lernen* (Tech. Rep. LSA-95-01, pp. 37–44). Zentrum für Lernende Systeme und Anwendungen, University of Kaiserslautern. /128/

Ferri, C., Flach, P., & Hernández, J. (2002). Learning decision trees using the area under the ROC curve. In C.Sammut & A. Hoffmann (Eds.), *Proceedings of the 19th International Conference on Machine Learning (ICML-02)*, Sydney, NSW (pp. 139–146). San Francisco: Morgan Kaufmann. /214/

Fisher, D. H. (1987). Knowledge acquisition via incremental conceptual clustering. *Machine Learning, 2*(2), 139–172. /245/

Flach, P. (1993). Predicate invention in inductive data engineering. In P. B. Brazdil (Ed.), *Proceedings of the 6th European Conference on Machine Learning (ECML-93)*, Vienna pp. 83–94. Berlin, Germany/New York: Springer. /97/

Flach, P. (1994). *Simply logical – Intelligent reasoning by example*. Chichester, UK/New York: Wiley. /104/

Flach, P. (1997). Normal forms for inductive logic programming. In N. Lavrač & S. Džeroski (Eds.), *Proceedings of the 7th International Workshop on Inductive Logic Programming (ILP-97)*, Prague, Czech Republic (pp. 149–156). Berlin, Germany/New York: Springer. /174/

Flach, P. (2003). The geometry of ROC space: Using ROC isometrics to understand machine learning metrics. In T. Fawcett & N. Mishra (Eds.), *Proceedings of the 20th International Conference on Machine Learning (ICML-03)*, Washington, DC (pp. 194–201). Menlo Park, CA: AAAI. /148, 177/

Flach, P., Giraud-Carrier, C., & Lloyd, J. (1998). Strongly typed inductive concept learning. In *Proceedings of the 8th International Conference on Inductive Logic Programming (ILP-98)*, Madison, WI (pp. 185–194). Berlin, Germany/New York: Springer. /96/

Flach, P., & Lachiche, N. (1999). 1BC: A first-order Bayesian classifier. In *Proceedings of the 9th International Workshop on Inductive Logic Programming (ILP-99)*, Bled, Slovenia (pp. 92–103). Berlin, Germany/New York: Springer. /102/

Flach, P., & Lachiche, N. (2001). Confirmation-guided discovery of first-order rules with Tertius. *Machine Learning, 42*(1/2), 61–95. /97/

Flach, P., & Lavrač, N. (2003). Rule induction. In M. Berthold & D. J. Hand (Eds.), *Intelligent data analysis* (2nd ed., pp. 229–267). Berlin, Germany/New York: Springer. /19, 95/

Flach, P., & Wu, S. (2005). Repairing concavities in ROC curves. In L. P. Kaelbling & A. Saffiotti (Eds.), *Proceedings of the 19th International Joint Conference on Artificial Intelligence (IJCAI-05)*, Edinburgh, UK (pp. 702–707). Professional Book Center. /214/

Fodor, J., & Roubens, M. (1994). *Fuzzy preference modelling and multicriteria decision support*. Dordrecht, The Netherlands/Boston: Kluwer. /223/

Frank, A., & Asuncion, A. (2010). *UCI machine learning repository*. Irvine, CA: University of California, School of Information and Computer Science. /48, 219/

Frank, E., & Hall, M. (2001). A simple approach to ordinal classification. In L. D. Raedt & P. Flach (Eds.), *Proceedings of the 12th European Conference on Machine Learning (ECML-01)*, Freiburg, Germany (pp. 145–156). Berlin, Germany/New York: Springer. /241, 243/

Frank, E., & Witten, I. H. (1998). Generating accurate rule sets without global optimization. In J. Shavlik (Ed.), *Proceedings of the 15th International Conference on Machine Learning (ICML-98)*, Madison, WI (pp. 144–151). San Francisco: Morgan Kaufmann. /55, 169, 183, 244/

Freund, Y., & Schapire, R. E. (1997). A decision-theoretic generalization of on-line learning and an application to boosting. *Journal of Computer and System Sciences, 55*(1), 119–139. /2/

Friedman, M. (1937). The use of ranks to avoid the assumption of normality implicit in the analysis of variance. *Journal of the American Statistical Association, 32*, 675–701. /90/

Friedman, A. (1986). *Fundamentals of logic design and switching theory*. Rockville, MD: Computer Science Press. /174/

Friedman, J. H. (1996). *Another approach to polychotomous classification* (Tech. rep.). Stanford, CA: Department of Statistics, Stanford University. /230/

Friedman, J. H. (1998). Data mining and statistics: What's the connection? In *Computing Science and Statistics: Proceedings of the 29th Symposium on the Interface*, Houston, TX. Fairfax Station, VA: Interface Foundation of North America. /3, 17/

Friedman, J. H., & Fisher, N. I. (1999). Bump hunting in high-dimensional data. *Statistics and Computing, 9*(2), 123–143. /2, 10, 21, 161, 261/

Friedman, N., Geiger, D., & Goldszmidt, M. (1997). Bayesian networks classifiers. *Machine Learning, 29*, 131–161. /222/

Friedman, J. H., & Popescu, B. E. (2008). Predictive learning via rule ensembles. *Annals of Applied Statistics, 2*, 916–954. /215, 245/

Fürnkranz, J. (1994a). FOSSIL: A robust relational learner. In F. Bergadano & L. De Raedt (Eds.), *Proceedings of the 7th European Conference on Machine Learning (ECML-94)*, Catania, Italy (pp. 122–137). Berlin, Germany/New York: Springer. /51, 154, 197, 198/

Fürnkranz, J. (1994b). Top-down pruning in relational learning. In A. G. Cohn (Ed.), *Proceedings of the 11th European Conference on Artificial Intelligence (ECAI-94)*, Amsterdam (pp. 453–457). Chichester, UK/New York: Wiley. /211/

Fürnkranz, J. (1995). A tight integration of pruning and learning (extended abstract). In N. Lavrač & S. Wrobel (Eds.), *Proceedings of the 8th European Conference on Machine Learning (ECML-95)*, Heraclion, Greece (pp. 291–294). Berlin, Germany/New York: Springer. /212/

Fürnkranz, J. (1997). Pruning algorithms for rule learning. *Machine Learning, 27*(2), 139–171. /12, 187, 197, 210, 211/

Fürnkranz, J. (2002a). A pathology of bottom-up hill-climbing in inductive rule learning. In N. Cesa-Bianchi, M. Numao, & R. Reischuk (Eds.), *Proceedings of the 13th European Conference on Algorithmic Learning Theory (ALT-02)*, Lübeck, Germany (pp. 263–277). Berlin, Germany/New York: Springer. /125, 126/

Fürnkranz, J. (2002b). Round robin classification. *Journal of Machine Learning Research, 2*, 721–747. /217, 230, 232/

Fürnkranz, J. (2003). Round robin ensembles. *Intelligent Data Analysis, 7*(5), 385–404. /233, 241/

Fürnkranz, J. (2005). From local to global patterns: Evaluation issues in rule learning algorithms. In K. Morik, J.-F. Boulicaut, & A. Siebes (Eds.), *Local pattern detection* (pp. 20–38). Berlin, Germany/New York: Springer. */185/*

Fürnkranz, J., & Flach, P. (2003). An analysis of rule evaluation metrics. In T. Fawcett & N. Mishra (Eds.), *Proceedings of the 20th International Conference on Machine Learning (ICML-03)*, Washington, DC (pp. 202–209). Menlo Park, CA: AAAI. */60/*

Fürnkranz, J., & Flach, P. (2004). An analysis of stopping and filtering criteria for rule learning. In J.-F. Boulicaut, F. Esposito, F. Giannotti, & D. Pedreschi (Eds.), *Proceedings of the 15th European Conference on Machine Learning (ECML-04)*, Pisa, Italy (pp. 123–133). Berlin, Germany/Heidelberg, Germany/New York: Springer. */60, 167/*

Fürnkranz, J., & Flach, P. (2005). ROC 'n' rule learning – Towards a better understanding of covering algorithms. *Machine Learning, 58*(1), 39–77. */57, 58, 60, 135, 148, 151, 154, 187, 194, 242/*

Fürnkranz, J., & Hüllermeier, E. (2003). Pairwise preference learning and ranking. In N. Lavrač, D. Gamberger, H. Blockeel, & L. Todorovski (Eds.), *Proceedings of the 14th European Conference on Machine Learning (ECML-03)*, Cavtat, Croatia (pp. 145–156). Berlin, Germany/New York: Springer. */238, 239/*

Fürnkranz, J., & Hüllermeier, E. (Eds.). (2010a). *Preference learning*. Heidelberg, Germany/New York: Springer. */237, 241/*

Fürnkranz, J., & Hüllermeier, E. (2010b). Preference learning and ranking by pairwise comparison. In J. Fürnkranz & E. Hüllermeier (Eds.), *Preference learning* (pp. 65–82). Heidelberg, Germany/New York: Springer. */217/*

Fürnkranz, J., Hüllermeier, E., Loza Mencía, E., & Brinker, K. (2008). Multilabel classification via calibrated label ranking. *Machine Learning, 73*(2), 133–153. */240/*

Fürnkranz, J., Hüllermeier, E., & Vanderlooy, S. (2009). Binary decomposition methods for multipartite ranking. In W. L. Buntine, M. Grobelnik, D. Mladenić, & J. Shawe-Taylor (Eds.), *Proceedings of the European Conference on Machine Learning and Principles and Practice of Knowledge Discovery in Databases (ECML/PKDD-09)*, Bled, Slovenia (Vol. Part I, pp. 359–374). Berlin, Germany: Springer. */243/*

Fürnkranz, J., & Knobbe, A. (2010). Special issue on global modeling using local patterns. *Data Mining and Knowledge Discovery, 21*(1), 1–8. */184/*

Fürnkranz, J., & Sima, J. F. (2010). On exploiting hierarchical label structure with pairwise classifiers. *SIGKDD Explorations, 12*(2), 21–25. Special Issue on Mining Unexpected Results. */231, 241/*

Fürnkranz, J., & Widmer, G. (1994). Incremental Reduced Error Pruning. In W. W. Cohen & H. Hirsh (Eds.), *Proceedings of the 11th International Conference on Machine Learning (ML-94)*, New Brunswick, NJ (pp. 70–77). San Francisco: Morgan Kaufmann. */142, 207, 210/*

Gamberger, D., & Lavrač, N. (2000). Confirmation rule sets. In D. A. Zighed, J. Komorowski, & J. Żytkow (Eds.), *Proceedings of the 4th European Conference on Principles of Data Mining and Knowledge Discovery (PKDD-00)*, Lyon, France (pp. 34–43). Berlin, Germany: Springer. */178, 186/*

Gamberger, D., & Lavrač, N. (2002). Expert-guided subgroup discovery: Methodology and application. *Journal of Artificial Intelligence Research, 17*, 501–527. */60, 73, 148, 149, 180, 249, 250, 259, 263, 268/*

Gamberger, D., Lavrač, N., & Fürnkranz, J. (2008). Handling unknown and imprecise attribute values in propositional rule learning: A feature-based approach. In T.-B. Ho & Z.-H. Zhou (Eds.), *Proceedings of the 10th Pacific Rim International Conference on Artificial Intelligence (PRICAI-08)*, Hanoi, Vietnam (pp. 636–645). Berlin, Germany/Heidelberg, Germany: Springer. */87/*

Gamberger, D., Lavrač, N., & Krstačić, G. (2002). Confirmation rule induction and its applications to coronary heart disease diagnosis and risk group discovery. *Journal of Intelligent and Fuzzy Systems, 12*(1), 35–48. */223/*

Gamberger, D., Lavrač, N., & Wettschereck., D. (2002). Subgroup visualization: A method and application in population screening. In *Proceedings of the 7th International Workshop on Intelligent Data Analysis in Medicine and Pharmacology (IDAMAP-02)*, Lyon, France (pp. 31–35). Lyon, France: ECAI /261, 263/

Gamberger, D., Lavrač, N., Zelezny, F., & Tolar, J. (2004). Induction of comprehensible models for gene expression datasets by subgroup discovery methodology. *Journal of Biomedical Informatics, 37*(4), 269–284. /93, 279/

Garriga, G. C., Kralj, P., & Lavrač, N. (2006). Closed sets for labeled data. In *Proceedings of the 10th European Conference on Principles and Practice of Knowledge Discovery in Databases (PKDD-06)*, Berlin, Germany (pp. 163 – 174). Berlin, Germany/New York: Springer /260/

Geibel, P., & Wysotzki, F. (1996). Learning relational concepts with decision trees. In L. Saitta (Ed.), *Proceedings of the 13th International Conference on Machine Learning (ICML-96)* (pp. 166–174). San Francisco: Morgan Kaufmann Publishers. /103/

Gelfand, S., Ravishankar, C., & Delp, E. (1991). An iterative growing and pruning algorithm for classification tree design. *IEEE Transactions on Pattern Analysis and Machine Intelligence, 13*(2), 163–174. /199/

Geng, L., & Hamilton, H. J. (2006). Interestingness measures for data mining: A survey. *ACM Computing Surveys, 38*(3), 37–68. /137/

Georgeff, M. P., & Wallace, C. S. (1984). A general criterion for inductive inference. In T. O'Shea (Ed.), *Proceedings of the Sixth European Conference on Artificial Intelligence (ECAI-84)*, Pisa, Italy (pp. 473–482). Amsterdam: Elsevier. /162/

Ghani, R. (2000). Using error-correcting codes for text classification. In *Proceedings of the 17th International Conference on Machine Learning (ICML-00)* (pp. 303–310). San Francisco: Morgan Kaufmann Publishers. /234/

Giordana, A., & Sale, C. (1992). Learning structured concepts using genetic algorithms. In Sleeman, D., Edwards, P. (eds.) *Proceedings of the 9th International Workshop on Machine Learning (ML-92)*, Edinburgh, UK (pp. 169–178). San Mateo, CA: Morgan Kaufmann. /123, 163/

Goethals, B. (2005). Frequent set mining. In O. Maimon & L. Rokach (Eds.), *The data mining and knowledge discovery handbook* (pp. 377–397). New York: Springer. /122/

Goldberg, D. E. (1989). *Genetic algorithms in search, optimization and machine learning*. Reading, MA: Addison-Wesley. /123/

Golub, T., Slonim, D., Tamayo, P., Huard, C., Gaaseenbeek, M., & Mesirov, J., et al. (1999). Molecular classification of cancer: Class discovery and class prediction by gene expression monitoring. *Science, 286*, 531–537. /274/

Gönen, M., & Heller, G. (2005). Concordance probability and discriminatory power in proportional hazards regression. *Biometrika, 92*(4), 965–970. /242/

Grant, J., & Minker, J. (1992). The impact of logic programming on databases. *Communications of the ACM, 35*(3), 66–81. /109/

Groff, J. R., & Weinberg, P. N. (2002). *SQL, the complete reference* (2nd ed.). New York: McGraw-Hill Osborne Media. /107/

Hall, M., Frank, E., Holmes, G., Pfahringer, B., Reutemann, P., & Witten, I. H. (2009). The WEKA data mining software: An update. *SIGKDD Explorations, 11*(1), 10–18. /5/

Han, J., Cai, Y., & Cercone, N. (1992). Knowledge discovery in databases: An attribute-oriented approach. In *Proceedings of the 18th Conference on Very Large Data Bases (VLDB-92)*, Vancouver, BC (pp. 547–559). San Mateo, CA: Morgan Kaufmann Publishers. /70/

Han, J., & Kamber, M. (2001). *Data mining: Concepts and techniques*. San Francisco: Morgan Kaufmann Publishers. /2, 70/

Han, J., Pei, J., Yin, Y., & Mao, R. (2004). Mining frequent patterns without candidate generation: A frequent-pattern tree approach. *Data Mining and Knowledge Discovery, 8*(1), 53–87. /122/

Hand, D. J. (2002). Pattern detection and discovery. In D. J. Hand, N. M. Adams, & R. J. Bolton (Eds.), *Pattern Detection and Discovery: Proceedings of the ESF Exploratory Workshop*, London (pp. 1–12). Berlin, Germany/New York: Springer. /184/

Har-Peled, S., Roth, D., & Zimak, D. (2002). Constraint classification: A new approach to multiclass classification. In N. Cesa-Bianchi, M. Numao, & R. Reischuk (Eds.), *Proceedings of the 13th International Conference on Algorithmic Learning Theory (ALT-02)*, Lübeck, Germany (pp. 365–379). Berlin, Germany/New York: Springer. /239/

Hart, P. E., Nilsson, N. J., & Raphael, B. (1968). A formal basis for the heuristic determination of minimum cost paths. *IEEE Transactions on Systems Science and Cybernetics, 4*(2), 100–107. /120/

Hastie, T., & Tibshirani, R. (1998). Classification by pairwise coupling. In M. Jordan, M. Kearns, & S. Solla (Eds.), *Advances in neural information processing systems 10 (NIPS-97)* (pp. 507–513). Cambridge, MA: MIT. /230, 239/

Hastie, T., Tibshirani, R., & Friedman, J. H. (2001). *The elements of statistical learning*. New York: Springer. /2/

Helft, N. (1989). Induction as nonmonotonic inference. In R. J. Brachman, H. J. Levesque, & R. Reiter (Eds.), *Proceedings of the 1st International Conference on Principles of Knowledge Representation and Reasoning (KR-89)*, Toronto, ON (pp. 149–156). San Mateo, CA: Morgan Kaufmann. /111/

Hernández-Orallo, J., & Ramírez-Quintana, M. (1999). A complete schema for inductive functional logic programming. In S. Džeroski & P. Flach (Eds.), *Proceedings of the 9th International Workshop on Inductive Logic Programming (ILP-99)*, Bled, Slovenia (pp. 116–127). Berlin, Germany/New York: Springer. /96/

Hilderman, R. J., & Peckham, T. (2005). A statistically sound alternative approach to mining contrast sets. In *Proceedings of the 4th Australia Data Mining Conference (AusDM-05)*, Sydney, NSW (pp. 157–172). /252/

Hipp, J., Güntzer, U., & Nakhaeizadeh, G. (2000). Algorithms for association rule mining – A general survey and comparison. *SIGKDD Explorations, 2*(1), 58–64. /122/

Hocquenghem, A. (1959). Codes correcteurs d'erreurs. *Chiffres, 2*, 147–156. In French. /235/

Holmes, G., Hall, M., & Frank, E. (1999). Generating rule sets from model trees. In N. Y. Foo (Ed.), *Proceedings of the 12th Australian Joint Conference on Artificial Intelligence (AI-99)*, Sydney, Australia (pp. 1–12). Berlin, Germany/New York: Springer. /244/

Holte, R., Acker, L., & Porter, B. (1989). Concept learning and the problem of small disjuncts. In *Proceedings of the 11th International Joint Conference on Artificial Intelligence (IJCAI-89)*, Detroit, MI (pp. 813–818). San Mateo, CA: Morgan Kaufmann. /159, 210/

Hoos, H. H., & Stützle, T. (2004). *Stochastic local search: Foundations and applications*. San Francisco: Morgan Kaufmann. /183/

Hsu, C.-W., & Lin, C.-J. (2002). A comparison of methods for multi-class support vector machines. *IEEE Transactions on Neural Networks, 13*(2), 415–425. /230, 231/

Hühn, J., & Hüllermeier, E. (2009a). FR3: A fuzzy rule learner for inducing reliable classifiers. *IEEE Transactions on Fuzzy Systems, 17*(1), 138–149. /223, 230/

Hühn, J., & Hüllermeier, E. (2009b). Furia: An algorithm for unordered fuzzy rule induction. *Data Mining and Knowledge Discovery, 19*(3), 293–319. /52, 92/

Hüllermeier, E. (2011). Fuzzy sets in machine learning and data mining. *Applied Soft Computing, 11*(2), 1493–1505. /92/

Hüllermeier, E., & Fürnkranz, J. (2010). On predictive accuracy and risk minimization in pairwise label ranking. *Journal of Computer and System Sciences, 76*(1), 49–62. /239/

Hüllermeier, E., Fürnkranz, J., Cheng, W., & Brinker, K. (2008). Label ranking by learning pairwise preferences. *Artificial Intelligence, 172*, 1897–1916. /238, 239/

Hüllermeier, E., & Vanderlooy, S. (2009). Why fuzzy decision trees are good rankers. *IEEE Transactions on Fuzzy Systems, 17*(6), 1233–1244. /215/

Janssen, F., & Fürnkranz, J. (2009). A re-evaluation of the over-searching phenomenon in inductive rule learning. In H. Park, S. Parthasarathy, H. Liu, & Z. Obradovic (Eds.), *Proceedings of the SIAM International Conference on Data Mining (SDM-09)*, Sparks, NV (pp. 329–340). Philadelphia: SIAM. /119, 155/

Janssen, F., & Fürnkranz, J. (2010). On the quest for optimal rule learning heuristics. *Machine Learning, 78*(3), 343–379. /144, 158, 164, 165/

Janssen, F., & Fürnkranz, J. (2011). Heuristic rule-based regression via dynamic reduction to classification. In T. Walsh (Ed.), *Proceedings of the 22nd International Joint Conference* on *Artificial Intelligence (IJCAI-11)*, Barcelona, Spain (pp. 1330–1335). Menlo Park, CA: AAAI. /244/

Jenkole, J., Kralj, P., Lavrač, N., & Sluga, A. (2007). A data mining experiment on manufacturing shop floor data. In *Proceedings of the 40th CIRP International Seminar on Manufacturing Systems*. Liverpool, UK: University of Liverpool /251/

Joachims, T. (2002). Optimizing search engines using clickthrough data. In *Proceedings of the 8th ACM SIGKDD International Conference on Knowledge Discovery and Data Mining (KDD-02)*, Edmonton, AB (pp. 133–142). New York: ACM. /243/

Joachims, T. (2006). Training linear SVMs in linear time. In T. Eliassi-Rad, L. H. Ungar, M. Craven, & D. Gunopulos (Eds.), *Proceedings of the 12th ACM SIGKDD International Conference on Knowledge Discovery and Data Mining (KDD-06)*, Philadelphia (pp. 217–226). New York: ACM. /243/

Joshi, S., Ramakrishnan, G., & Srinivasan, A. (2008). Feature construction using theory-guided sampling and randomised search. In F. Zelezný & N. Lavrac (Eds.), *Proceedings of the 18th International Conference on Inductive Logic Programming (ILP-08)*, Prague, Czech Republic (pp. 140–157). Berlin, Germany/New York: Springer. /53/

Jovanoski, V., & Lavrač, N. (2001). Classification rule learning with APRIORI-C. In P. Brazdil & A. Jorge (Eds.), *Proceedings of the 10th Portuguese Conference on Artificial Intelligence (EPIA 2001)*, Porto, Portugal (pp. 44–51). Berlin, Germany/New York: Springer. /54, 122, 185, 192/

Karalič, A., & Bratko, I. (1997). First order regression. *Machine Learning, 26*(2/3), 147–176. Special Issue on *Inductive Logic Programming*. /244/

Kaufman, K. A., & Michalski, R. S. (2000). An adjustable rule learner for pattern discovery using the AQ methodology. *Journal of Intelligent Information Systems, 14*, 199–216. /48/

Kavšek, B., & Lavrač, N. (2006). Apriori-SD: Adapting association rule learning to subgroup discovery. *Applied Artificial Intelligence, 20*(7), 543–583. /122, 249, 259/

King, R. D., Whelan, K. E., Jones, F. M., Reiser, P., Bryant, C., & Muggleton, S., et al. (2004). Functional genomic hypothesis generation and experimentation by a robot. *Nature, 427*, 247–252. /52/

Kirkpatrick, S., Gelatt, C., & Vecchi, M. (1983). Optimization by simulated annealing. *Science, 220*, 671–680. /123/

Kirsten, M., Wrobel, S., & Horvath, T. (2001). Distance based approaches to relational learning and clustering. In S. Džeroski & N. Lavrač (Eds.), *Relational data mining* (pp. 213–232). Berlin, Germany/New York: Springer. /97/

Kittler, J., Ghaderi, R., Windeatt, T., & Matas, J. (2003). Face verification via error correcting output codes. *Image and Vision Computing, 21*(13–14), 1163–1169. /234/

Klösgen, W. (1992). Problems for knowledge discovery in databases and their treatment in the statistics interpreter EXPLORA. *International Journal of Intelligent Systems, 7*(7), 649–673. /157, 158/

Klösgen, W. (1996). Explora: A multipattern and multistrategy discovery assistant. In U. M. Fayyad, G. Piatetsky-Shapiro, P. Smyth, & R. Uthurusamy (Eds.), *Advances in knowledge discovery and data mining* (pp. 249–271). Menlo Park, CA: AAAI. Chap. 10. /143, 157, 167, 248, 249, 250/

Klösgen, W., & May, M. (2002). Spatial subgroup mining integrated in an object-relational spatial database. In *Proceedings of the 6th European Conference on Principles and Practice of Knowledge Discovery in Databases (PKDD-02)* (pp. 275–286). Berlin, Germany/New York: Springer. /249/

Klösgen, W., May, M., & Petch, J. (2003). Mining census data for spatial effects on mortality. *Intelligent Data Analysis*, 7(6):521–540. /250/

Knerr, S., Personnaz, L., & Dreyfus, G. (1990). Single-layer learning revisited: A stepwise procedure for building and training a neural network. In F. Fogelman Soulié & J. Hérault (Eds.), *Neurocomputing: Algorithms, architectures and applications* (NATO ASI Series, Vol. F68, pp. 41–50). Berlin, Germany/New York: Springer. /230/

Knerr, S., Personnaz, L., & Dreyfus, G. (1992). Handwritten digit recognition by neural networks with single-layer training. *IEEE Transactions on Neural Networks, 3*(6), 962–968. /230, 231/

Knobbe, A. J., Crémilleux, B., Fürnkranz, J., & Scholz, M. (2008). From local patterns to global models: The LeGo approach to data mining. In J. Fürnkranz & A. J. Knobbe (Eds.), *From Local Patterns to Global Models: Proceedings of the ECML/PKDD-08 Workshop (LeGo-08)*, Antwerp, Belgium (pp. 1–16). /184, 185/

Koller, D., & Sahami, M. (1997). Hierarchically classifying documents using very few words. In *Proceedings of the 14th International Conference on Machine Learning (ICML-97)*, Nashville, TN (pp. 170–178). San Francisco: Morgan Kaufmann Publishers /241/

Kong, E. B., & Dietterich, T. G. (1995). Error-correcting output coding corrects bias and variance. In *Proceedings of the 12th International Conference on Machine Learning (ICML-95)* (pp. 313–321). San Mateo, CA: Morgan Kaufmann. /234/

Kononenko, I., & Kovačič, M. (1992). Learning as optimization: Stochastic generation of multiple knowledge. In D. Sleeman & P. Edwards (Eds.), *Proceedings of the 9th International Workshop on Machine Learning (ML-92)* (pp. 257–262). San Mateo, CA: Morgan Kaufmann. /123, 128/

Kotsiantis, S., Zaharakis, I., & Pintelas, P. (2006). Supervised machine learning: A review of classification techniques. *Artificial Intelligence Review, 26*, 159–190. /6/

Kovačič, M. (1991). Markovian neural networks. *Biological Cybernetics, 64*, 337–342. /123/

Kovačič, M. (1994a). MDL-heuristics in ILP revised. In *Proceedings of the ML-COLT-94 Workshop on Applications of Descriptional Complexity to Inductive, Statistical, and Visual Inference*. New Brunswick, NJ. /162/

Kovačič, M. (1994b). *Stochastic inductive logic programming*. Ph.D. thesis, Department of Computer and Information Science, University of Ljubljana. /123, 162/

Kralj, P., Grubešič, A., Toplak, N., Gruden, K., Lavrač, N., & Garriga, G. C. (2006). Application of closed itemset mining for class labeled data in functional genomics. *Informatica Medica Slovenica, 11*(1), 40–45. /260/

Kralj, P., Lavrač, N., Gamberger, D., & Krstačić, A. (2007a). Contrast set mining for distinguishing between similar diseases. In *Proceedings of the 11th Conference on Artificial Intelligence in Medicine (AIME-07)*, Amsterdam (pp. 109–118). Berlin, Germany: Springer /250/

Kralj, P., Lavrač, N., Gamberger, D., & Krstačić, A. (2007b). Contrast set mining through subgroup discovery applied to brain ischaemia data. In *Proceedings of the 11th Pacific-Asia Conference on Advances in Knowledge Discovery and Data Mining (PAKDD-07)*, Nanjing, China (pp. 579–586). Berlin, Germany/New York: Springer /250, 252/

Kralj, P., Lavrač, N., & Zupan, B. (2005). Subgroup visualization. In *Proceedings of the 8th International Multiconference Information Society (IS-05)*, Ljubljana, Slovenia (pp. 228–231). Ljubljana, Slovenia: Institut Jožef Stefan. /261, 264/

Kralj Novak, P., Lavrač, N., & Webb, G. I. (2009). Supervised descriptive rule discovery: A unifying survey of contrast set, emerging pattern and subgroup mining. *Journal of Machine Learning Research, 10*, 377–403. /15, 167, 247, 258/

Kramer, S. (1996). Structural regression trees. In *Proceedings of the 13th National Conference on Artificial Intelligence (AAAI-96)* (pp. 812–819). Menlo Park, CA: AAAI. /13, 244/

Kramer, S., & Frank, E. (2000). Bottom-up propositionalization. In *Proceedings of the ILP-2000 Work-In-Progress Track* (pp. 156–162). London: Imperial College. /103/

Kramer, S., Lavrač, N., & Flach, P. (2001). Propositionalization approaches to relational data mining. In S. Džeroski & N. Lavrač (Eds.), *Relational data mining* (pp. 262–291). Berlin, Germany: Springer. /53, 101/

Kramer, S., Pfahringer, B., & Helma, C. (2000). Stochastic propositionalization of non-determinate background knowledge. In *Proceedings of the 8th International Conference on Inductive Logic Programming (ILP-2000)*, Madison, WI (pp. 80–94). Berlin, Germany/New York: Springer. /103/

Kreßel, U. H.-G. (1999). Pairwise classification and support vector machines. In B. Schölkopf, C. Burges, & A. Smola (Eds.), *Advances in Kernel methods: Support vector learning* (pp. 255–268). Cambridge, MA: MIT. Chap. 15. /230/

Krogel, M. A., Rawles, S., Železný, F., Flach, P., Lavrač, N., & Wrobel, S. (2003). Comparative evaluation of approaches to propositionalization. In T. Horvath & A. Yamamoto (Eds.),

Proceedings of the 13th International Conference on Inductive Logic Programming (ILP-2003), Szeged, Hungary (pp. 197–214). Berlin, Germany/New York: Springer. /102/

Kullback, S., & Leibler, R. (1951). On information and sufficiency. *Annals of Mathematical Statistics, 22*(1), 79–86. /147/

Landwehr, N., Kersting, K., & De Raedt, L. (2007). Integrating Naive Bayes and FOIL. *Journal of Machine Learning Research, 8*, 481–507. /222/

Langford, J., Oliveira, R., & Zadrozny, B. (2006). Predicting conditional quantiles via reduction to classification. In *Proceedings of the 22nd Conference Annual Conference on Uncertainty in Artificial Intelligence (UAI-06)*, Cambridge, MA (pp. 257–264). Arlington, VA: AUAI. /244/

Langley, P. (1996). *Elements of machine learning*. San Francisco: Morgan Kaufmann. /2/

Langley, P., & Simon, H. (1995). Applications of machine learning and rule induction. *Communications of the ACM, 38*(11), 54–64. /267/

Lavrač, N., Cestnik, B., & Džeroski, S. (1992a). Search heuristics in empirical Inductive Logic Programming. In *Logical Approaches to Machine Learning, Workshop Notes of the 10th European Conference on AI*, Vienna. /164/

Lavrač, N., Cestnik, B., & Džeroski, S. (1992b). Use of heuristics in empirical Inductive Logic Programming. In S. H. Muggleton & K. Furukawa (Eds.), *Proceedings of the 2nd International Workshop on Inductive Logic Programming (ILP-92)*. No. TM-1182 in ICOT Technical Memorandum, Institute for New Generation Computer Technology, Tokyo, Japan. /164/

Lavrač, N., Cestnik, B., Gamberger, D., & Flach, P. A. (2004). Decision support through subgroup discovery: Three case studies and the lessons learned. *Machine Learning, 57*(1–2):115–143. Special issue on Data Mining Lessons Learned. /251/

Lavrač, N., & Džeroski, S. (1994a). *Inductive logic programming: Techniques and applications*. New York: Ellis Horwood. /2, 7, 53, 67, 76, 95, 96, 101, 129/

Lavrač, N., & Džeroski, S. (1994b). Weakening the language bias in LINUS. *Journal of Experimental and Theoretical Artificial Intelligence, 6*, 95–119. /110/

Lavrač, N., Džeroski, S., & Grobelnik, M. (1991). Learning nonrecursive definitions of relations with LINUS. In *Proceedings of the 5th European Working Session on Learning (EWSL-91)*, Porto, Portugal (pp. 265–281). Berlin, Germany: Springer. /13, 53, 67, 76, 101/

Lavrač, N., & Flach, P. (2001). An extended transformation approach to inductive logic programming. *ACM Transactions on Computational Logic, 2*(4), 458–494. /67, 102/

Lavrač, N., Flach, P., & Zupan, B. (1999). Rule evaluation measures: A unifying view. In S. Džeroski & P. Flach (Eds.), *Proceedings of the 9th International Workshop on Inductive Logic Programming (ILP-99)*, Bled, Slovenia (pp. 174–185). Berlin, Germany/New York: Springer. /143, 167/

Lavrač, N., Fürnkranz, J., & Gamberger, D. (2010). Explicit feature construction and manipulation for covering rule learning algorithms. In J. Koronacki, Z. Ras, S. T. Wierzchon, & J. Kacprzyk (Eds.), *advances in machine learning II—Dedicated to the memory of Professor Ryszard S. Michalski* (pp. 121–146). Berlin, Germany/Heidelberg, Germany: Springer. /65, 87/

Lavrač, N., Gamberger, D., & Jovanoski, V. (1999). A sudy of relevance for learning in deductive databases. *The Journal of Logic Programming, 40*(2/3), 215–249. /77/

Lavrač, N., & Grobelnik, M. (2003). Data mining. In D. Mladenić, N. Lavrač, M. Bohanec, & S. Moyle (Eds.), *Data mining and decision support: Integration and collaboration* (pp. 3–14). Boston: Kluwer. /1/

Lavrač, N., Kavšek, B., Flach, P., & Todorovski, L. (2004). Subgroup discovery with CN2-SD. *Journal of Machine Learning Research, 5*, 153–188. /143, 167, 178, 179, 180, 249, 250, 259/

Lavrač, N., Kok, J., de Bruin, J., & Podpečan, V. (Eds.). (2008). *Proceedings of the ECML-PKDD-08 Workshop on Third Generation Generation Data Mining: Towards Service-Oriented Knowledge Discovery (SoKD-08)*, Antwerp, Belgium. /17/

Lavrač, N., Kralj, P., Gamberger, D., & Krstačić, A. (2007). Supporting factors to improve the explanatory potential of contrast set mining: Analyzing brain ischaemia data. In *Proceedings of the 11th Mediterranean Conference on Medical and Biological Engineering and Computing (MEDICON-07)*, Ljubljana, Slovenia (pp. 157–161). Berlin, Germany: Springer. /250/

Lavrač, N., Podpečan, V., Kok, J., & de Bruin, J. (Eds.). (2009). *Proceedings of the ECML-PKDD-09 Workshop on Service-Oriented Knowledge Discovery (SoKD-09)*, Bled, Slovenia. /17/

Li, J., Dong, G., & Ramamohanarao, K. (2000). Instance-based classification by emerging patterns. In *Proceedings of the 14th European Conference on Principles and Practice of Knowledge Discovery in Databases (PKDD-2000)*, Lyon, France (pp. 191–200). Berlin, Germany/New York: Springer. /253/

Li, J., Dong, G., & Ramamohanarao, K. (2001). Making use of the most expressive jumping emerging patterns for classification. *Knowledge and Information Systems, 3*(2), 1–29. /253/

Li, J., Liu, H., Downing, J. R., Yeoh, A. E.-J., & Wong, L. (2003). Simple rules underlying gene expression profiles of more than six subtypes of acute lymphoblastic leukemia (ALL) patients. *Bioinformatics, 19*(1), 71–78. /254/

Li, J., & Wong, L. (2002a). Geography of differences between two classes of data. In *Proceedings of the 6th European Conference on Principles of Data Mining and Knowledge Discovery (PKDD-02)*, Helsinki, Finland (pp. 325–337). Berlin, Germany/New York: Springer. /274, 275/

Li, J., & Wong, L. (2002b). Identifying good diagnostic gene groups from gene expression profiles using the concept of emerging patterns. *Bioinformatics, 18*(10), 1406–1407. /255/

Li, W., Han, J., & Pei, J. (2001). CMAR: Accurate and efficient classification based on multiple class-association rules. In *Proceedings of the IEEE Conference on Data Mining (ICDM-01)*, San Jose, CA (pp. 369–376). Los Alamitos, CA: IEEE. /54, 185/

Lin, J., & Keogh, E. (2006). Group SAX: Extending the notion of contrast sets to time series and multimedia data. In *Proceedings of the 10th European Conference on Principles and Practice of Knowledge Discovery in Databases (PKDD-06)*, Berlin, Germany (pp. 284–296). Berlin, Germany/New York: Springer. /252/

Lin, H.-T., Lin, C.-J., & Weng, R. C. (2007). A note on Platt's probabilistic outputs for support vector machines. *Machine Learning, 68*(3), 267–276. /215/

Lindgren, T., & Boström, H. (2004). Resolving rule conflicts with double induction. *Intelligent Data Analysis, 8*(5), 457–468. /222/

Liu, B., Hsu, W., Han, H.-S., & Xia, Y. (2000). Mining changes for real-life applications. In *Proceedings of the 2nd International Conference on Data Warehousing and Knowledge Discovery (DaWaK-2000)*, London (pp. 337–346). Berlin, Germany: Springer. /260/

Liu, B., Hsu, W., & Ma, Y. (1998). Integrating classification and association rule mining. In R. Agrawal, P. Stolorz, & G. Piatetsky-Shapiro (Eds.), *Proceedings of the 4th International Conference on Knowledge Discovery and Data Mining (KDD-98)* (pp. 80–86). Menlo Park, CA: AAAI. /54,122, 184, 192/

Liu, B., Hsu, W., & Ma, Y. (2001). Discovering the set of fundamental rule changes. In *Proceedings of the 7th ACM SIGKDD International Conference on Knowledge Discovery and Data Mining (KDD-01)*, San Francisco (pp. 335–340). New York: ACM. /260/

Liu, B., Ma, Y., & Wong, C.-K. (2000). Improving an exhaustive search based rule learner. In D. A. Zighed, H. J. Komorowski, & J. M. Zytkow (Eds.), *Proceedings of the 4th European Conference on Principles and Practice of Knowledge Discovery in Databases (PKDD-2000)*, Lyon, France (pp. 504–509). Berlin, Germany: Springer. /54, 184, 192/

Lloyd, J. W. (1987). *Foundations of logic programming* (2nd extended ed.). Berlin, Germany: Springer. /13, 104, 106/

Loza Mencía, E., Park, S.-H., & Fürnkranz, J. (2009). Efficient voting prediction for pairwise multilabel classification. In *Proceedings of the 17th European Symposium on Artificial Neural Networks (ESANN-09)*, Bruges, Belgium (pp. 117–122). Evere, Belgium: d-side publications. /240/

Lu, B.-L., & Ito, M. (1999). Task decomposition and module combination based on class relations: A modular neural network for pattern classification. *IEEE Transactions on Neural Networks, 10*(5), 1244–1256. /230/

Macskassy, S. A., Provost, F., & Rosset, S. (2005). ROC confidence bands: An empirical evaluation. In *Proceedings of the 22nd International Conference on Machine Learning (ICML-05)*, Bonn, Germany (pp. 537–544). New York: ACM. /59/

MacWilliams, F. J., & Sloane, N. J. A. (1983). *The theory of error-correcting codes*. North Holland, The Netherlands: North-Holland Mathematical Library. /233/

Major, J. A., & Mangano, J. J. (1995). Selecting among rules induced from a hurricane database. *Journal of Intelligent Information Systems, 4*(1), 39–52. /157, 166/

Manning, C. D., & Schütze, H. (1999). *Foundations of statistical natural language processing*. Cambridge, MA: MIT. /215/

May, M., & Ragia, L. (2002). Spatial subgroup discovery applied to the analysis of vegetation data. In *Proceedings of the 4th International Conference on Practical Aspects of Knowledge Management (PAKM-2002)*, Vienna (pp. 49–61). Berlin, Germany/New York: Springer. /250/

Mease, D., & Wyner, A. (2008). Evidence contrary to the statistical view of boosting. *Journal of Machine Learning Research, 9*, 131–156. /2/

Melvin, I., Ie, E., Weston, J., Noble, W. S., & Leslie, C. (2007). Multi-class protein classification using adaptive codes. *Journal of Machine Learning Research, 8*, 1557–1581. /234, 235/

Michalski, R. S. (1969). On the quasi-minimal solution of the covering problem. In *Proceedings of the 5th International Symposium on Information Processing (FCIP-69)*, Bled, Yugoslavia (Switching circuits, Vol. A3, pp. 125–128). /3, 40, 48, 245/

Michalski, R. S. (1973). AQVAL/1—Computer implementation of a variable-valued logic system VL_1 and examples of its application to pattern recognition. In *Proceedings of the 1st International Joint Conference on Pattern Recognition*, Washington, DC (pp. 3–17). Northridge, CA: IEEE /68/

Michalski, R. S. (1980). Pattern recognition and rule-guided inference. *IEEE Transactions on Pattern Analysis and Machine Intelligence, 2*, 349–361. /3, 48, 68, 245/

Michalski, R. S. (1983). A theory and methodology of inductive learning. *Artificial Intelligence, 20*(2), 111–162. /163/

Michalski, R. S., Carbonell, J. G., & Mitchell, T. M. (Eds.). (1983). *Machine learning: An artificial intelligence approach* (Vol. I). Palo Alto, CA: Tioga. /3/

Michalski, R. S., Carbonell, J. G., & Mitchell, T. M. (Eds.). (1986). *Machine learning: An artificial intelligence approach* (Vol. II). Los Altos, CA: Morgan Kaufmann. /3/

Michalski, R. S., & Larson, J. B. (1978). *Selection of most representative training examples and incremental generation of VL1 hypotheses: the underlying methodology and the description of programs ESEL and AQ11* (Tech. Rep. 78-867). Department of Computer Science, University of Illinois at Urbana-Champaign. /48/

Michalski, R. S., Mozetič, I., Hong, J., & Lavrač, N. (1986). The multi-purpose incremental learning system AQ15 and its testing application to three medical domains. In *Proceedings of the 5th National Conference on Artificial Intelligence (AAAI-86)*, Philadelphia (pp. 1041–1045). Menlo Park, CA: AAAI. /1, 3, 48, 68, 118, 200/

Michalski, R. S., & Stepp, R. E. (1983). Learning from observation: Conceptual clustering. In R. Michalski, J. Carbonell, & T. Mitchell (Eds.), *Machine learning: An artificial intelligence approach*. Palo Alto, CA: Tioga. /245/

Michie, D., Muggleton, S. H., Page, D., & Srinivasan, A. (1994). *To the international computing community: A New East-West challenge* (Tech. Rep.). Oxford, UK: Oxford University Computing laboratory. /98/

Michie, D., Spiegelhalter, D., & Taylor, C. C. (Eds.). (1994). *Machine learning, neural and statistical classification*. New York: Ellis Horwood. /2/

Mierswa, I., Wurst, M., Klinkenberg, R., Scholz, M., & Euler, T. (2006). Yale: Rapid prototyping for complex data mining tasks. In L. Ungar, M. Craven, D. Gunopulos, & T. Eliassi-Rad (Eds.), *KDD '06: Proceedings of the 12th ACM SIGKDD International Conference on Knowledge Discovery and Data Mining*, Philadelphia (pp. 935–940). New York: ACM. /5/

Mingers, J. (1989a). An empirical comparison of pruning methods for decision tree induction. *Machine Learning, 4*, 227–243. /200/

Mingers, J. (1989b). An empirical comparison of selection measures for decision-tree induction. *Machine Learning, 3*, 319–342. /136/

Mitchell, T. M. (1982). Generalization as search. *Artificial Intelligence, 18*(2), 203–226. /29, 114/

Mitchell, T. M. (1997). *Machine learning*. New York: McGraw Hill. /1, 2, 30/

Mladenić, D. (1993). Combinatorial optimization in inductive concept learning. In *Proceedings of the 10th International Conference on Machine Learning (ML-93)*, Amherst, MA (pp. 205–211). San Mateo, CA: Morgan Kaufmann. /117, 123, 128/

Mooney, R. J. (1995). Encouraging experimental results on learning CNF. *Machine Learning, 19*, 79–92. /51, 174/

Mooney, R. J., & Califf, M. E. (1995). Induction of first-order decision lists: Results on learning the past tense of English verbs. *Journal of Artificial Intelligence Research, 3*, 1–24. /226/

Morik, K., Boulicaut, J.-F., & Siebes, A. (Eds.). (2005). *Local pattern detection*. Berlin, Germany/New York: Springer. /184/

Muggleton, S. H. (1987). Structuring knowledge by asking questions. In Bratko, I., & Lavrač, N. (Eds.), *Progress in machine learning* (pp. 218–229). Wilmslow, England: Sigma Press. /127/

Muggleton, S. H. (1988). A strategy for constructing new predicates in first order logic. In *Proceedings of the 3rd European Working Session on Learning (EWSL-88)* (pp. 123–130). London: Pitman. /127/

Muggleton, S. H. (1991). Inverting the resolution principle. In J. E. Hayes, D. Michie, & E. Tyugu (Eds.), *Machine intelligence 12* (pp. 93–103). Oxford, UK: Clarendon. Chap. 7 /127/

Muggleton, S. H. (Ed.). (1992). *Inductive logic programming*. London: Academic. /2, 7, 96/

Muggleton, S. H. (1995). Inverse entailment and Progol. *New Generation Computing, 13*(3,4), 245–286. Special Issue on Inductive Logic Programming. /52, 53, 120, 141/

Muggleton, S. H., & Buntine, W. L. (1988). Machine invention of first-order predicates by inverting resolution. In *Proceedings of the 5th International Conference on Machine Learning (ML-88)*, Ann Arbor, MI (pp. 339–352). San Mateo, CA: Morgan Kaufmann /127/

Muggleton, S. H., & Feng, C. (1990). Efficient induction of logic programs. In *Proceedings of the 1st Conference on Algorithmic Learning Theory*, Tokyo (pp. 1–14). Tokyo: Japanese Society for Artificial Intelligence. /110, 131, 133/

Muggleton, S. H., & Firth, J. (2001). Relational rule induction with CProgol4.4: A tutorial introduction. In S. Džeroski & N. Lavrač (Eds.), *Relational data mining* (pp. 160–188). Berlin, Germany: Springer. Chap. 7. /52/

Muggleton, S. H., Santos, J. C. A., & Tamaddoni-Nezhad, A. (2009). ProGolem: A system based on relative minimal generalisation. In L. De Raedt (Ed.), *Proceedings of the 19th International Conference on Inductive Logic Programming (ILP-09)*, Leuven, Belgium (pp. 131–148). Springer. /133/

Mutter, S., Hall, M., & Frank, E. (2004). Using classification to evaluate the output of confidence-based association rule mining. In G. I. Webb & X. Yu (Eds.), *Proceedings of the Australian Joint Conference on Artificial Intelligence (AI-04)*, Cairns, QLD (pp. 538–549). Berlin, Germany: Springer. /54, 185/

Nemenyi, P. (1963). *Distribution-free multiple comparisons*. Ph.D. thesis, Princeton University. /90/

Niblett, T., & Bratko, I. (1987). Learning decision rules in noisy domains. In M. A. Bramer (Ed.), *Research and development in expert systems III* (pp. 25–34). Brighton, U.K.: Cambridge University Press. /149, 200/

Niculescu-Mizil, A., & Caruana, R. (2005a). Obtaining calibrated probabilities from boosting. In *Proceedings of the 21st Conference in Uncertainty in Artificial Intelligence (UAI-05)*, Edinburgh, UK (p. 413). Corvallis, OR: AUAI. /215/

Niculescu-Mizil, A., & Caruana, R. (2005b). Predicting good probabilities with supervised learning. In De Raedt, L. & S. Wrobel (Eds.), *Proceedings of the 22nd International Conference on Machine Learning (ICML 2005)*, Bonn, Germany (pp. 625–632). New York: ACM. /215/

Pagallo, G., & Haussler, D. (1990). Boolean feature discovery in empirical learning. *Machine Learning, 5*, 71–99. /13, 40, 145, 199, 204/

Park, S.-H., & Fürnkranz, J. (2007). Efficient pairwise classification. In J. N. Kok, J. Koronacki, R. López de Mántaras, S. Matwin, D. Mladenić, & A. Skowron (Eds.), *Proceedings of 18th European Conference on Machine Learning (ECML-07)*, Warsaw, Poland (pp. 658–665). Berlin, Germany/New York: Springer. /232/

Park, S.-H., & Fürnkranz, J. (2009). Efficient decoding of ternary error-correcting output codes for multiclass classification. In W. L. Buntine, M. Grobelnik, D. Mladenić, & J. Shawe-Taylor (Eds.), *Proceedings of the European Conference on Machine Learning and Principles and Practice of Knowledge Discovery in Databases (ECML/PKDD-09)*, Bled, Slovenia (Vol. Part II, pp. 189–204). Berlin, Germany: Springer. /217, 236/

Pazzani, M., Merz, C. J., Murphy, P., Ali, K., Hume, T., & Brunk, C. (1994). Reducing misclassification costs. In W. W. Cohen & H. Hirsh (Eds.), *Proceedings of the 11th International Conference on Machine Learning (ML-94)* (pp. 217–225). New Brunswick, NJ: Morgan Kaufmann. /221/

Pearl, J. (1988). *Probabilistic reasoning in intelligent systems: Networks of plausible inference*. San Mateo, CA: Morgan Kaufmann. /2/

Pechter, R. (2009). What's PMML and what's new in PMML 4.0. *SIGKDD Explorations, 11*, 19–25. /5/

Pelleg, D., & Moore, A. (2001). Mixtures of rectangles: Interpretable soft clustering. In C. E. Brodley & A. P. Danyluk (Eds.), *Proceedings of the 18th International Conference on Machine Learning (ICML-01)*, Williamstown, MA (pp. 401–408). San Francisco: Morgan Kaufmann. /245/

Peña Castillo, L., & Wrobel, S. (2004). A comparative study on methods for reducing myopia of hill-climbing search in multirelational learning. In C. E. Brodley (Ed.), *Proceedings of the 21st International Conference on Machine Learning (ICML-2004)*, Banff, AB. New York: ACM. /111, 117/

Pfahringer, B. (1995a). A new MDL measure for robust rule induction (extended abstract). In N. Lavrač & S. Wrobel (Eds.), *Proceedings of the 8th European Conference on Machine Learning (ECML-95)*, Heraclion, Greece (pp. 331–334). Berlin, Germany/New York: Springer. /162, 193, 206/

Pfahringer, B. (1995b). *Practical uses of the minimum description length principle in inductive learning*. Ph.D. thesis, Technische Universität Wien. /162/

Pfahringer, B., Holmes, G., & Wang, C. (2005). Millions of random rules. In J. Fürnkranz (Ed.), *Proceedings of the ECML/PKDD Workshop on Advances in Inductive Rule Learning*, Pisa, Italy. /183/

Piatetsky-Shapiro, G. (1991). Discovery, analysis, and presentation of strong rules. In G. Piatetsky-Shapiro & W. J. Frawley (Eds.), *Knowledge discovery in databases* (pp. 229–248). Menlo Park, CA: MIT. /143, 154, 156, 157, 165, 166, 167/

Piatetsky-Shapiro, G. & Frawley, W. J. (Eds.). (1991). *Knowledge discovery in databases*. Menlo Park, CA: MIT. /2, 3/

Pietraszek, T. (2007). On the use of ROC analysis for the optimization of abstaining classifiers. *Machine Learning, 68*(2), 137–169. /223/

Pimenta, E., Gama, J., & de Leon Ferreira de Carvalho, A. C. P. (2008). The dimension of ECOCs for multiclass classification problems. *International Journal on Artificial Intelligence Tools, 17*(3), 433–447. /234/

Platt, J. C. (1999). Probabilistic outputs for support vector machines and comparisons to regularized likelihood methods. In A. Smola, P. Bartlett, B. Schölkopf, & D. Schuurmans (Eds.), *Advances in large margin classifiers*. Cambridge, MA: MIT. /215/

Platt, J. C., Cristianini, N., & Shawe-Taylor, J. (2000). Large margin DAGs for multiclass classification. In S. A. Solla, T. K. Leen, & K.-R. Müller (Eds.), *Advances in neural information processing systems 12 (NIPS-99)* (pp. 547–553). Cambridge, MA/London: MIT. /243/

Plotkin, G. D. (1970). A note on inductive generalisation. In B. Meltzer & D. Michie (Eds.), *Machine intelligence 5* (pp. 153–163). New York: Elsevier/North-Holland. /132/

Plotkin, G. D. (1971). A further note on inductive generalisation. In B. Meltzer & D. Michie (Eds.), *Machine intelligence 6* (pp. 101–124). New York: Elsevier/North-Holland. /133/

Pompe, U., Kovačič, M., & Kononenko, I. (1993). SFOIL: Stochastic approach to inductive logic programming. In *Proceedings of the 2nd Slovenian Conference on Electrical Engineering and Computer Science (ERK-93)*, Portorož, Slovenia (Vol. B, pp. 189–192). /123, 191/

Prati, R. C., & Flach, P. A. (2005). Roccer: An algorithm for rule learning based on ROC analysis. In L. P. Kaelbling & A. Saffiotti (Eds.), *Proceedings of the 19th International Joint Conference on Artificial Intelligence (IJCAI-05)*, Edinburgh, UK (pp. 823–828). Professional Book Center. /242/

Price, D., Knerr, S., Personnaz, L., & Dreyfus, G. (1995). Pairwise neural network classifiers with probabilistic outputs. In G. Tesauro, D. Touretzky, & T. Leen (Eds.), *Advances in neural information processing systems 7 (NIPS-94)* (pp. 1109–1116). Cambridge, MA: MIT. /230/

Provost, F. J., & Domingos, P. (2003). Tree induction for probability-based ranking. *Machine Learning, 52*(3), 199–215. /215/

Provost, F., & Fawcett, T. (2001). Robust classification for imprecise environments. *Machine Learning, 42*(3), 203–231. /58, 59/

Pujol, O., Radeva, P., & Vitriá, J. (2006). Discriminant ECOC: A heuristic method for application dependent design of error correcting output codes. *IEEE Transactions on Pattern Analysis and Machine Intelligence, 28*(6), 1007–1012. /235/

Quevedo, J. R., Montañés, E., Luaces, O., & del Coz, J. J. (2010). Adapting decision DAGs for multipartite ranking. In J. L. Balcázar, F. Bonchi, A. Gionis, & M. Sebag (Eds.), *Proceedings of the European Conference on Machine Learning and Knowledge Discovery in Databases (ECML/PKDD-10)* Barcelona, Spain (Part III, pp. 115–130). Berlin, Germany/Heidelberg, Germany: Springer. /243/

Quinlan, J. R. (1979). Discovering rules by induction from large collections of examples. In D. Michie (Ed.), *Expert systems in the micro electronic age* (pp. 168–201). Edinburgh, UK: Edinburgh University Press. /3/

Quinlan, J. R. (1983). Learning efficient classification procedures and their application to chess end games. In R. S. Michalski, J. G. Carbonell, & T. M. Mitchell, (Eds.), *Machine learning. An artificial intelligence approach* (pp. 463–482). Palo Alto, CA: Tioga. /49, 146, 169/

Quinlan, J. R. (1986). Induction of decision trees. *Machine Learning, 1*, 81–106. /2, 3, 6, 9, 66, 146/

Quinlan, J. R. (1987a). Generating production rules from decision trees. In *Proceedings of the 10th International Joint Conference on Artificial Intelligence (IJCAI-87)* (pp. 304–307). Los Altos, CA: Morgan Kaufmann. /12, 55, 183, 211, 221/

Quinlan, J. R. (1987b). Simplifying decision trees. *International Journal of Man-Machine Studies, 27*, 221–234. /199, 200/

Quinlan, J. R. (1990). Learning logical definitions from relations. *Machine Learning, 5*, 239–266. /7, 50, 66, 96, 97, 120, 130, 159, 169, 191, 192, 193/

Quinlan, J. R. (1991). Determinate literals in inductive logic programming. In *Proceedings of the 8th International Workshop on Machine Learning (ML-91)* (pp. 442–446). San Mateo, CA: Morgan Kaufmann /51, 110/

Quinlan, J. R. (1992). Learning with continuous classes. In N. Adams & L. Sterling (Eds.), *Proceedings of the 5th Australian Joint Conference on Artificial Intelligence*, Hobart, TAS (pp. 343–348). Singapore: World Scientific. /244/

Quinlan, J. R. (1993). *C4.5: Programs for machine learning*. San Mateo, CA: Morgan Kaufmann. /v, 12, 52, 55, 183, 206, 221/

Quinlan, J. R. (1994). The minimum description length principle and categorical theories. In W. Cohen & H. Hirsh (Eds.), *Proceedings of the 11th International Conference on Machine Learning (ML-94)* (pp. 233–241). New Brunswick, NJ: Morgan Kaufmann. /162, 206/

Quinlan, J. R. (1995). MDL and categorical theories (continued). In A. Prieditis & S. J. Russell (Eds.), *Proceedings of the 12th International Conference on Machine Learning (ICML-95)*, Tahoe City, CA (pp. 464–470). San Francisco: Morgan Kaufmann. /206, 211/

Quinlan, J. R., & Cameron-Jones, R. M. (1995a). Induction of logic programs: FOIL and related systems. *New Generation Computing, 13*(3,4), 287–312. Special Issue on Inductive Logic Programming. /51, 111/

Quinlan, J. R., & Cameron-Jones, R. M. (1995b). Oversearching and layered search in empirical learning. In C. Mellish (Ed.), *Proceedings of the 14th International Joint Conference on Artificial Intelligence (IJCAI-95)*, Montréal, QC (pp. 1019–1024). San Mateo, CA: Morgan Kaufmann. /119/

Ramakrishnan, G., Joshi, S., Balakrishnan, S., & Srinivasan, A. (2008). Using ILP to construct features for information extraction from semi-structured text. In H. Blockeel, J. Ramon, J. W. Shavlik, & P. Tadepalli (Eds.), *Proceedings of the 17th International Conference on Inductive Logic Programming (ILP-07)*, Corvallis, OR (pp. 211–224). Springer. /53/

Ramaswamy, S., Tamayo, P., Rifkin, R., Mukherjee, S., Yeang, C.-H., & Angelo, M., et al. (2001). Multiclass cancer diagnosis using tumor gene expression signatures. *Proceedings of the National Academy of Sciences, 98*(26), 15149–15154. /274, 280/

Rijnbeek, P. R., & Kors, J. A. (2010). Finding a short and accurate decision rule in disjunctive normal form by exhaustive search. *Machine Learning, 80*(1), 33–62. /183/

Ripley, B. D. (1996). *Pattern recognition and neural networks*. Cambridge, MA/New York Cambridge University Press. /2/

Rissanen, J. (1978). Modeling by shortest data description. *Automatica, 14*, 465–471. /162, 192, 221/

Rivest, R. L. (1987). Learning decision lists. *Machine Learning, 2*, 229–246. /12, 50, 116, 117/

Rouveirol, C. (1992). Extensions of inversion of resolution applied to theory completion. In S. H. Muggleton (Ed.), *Inductive logic programming* (pp. 63–92). London: Academic. /127, 133/

Rouveirol, C. (1994). Flattening and saturation: Two representation changes for generalization. *Machine Learning, 14*, 219–232. Special issue on Evaluating and Changing Representation. /105, 133/

Rouveirol, C., & Puget, J. F. (1990). Beyond inversion of resolution. In *Proceedings of the 7th International Conference on Machine Learning (ML-90)*, Austin, TX (pp. 122–130). San Mateo, CA: Morgan Kaufmann. /127/

Rückert, U., & De Raedt, L. (2008). An experimental evaluation of simplicity in rule learning. *Artificial Intelligence, 172*(1), 19–28. /183/

Rückert, U., & Kramer, S. (2003). Stochastic local search in k-term DNF learning. In T. Fawcett & N. Mishra (Eds.), *Proceedings of the 20th International Conference on Machine Learning (ICML-03)*, Washington, DC (pp. 648–655). Menlo Park, CA: AAAI. /183/

Rückert, U., & Kramer, S. (2008). Margin-based first-order rule learning. *Machine Learning, 70*(2–3), 189–206. /215/

Rumelhart, D. E., & McClelland, J. L. (Eds.). (1986). *Parallel distributed processing: explorations in the microstructure of cognition, vol. 1: Foundations*. Cambridge, MA: MIT. /2, 3/

Rüping, S. (2006). Robust probabilistic calibration. In J. Fürnkranz, T. Scheffer, & M. Spiliopoulou (Eds.), *Proceedings of the 17th European Conference on Machine Learning (ECML/PKDD-06)*, Berlin, Germany (pp. 743–750). Berlin, Germany/New York: Springer. /215/

Salzberg, S. (1991). A nearest hyperrectangle learning method. *Machine Learning, 6*, 251–276. /181, 223/

Schapire, R. E., Freund, Y., Bartlett, P., & Lee, W. S. (1998). Boosting the margin: A new explanation for the effectiveness of voting methods. *The Annals of Statistics, 26*(5), 1651–1686. /2/

Scheffer, T., & Wrobel, S. (2002). Finding the most interesting patterns in a database quickly by using sequential sampling. *Journal of Machine Learning Research, 3*, 833–862. /186/

Schmidt, M. S., & Gish, H. (1996). Speaker identification via support vector classifiers. In *Proceedings of the 21st IEEE International Conference on Acoustics, Speech, and Signal Processing (ICASSP-96)*, Atlanta, GA (pp. 105–108). Piscataway, NJ: IEEE. /230/

Schölkopf, B., & Smola, A. J. (2001). *Learning with kernels: Support vector machines, regularization, optimization, and beyond*. Cambridge, MA: MIT. /2/

Shapiro, E. Y. (1981). An algorithm that infers theories from facts. In *Proceedings of the 7th International Joint Conference on Artificial Intelligence (IJCAI-81)* (pp. 446–451). Los Altos, CA: William Kaufmann. /110/

Shapiro, E. Y. (1982). *Algorithmic program debugging*. Cambridge, MA: MIT. /97/

Shapiro, E. Y. (1991). Inductive inference of theories from facts. In J. L. Lassez & G. D. Plotkin (Eds.), *Computational logic: Essays in honor of Alan Robinson* (pp. 199–255). Cambridge, MA: MIT. /97/

Silberschatz, A., & Tuzhilin, A. (1995). On subjective measure of interestingness in knowledge discovery. In *Proceedings of the 1st International Conference on Knowledge Discovery and Data Mining (KDD-95)*, Montréal, QC (pp. 275–281). Menlo Park, CA: AAAI /287/

Simeon, M., & Hilderman, R. J. (2007). Exploratory quantitative contrast set mining: A discretization approach. In *Proceedings of the 19th IEEE International Conference on Tools with Artificial Intelligence (ICTAI-07)*, Patras, Greece (Vol.2, pp. 124–131). Los Alamitos, CA: IEEE. /252/

Siu, K., Butler, S., Beveridge, T., Gillam, J., Hall, C., & Kaye, A., et al. (2005). Identifying markers of pathology in SAXS data of malignant tissues of the brain. *Nuclear Instruments and Methods in Physics Research A, 548*, 140–146. /252/

Smola, A. J., & Schölkopf, B. (2004). A tutorial on support vector regression. *Statistics and Computing, 14*, 199–222. /244/

Smyth, P., & Goodman, R. M. (1991). Rule induction using information theory. In G. Piatetsky-Shapiro & W. J. Frawley (Eds.), *Knowledge discovery in databases* (pp. 159–176). London: MIT. /153, 183/

Soares, C. (2003). Is the UCI repository useful for data mining? In F. Moura-Pires & S. Abreu (Eds.), *Proceedings of the 11th Portuguese Conference on Artificial Intelligence (EPIA-03)*, Beja, Portugal (pp. 209–223). Berlin, Germany/Heidelberg, Germany: Springer. /48/

Song, H. S., Kimb, J. K., & Kima, S. H. (2001). Mining the change of customer behavior in an internet shopping mall. *Expert Systems with Applications, 21*(3), 157–168. /255/

Soulet, A., Crémilleux, B., & Rioult, F. (2004). Condensed representation of emerging patterns. In *Proceedings of the 8th Pacific-Asia Conference on Knowledge Discovery and Data Mining (PAKDD-04)*, Sydney, NSW (pp. 127–132). Berlin, Germany/New York: Springer. /254/

Specia, L., Srinivasan, A., Joshi, S., Ramakrishnan, G., & das Graças Volpe Nunes, M. (2009). An investigation into feature construction to assist word sense disambiguation. *Machine Learning, 76*(1), 109–136. /53/

Srinivasan, A. (1999). The Aleph manual. http://web.comlab.ox.ac.uk/oucl/research/areas/machlearn/Aleph/. /53/

Srinivasan, A., & King, R. D. (1997). Feature construction with inductive logic programming: A study of quantitative predictions of biological activity by structural attributes. In S. Muggleton (Ed.), *Proceedings of the 6th International Workshop, on Inductive Logic Programming (ILP-96)*, Stockholm (pp. 89–104). Berlin, Germany/New York: Springer. /53/

Stepp, R. E., & Michalski, R. S. (1986). Conceptual clustering of structured objects: A goal-oriented approach. *Artificial Intelligence, 28*(1), 43–69. /245/

Sterling, L., & Shapiro, E. (1994). *The art of prolog—Advanced programming techniques* (2nd ed.). Cambridge, MA: MIT. /13, 104/

Sternberg, M. J., & Muggleton, S. H. (2003). Structure activity relationships (SAR) and pharmacophore discovery using inductive logic programming (ILP). *QSAR and Combinatorial Science, 22*(5), 527–532. /52/

Sulzmann, J.-N., & Fürnkranz, J. (2008). A comparison of techniques for selecting and combining class association rules. In J. Fürnkranz & A. J. Knobbe (Eds.), *From Local Patterns to Global Models: Proceedings of the ECML/PKDD-08 Workshop (LeGo-08)*, Antwerp, Belgium (pp. 154–168). /185/

Sulzmann, J.-N., & Fürnkranz, J. (2009). An empirical comparison of probability estimation techniques for probabilistic rules. In J.Gama, V. Santos Costa, A. Jorge, & P. B. Brazdil (Eds.), *Proceedings of the 12th International Conference on Discovery Science (DS-09)* (pp. 317–331). Berlin, Germany/New York: Springer. Winner of Carl Smith Award for Best Student Paper. /215/

Sulzmann, J.-N., & Fürnkranz, J. (2011). Rule stacking: An approach for compressing an ensemble of rule sets into a single classifier. In T. Elomaa, J. Hollmèn, & H. Mannila (Eds.), *Proceedings of the 14th International Conference on Discovery Science (DS-11)*, Espoo, Finland (pp. 323–334). Berlin, Germany/New York: Springer. /232/

Suzuki, E. (2006). Data mining methods for discovering interesting exceptions from an unsupervised table. *Journal of Universal Computer Science, 12*(6), 627–653. /260/

Tan, P.-N., Kumar, V., & Srivastava, J. (2002). Selecting the right interestingness measure for association patterns. In *Proceedings of the 8th ACM SIGKDD International Conference on Knowledge Discovery and Data Mining (KDD-02)*, Edmonton, AB (pp. 32–41). New York: ACM. */137, 167/*

Tan, P.-N., Kumar, V., & Srivastava, J. (2004). Selecting the right objective measure for association analysis. *Information Systems, 29*(4), 293–313. */137/*

Theron, H., & Cloete, I. (1996). BEXA: A covering algorithm for learning propositional concept descriptions. *Machine Learning, 24*, 5–40. */50, 68, 118, 150, 196, 211/*

Todorovski, L., Flach, P., & Lavrač, N. (2000). Predictive performance of weighted relative accuracy. In D. Zighed, J. Komorowski, & J. Zytkow (Eds.), *Proceedings of the 4th European Symposium on Principles of Data Mining and Knowledge Discovery (PKDD-2000)*, Lyon, France (pp. 255–264). Berlin, Germany/Springer. */143, 164, 198/*

Torgo, L. (1995). Data fitting with rule-based regression. In J. Zizka & P. B. Brazdil (Eds.), *Proceedings of the 2nd International Workshop on Artificial Intelligence Techniques (AIT-95)*. Brno, Czech Republic: Springer. */244/*

Torgo, L. (2010). *Data mining with R: Learning with case studies* (Data mining and knowledge discovery series). Boca Raton: Chapman & Hall/CRC. */5/*

Torgo, L., & Gama, J. (1997). Regression using classification algorithms. *Intelligent Data Analysis, 1*(4), 275-292. */244/*

Ullman, J. D. (1988). *Principles of database and knowledge base systems* (Vol. I). Rockville, MA: Computer Science Press. */107, 108, 109/*

Van Horn, K. S., & Martinez, T. R. (1993). The BBG rule induction algorithm. In *Proceedings of the 6th Australian Joint Conference on Artificial Intelligence (AI-93)*, Melbourne, VIC (pp. 348–355). Singapore: World Scientific. */226/*

van Rijsbergen, C. J. (1979). *Information retrieval* (2nd ed.). London: Butterworths. */148/*

Vapnik, V. (1995). *The nature of statististical learning theory*. New York: Springer. */2/*

Venturini, G. (1993). SIA: A supervised inductive algorithm with genetic search for learning attributes based concepts. In P. Brazdil (Ed.), *Proceedings of the 6th European Conference on Machine Learning (ECML-93)*, Vienna (pp. 280–296). Berlin, Germany: Springer. */123, 127, 162, 163/*

Wallace, C. S., & Boulton, D. M. (1968). An information measure for classification. *Computer Journal, 11*, 185–194. */162, 192/*

Wang, B., & Zhang, H. (2006). Improving the ranking performance of decision trees. In J. Fürnkranz, T. Scheffer, & M. Spiliopoulou (Eds.), *Proceedings of the 17th European Conference on Machine Learning (ECML-06)* (pp. 461–472). Berlin, Germany: Springer. */215/*

Wang, K., Zhou, S., Fu, A. W.-C., & Yu, J. X. (2003). Mining changes of classification by correspondence tracing. In *Proceedings of the 3rd SIAM International Conference on Data Mining (SDM-03)* (pp. 95–106). Philadelphia: SIAM */260/*

Watanabe, L., & Rendell, L. (1991). Learning structural decision trees from examples. In *Proceedings of the 12th International Joint Conference on Artificial Intelligence (IJCAI-91)*, Sydney, NSW (pp. 770–776). San Mateo, CA: Morgan Kaufmann. */13/*

Webb, G. I. (1992). *Learning disjunctive class descriptions by least generalisation* (Tech. Rep. TR C92/9). Geelong, VIC: Deakin University, School of Computing & Mathematics. */128, 140/*

Webb, G. I. (1993). Systematic search for categorical attribute-value data-driven machine learning. In C. Rowles, H. Liu, & N. Foo (Eds.), *Proceedings of the 6th Australian Joint Conference of Artificial Intelligence (AI'93)*, Melbourne, VIC (pp. 342–347). Singapore: World Scientific. */119, 150/*

Webb, G. I. (1994). Recent progress in learning decision lists by prepending inferred rules. In *Proceedings of the 2nd Singapore International Conference on Intelligent Systems* (pp. B280–B285). Singapore: World Scientific. */226/*

Webb, G. I. (1995). OPUS: An efficient admissible algorithm for unordered search. *Journal of Artificial Intelligence Research, 5*, 431–465. */15, 54, 120, 252/*

Webb, G. I. (1996). Further experimental evidence aganst the utility of Occam's razor. *Journal of Artificial Intelligence Research, 4*, 397–417. */9/*

Webb, G. I. (2000). Efficient search for association rules. In *Proceedings of the 6th ACM SIGKDD International Conference on Knowledge Discovery and Data Mining (KDD-2000)*, Boston (pp. 99–107). New York: ACM. /54, 122/

Webb, G. I. (2001). Discovering associations with numeric variables. In *Proceedings of the 7th ACM SIGKDD International Conference on Knowledge Discovery and Data Mining (KDD-01)*, San Francisco (pp. 383–388). New York: ACM. /261/

Webb, G. I. (2007). Discovering significant patterns. *Machine Learning, 68*(1), 1–33. /251, 252, 259/

Webb, G. I., & Brkič, N. (1993). Learning decision lists by prepending inferred rules. In *Proceedings of the AI'93 Workshop on Machine Learning and Hybrid Systems*, Melbourne, VIC (pp. 6–10). Melbourne, Australia. /226/

Webb, G. I., Butler, S. M., & Newlands, D. (2003). On detecting differences between groups. In *Proceedings of the 9th ACM SIGKDD International Conference on Knowledge Discovery and Data Mining (KDD-03)*, Washington, DC (pp. 256–265). New York: ACM. /251, 252, 259/

Webb, G. I., & Zhang, S. (2005). *k*-optimal rule discovery. *Data Mining and Knowledge Discovery, 10*(1), 39–79. /54/

Weiss, S. M., & Indurkhya, N. (1991). Reduced complexity rule induction. In *Proceedings of the 12th International Joint Conference on Artificial Intelligence (IJCAI-91)*, Sydney, NSW (pp. 678–684). San Mateo, CA: Morgan Kaufmann. /128, 145, 202/

Weiss, S. M., & Indurkhya, N. (1994). Small sample decision tree pruning. In *Proceedings of the 11th Conference on Machine Learning*, (pp. 335–342). New Brunswick, NJ: Rutgers University. /212/

Weiss, S. M., & Indurkhya, N. (1995). Rule-based machine learning methods for functional prediction. *Journal of Artificial Intelligence Research, 3*, 383–403. /244/

Weiss, S. M., & Indurkhya, N. (2000). Lightweight rule induction. In P. Langley (Ed.), *Proceedings of the 17th International Conference on Machine Learning (ICML-2000)*, Stanford, CA (pp. 1135–1142). San Francisco, CA: Morgan Kaufmann. /178, 186/

Wettschereck, D. (2002). A KDDSE-independent PMML visualizer. In *Proceedings of 2nd Workshop on Integration Aspects of Data Mining, Decision Support and Meta-Learning (IDDM-02)* (pp. 150–155). Helsinki, Finland: Helsinki University /261/

Widmer, G. (1993). Combining knowledge-based and instance-based learning to exploit qualitative knowledge. *Informatica, 17*, 371–385. Special Issue on Multistrategy Learning. /128/

Widmer, G. (2003). Discovering simple rules in complex data: A meta-learning algorithm and some surprising musical discoveries. *Artificial Intelligence, 146*(2), 129–148. /182/

Wiese, M. (1996). A bidirectional ILP algorithm. In *Proceedings of the MLnet Familiarization Workshop on Data Mining with Inductive Logic Programming (ILP for KDD)*, Bari, Italy (pp. 61–72). /128/

Windeatt, T., & Ghaderi, R. (2003). Coding and decoding strategies for multi-class learning problems. *Information Fusion, 4*(1), 11–21. /235/

Wirth, R. (1988). Learning by failure to prove. In *Proceedings of the Third European Working Session on Learning*, Glasgow, UK (pp. 237–251). London: Pitman. /127/

Witten, I. H., & Frank, E. (2005). *Data mining: Practical machine learning tools and techniques with Java implementations* (2nd ed.). Amsterdam/Boston: Morgan Kaufmann Publishers. /2, 5, 10, 23, 52/

Wohlrab, L., & Fürnkranz, J. (2011). A review and comparison of strategies for handling missing values in separate-and-conquer rule learning. *Journal of Intelligent Information Systems, 36*(1), 73–98. /87, 89, 90/

Wolpert, D. H. (1992). Stacked generalization. *Neural Networks, 5*(2), 241–260. /232/

Wong, T.-T., & Tseng, K.-L. (2005). Mining negative contrast sets from data with discrete attributes. *Expert Systems with Applications, 29*(2), 401–407. /252/

Wrobel, S. (1996). First order theory refinement. In L. De Raedt (Ed.), *Advances in inductive logic programming* (pp. 14–33). Amsterdam: IOS Press. /183/

Wrobel, S. (1997). An algorithm for multi-relational discovery of subgroups. In *Proceedings of the 1st European Symposium on Principles of Data Mining and Knowledge Discovery (PKDD-97)* (pp. 78–87). Berlin, Germany: Springer. */157, 158, 248, 249, 256, 258/*

Wrobel, S. (2001). Inductive logic programming for knowledge discovery in databases. In S. Džeroski & N. Lavrač (Eds.), *Relational data mining* (pp. 74–101). Berlin, Germany/New York: Springer. */97, 249, 261, 262/*

Wu, T., Chen, Y., & Han, J. (2007). Association mining in large databases: A re-examination of its measures. In *Proceedings of the 11th European Symposium on Principles of Data Mining and Knowledge Discovery (PKDD-07)*, Warsaw, Poland (pp. 621–628). Springer. */167/*

Wu, T.-F., Lin, C.-J., & Weng, R. C. (2004). Probability estimates for multi-class classification by pairwise coupling. *Journal of Machine Learning Research, 5*, 975–1005. */239/*

Xiong, H., Shekhar, S., Tan, P.-N., & Kumar, V. (2004). Exploiting a support-based upper bound of Pearson's correlation coefficient for efficiently identifying strongly correlated pairs. In *Proceedings of the 10th ACM SIGKDD International Conference on Knowledge Discovery and Data Mining (KDD-04)*, Seattle, WA (pp. 334–343). New York: ACM. */154/*

Yang, Y., Webb, G. I., & Wu, X. (2005). Discretization methods. In O. Maimon & L. Rokach (Eds.), *The data mining and knowledge discovery handbook* (pp. 113–130). New York: Springer. */75/*

Yin, X., & Han, J. (2003). CPAR: Classification based on predictive association rules. In D. Barbará & C. Kamath (Eds.) *Proceedings of the SIAM Conference on Data Mining (SDM-03)* (pp. 331–335). Philadelphia: SIAM. */54, 185/*

Zadeh, L. A. (1965). Fuzzy sets. *Information and Control, 8*(3), 338–353. */92/*

Zadrozny, B., & Elkan, C. (2001). Obtaining calibrated probability estimates from decision trees and naive bayesian classifiers. In C. E. Brodley & A. P. Danyluk (Eds.), *Proceedings of the 18th International Conference on Machine Learning (ICML 2001)*, Williams College, Williamstown, MA (pp. 609–616). San Francisco: Morgan Kaufmann. */215/*

Zadrozny, B., & Elkan, C. (2002). Transforming classifier scores into accurate multiclass probability estimates. In *Proceedings of the 8th ACM SIGKDD International Conference on Knowledge Discovery and Data Mining (KDD-02)*, Edmonton, AB (pp. 694–699). New York: ACM. */215/*

Zaki, M. J., Parthasarathy, S., Ogihara, M., & Li, W. (1997). New algorithms for fast discovery of association rules. In *Proceedings of the 3rd International Conference on Knowledge Discovery and Data Mining (KDD-97)*, Newport, CA (pp. 283–286). Menlo Park, CA: AAAI. */122/*

Zelezný, F., & Lavrač, N. (2006). Propositionalization-based relational subgroup discovery with RSD. *Machine Learning, 62*, 33–63. */102, 249/*

Zelle, J. M., Mooney, R. J., & Konvisser, J. B. (1994). Combining top-down and bottom-up techniques in inductive logic programming. In W. Cohen & H. Hirsh (Eds.), *Proceedings of the 11th International Conference on Machine Learning (ML-94)*, New Brunswick, NJ (pp. 343–351). San Francisco: Morgan Kaufmann. */182/*

Zenko, B. (2007). *Learning Predictive Clustering Rules*. Ph.D. thesis, University of Ljubljana, Faculty of Computer and Information Science, Ljubljana, Slovenia. */245/*

Zenko, B., Džeroski, S., & Struyf, J. (2006). Learning predictive clustering rules. In F. Bonchi & J.-F. Boulicaut (Eds.), *Proceedings of the 4th International Workshop on Knowledge Discovery in Inductive Databases (KDID-05)*, Porto, Portugal (pp. 234–250). Berlin, Germany/New York: Springer. */245/*

Zhang, C., & Zhang, S. (2002). *Association rule mining: Models and algorithms*. Berlin, Germany/New York: Springer. */15, 122/*

Zhou, W., & Xiong, H. (2011). Checkpoint evolution for volatile correlation computing. *Machine Learning, 83*(1), 103–131. */154/*

Zimmermann, A., & De Raedt, L. (2009). Cluster-grouping: From subgroup discovery to clustering. *Machine Learning, 77*(1), 125–159. */245/*

Index

J. Fürnkranz et al., *Foundations of Rule Learning*, Cognitive Technologies,
DOI 10.1007/978-3-540-75197-7, © Springer-Verlag Berlin Heidelberg 2012